Genes and Behaviour: Beyond Nature-Nurture

Genes and Behaviour: Beyond Nature-Nurture

Edited by
David J. Hosken
John Hunt
Nina Wedell

Registered Offices
John Wiley & Sons, Inc., 111 River Street, Hoboken, NJ 07030, USA
John Wiley & Sons Ltd, The Atrium, Southern Gate, Chichester, West Sussex, PO19 8SQ, UK

Editorial Office
The Atrium, Southern Gate, Chichester, West Sussex, PO19 8SQ, UK

For details of our global editorial offices, customer services, and more information about Wiley products visit us at www.wiley.com.

Library of Congress Cataloging-in-Publication Data has been Applied for
9781119313427

Cover Design: Wiley
Cover Image: © Pixtum/iStock.com

Set in 10/12pt WarnockPro by SPi Global, Chennai, India
Printed in Singapore by C.O.S. Printers Pte Ltd

10 9 8 7 6 5 4 3 2 1

To our children, Rebekka, Noam, Nils, Jackson, Matilda & Jake.

Contents

List of Contributors

C. Ruth Archer
Centre for Ecology & Conservation
University of Exeter
Penryn Campus
Penryn
Cornwall
UK

Louise Barrett
Department of Psychology
University of Lethbridge
Calgary
Canada

Neeltje J. Boogert
Centre for Ecology & Conservation
University of Exeter
Penryn Campus
Penryn
Cornwall
UK

Amanda Bretman
School of Biology
Faculty of Biological Sciences
University of Leeds
Leeds
UK

Thomas E. Currie
Centre for Ecology & Conservation
University of Exeter
Penryn Campus
Penryn
Cornwall
UK

Sasha R.X. Dall
Centre for Ecology & Conservation
University of Exeter
Penryn Campus
Penryn
Cornwall
UK

Stephen F. Goodwin
Centre for Neural Circuits & Behaviour
University of Oxford
Oxford
UK

David J. Hosken
Centre for Ecology & Conservation
University of Exeter
Penryn Campus
Penryn
Cornwall
UK

Clarissa M. House
School of Science and Health and
The Hawkesbury Institute for the
Environment
Western Sydney University
Hawkesbury
NSW
Australia

John Hunt
Centre for Ecology & Conservation
University of Exeter
Penryn Campus
Penryn
Cornwall
UK

and

School of Science and Health and
The Hawkesbury Institute for the
Environment
Western Sydney University
Hawkesbury
NSW
Australia

Simone Immler
School of Biological Sciences
University of East Anglia
Norwich
UK

Laurent Keller
Department of Ecology and Evolution
University of Lausanne
Lausanne
Switzerland

John M. McNamara
School of Mathematics
University of Bristol
Bristol
UK

Allen J. Moore
Department of Genetics
University of Georgia
Athens
USA

James Rapkin
Centre for Ecology & Conservation
University of Exeter
Penryn Campus
Penryn
Cornwall
UK

Nick J. Royle
Centre for Ecology & Conservation
University of Exeter
Penryn Campus
Penryn
Cornwall
UK

Gert Stulp
Department of Sociology
University of Groningen/Inter-university
Center for Social Science Theory and
Methodology (ICS)
Groningen
The Netherlands

Alex Thornton
Centre for Ecology & Conservation
University of Exeter
Penryn Campus
Penryn
Cornwall
UK

Nina Wedell
Centre for Ecology & Conservation
University of Exeter
Penryn Campus
Penryn
Cornwall
UK

Chelsea A. Weitekamp
Department of Ecology and Evolution
University of Lausanne
Lausanne
Switzerland

Alastair J. Wilson
Centre for Ecology & Conservation
University of Exeter
Penryn Campus
Penryn
Cornwall
UK

Preface

Our desire to edit a volume on how nature and nurture affect behavior comes largely from frustration, which may not be the best motivation. As we outline in Chapter 1, we simply got bored with hearing (reading) the false dichotomy – it's either nature or nurture – and wanted to highlight the much more subtle but easy to understand interplay between the two and to discuss some of the fascinating findings that have emerged in the study of behavior. And to ensure we did not have to deal with the equally boring 'humans are fundamentally different', we also invited authors to contribute specific chapters that focus on our species and on traits that are often seen as especially important to us, like our highly developed cognitive abilities. How well we have covered all these interesting bases we leave to the reader, but we hope some of the excitement of behavioural genetics (in the broadest sense) comes through in the chapters that follow.

The book starts with some general concepts, that after our introductory chapter on the demise of the nature versus nurture dichotomy is headed by a typically provocative chapter by Sasha Dall, John McNamara and Alastair Wilson, who explore whether we need to identify genes to understand behaviour and then dissect the difference between the phenotypic gambit and quantitative genetics approach to studying behaviour, all in the context of Tinbergen's '4 whys'. This is sure to clarify some issues and generate considerable debate. Tom Currie then explores the relationships between culture, genes and behaviour in humans. In many ways, this outlines the importance of social environments, an issue that permeates many of the chapters that follow. Interestingly, Tom suggests that thinking of culture as a distinct inheritance system can facilitate our understanding of behavioural evolution. This is followed by John Hunt and co-authors who explore similar issues through the lens of quantitative genetics to review how indirect genetic effects affect the evolution of behaviour. Here, they distinguish between trait-based and variance partitioning approaches and the chapter nicely contrasts with the two that precede it. Chapter 5 is a pragmatic contribution by Chelsea Weitekamp and Laurent Keller that embraces much of what the preceding chapters discuss. They review effects of single genes on complex behaviours, including many textbook examples like the *foraging* gene, while also noting that most traits, including behaviour, are underpinned by many genes of small effect. Their discussion should dispel any vestiges of doubt that genes affecting behaviour are either non-existent or impossible to identify.

The next cluster of chapters largely focuses on gene-by-environment interactions for specific behaviours. It starts with a chapter by David Hosken and co-authors on

Drosophila sexual interactions. *Drosophila* have been a workhorse for behavioural genetics and the depth and breadth of information that has accumulated are revealed in this contribution that moves from single genes of large effect to many genes of small effect and how they interact with the environment to produce statistical gene-by-environment interactions. It particularly focuses on social environments and reiterates an overall theme: establishing the genotype-phenotype link remains a holy grail of modern biology but one that will be greatly complicated by interactions. Nick Royle and Allen Moore follow with a review of gene and environmental effects in parental care. This includes discussion of genotype-by-family environmental effects and how social environments can be transmitted across generations, meaning that two elements determining phenotypes can be heritable. This group of chapters ends with a contribution by Nina Wedell reviewing the effects of non-self genes on self-behaviour. This includes recent work on microbiomes and 'selfish genes', and makes the obvious point (given the chapter subject matter) that not only do our own genes directly influence our behaviour, but the genes of other organisms that share our bodies can also have profound effects, which expands the realm of environment–gene impacts on focal genes.

Next follow two chapters focused on comparative cognition and human evolution-ary psychology. The first is by Alex Thornton and Neeltje Boogert and explores how knowledge is acquired and how intelligence evolves, from both a biological and psycho-logical perspective. They make the point that this approach may have applied implica-tions because understanding what they term the 'cognitive reaction norm' can inform us about animal welfare, including humans. Louise Barrett and Gert Stulp then focus on human behaviour and how evolutionary thinking has shaped understanding of our own behaviour. They note that for humans, genes and environment (biology and cul-ture) also includes the socio-economic environment, and make the interesting point that our attempts at improving our health and well-being have generated a whole new set of problems to address. Adopting an approach informed by evolutionary principles may offer new insight into how to solve the issues they raise. This is followed by Simone Immler's review of modern techniques that can be employed to uncover genes affecting behaviour and their limitations, as well as discussion of a few case studies of what and how. We (Wedell, Hunt and Hosken) close the book by pondering what we have learnt in the preceding chapters and what excites researchers interested in behavioural genetics. We discuss what we (think we) know, what we need to know, and what we should do next. We note here that the book largely ignores more mechanistic elements of nature-nurture and thus we did not include chapters on behavioural circuits, regulation and neurology, for example, except for small sections here and there.

We would like to thank all the contributors for their excellent chapters, for their gen-erous reviews, and for their (mostly!) timely responses. Sorry it took much longer than we anticipated. We would also like to thank wonderful colleagues, who are not authors, for their generosity in offering reviews that were insightful, collegial and super helpful: in no particular order, Matt Dean, Hanna Kokko, Rebecca Kilner, Alex Mesoudi, Judith Mank, Nathan Bailey, Barbara Taborsky, Marla Sokolowski, Stuart Wigby, Luc Bussiere, Erik Potsma, and Ben Longdon – without your collective help the book would be less clear and probably less interesting too. We would also very much like to thank all the wonderful colleagues and friends who have helped shape how we view these matters through endless discussion over the years. Finally we thank the publishing team for their patience and help.

July 2018

David J. Hosken
John Hunt
Nina Wedell

This is a welcome modern take on the nature-nurture debate. It provides a thoroughly comprehensive coverage while including many of the recently-recognised subtleties of the relationships between genes and behaviour: no matter the species, there is no dichotomy, and this book provides piles of clear, cogent evidence that would be accessible to senior undergraduate and graduate students. It's one for my bookshelf!

Professor Susan Healy, University of St Andrews

This excellent synthesis of our current understanding of genes and behaviour offers essential reading for students and established researchers across a wide range of disciplines, including the study of both animal and human behaviour. It offers an impressive collection of accessible contributions from leading researchers and will inspire the next generation to explore new horizons in this expanding field.

Professor Paula Stockley, University of Liverpool

1

Nature, Nurture, and Nature-by-Nurture – Killing the Dichotomy

David J. Hosken[1], John Hunt[1,2] and Nina Wedell[1]

[1] Centre for Ecology & Conservation, University of Exeter, Penryn Campus, Penryn, TR10 9EZ, UK
[2] School of Science and Health & Hawkesbury Institute for the Environment, Western Sydney University, Hawkesbury, NSW 2793, Australia

The primary purpose of this book is to provide a broad snapshot of recent findings showing how the environment and genes influence behaviour. At face value, this should be uncontroversial but unfortunately, the history of genetics includes eugenic movements and Lysenkoism. As a result, discussions of how nature and nurture affect behaviour have been dogged by polemic disputes because ideological views about their contributions have tended to cloud what is really an empirical question. This is in some ways exemplified by the book *Not in Our Genes* (Lewontin et al. 1984), which begins with a political confession from the authors – we are committed socialists – and starts with a chapter on right-wing politics and determinism. For us, the evidence, and not political or any other beliefs, is what counts and any 'belief' approach puts the desire for the world to be a certain way ahead of the evidence that it is not so, ultimately committing a version of the naturalistic fallacy – if something is 'natural', it is morally correct, which is clearly rubbish (also see Chapter 10). Infanticide, cannibalism, forced copulation (rape), and killing other members of your species (murder) are rife in nature, but it would be difficult to convince anyone of intelligence that these acts are moral because they are natural. Furthermore, 'politically' motivated arguments against 'reductionism', reducing complex behaviours to single causes, are frequently concocted to protect against a biological determinism that must be fought at all costs. However, as we hope to explain, acknowledging that there are genes underlying behaviour, even genes of large effect, is imperative if that is what the data tell us. After all, it is no use playing music to cows if milk yield is totally determined by genes and unaffected by the environment, and as we outline below, in a polygenic world that includes inevitable environmental effects and all manner of interactions, prediction is tricky and determinism dubious because of the probabilistic and complex nature of the gene–behaviour link. But again, even if single genes were completely responsible for single behaviours, which they cannot be in the strictest sense (see below), let us not fall into a naturalistic fallacy.

Rather than engage in further fruitless arguments about world-views, this book explores exciting new findings about behaviour and where we go from here. Before moving on to these new advances and the interesting questions that arise from them, we wish to make another – a final? – attempt to kill the nature versus nurture polarity

Genes and Behaviour: Beyond Nature-Nurture, First Edition.
Edited by David J. Hosken, John Hunt and Nina Wedell.
© 2019 John Wiley & Sons Ltd. Published 2019 by John Wiley & Sons Ltd.

that has plagued the study of behaviour. This dichotomy is largely, but not totally, dead in academic circles but still haunts many debates outside academia, from views on teaching and punishment to politics and the media more generally. It potentially has grave consequences and is a serious distraction to the much more fruitful and interesting discussion about the determinants and influences of behaviour.

Most behaviours, like any aspect of the phenotype, are not influenced by either nature or nurture but by both and by the statistical interaction between nature and nurture (see reviews in Boake 1994; Sokolowski 2001; Bucan and Abel 2002; van Oers et al. 2005; Hunt and Hosken 2014; Anholt and Mackay 2015) (see also Chapters 4, 6, and 7). To explain, starting with the genetic effects, behaviours (and other characters, for that matter) are typically polygenic (Anholt and Mackay 2004). That is, they have complicated genetic architecture that involves many segregating genes with pleiotropic effects and are characterized by complicated epistatic interactions (Anholt and Mackay 2004). In other words, there are lots of genes, each can affect many characters, and the effects of any one gene frequently depend on the other genes it is associated with. There are exceptions to some of this (see Chapter 5), with, for example, foraging movement in *Drosophila melanogaster* having two distinct behavioural phenotypes that are largely determined by a single gene (reviewed in Sokolowski 2001), and aggression being altered by transposon upregulation of a cytochrome P450 gene (Rostant et al. 2017). However, even these large single-gene effects can be complicated by epistasis (gene–gene interactions) (e.g. Smith et al. 2011; Rostant et al. 2015).

Nonetheless, most behaviours are influenced by many genes, often of small effect, and because of this, we may never uncover all the precise genes that influence a behavioural phenotype. As a result, a statistical approach is needed to describe the average effects of genes on a behaviour and, importantly, to show how genes affect the variation around the mean. The distinction between an average effect and the variation around it is crucial, because for the most part there is not a single gene for phenotype A or B; rather, there are many genes that alter the probability of expressing phenotype A or B. Thus, many interesting traits do not vary discretely but are continuous (Falconer 1981; Roff 1997; Lynch and Walsh 1998), and genes influence the likelihood that an individual will express more or less of the trait in question.

The simplest statistical approach to understanding these relationships involves partitioning the variation in the behaviour of interest into the sum of the genetic effects and the variance unexplained is then due to the environment (which includes maternal/paternal effects, indirect genetic effects, ecology and abiotic factors like temperature, food, and water), or alternatively, testing a range of genotypes across environments and then partitioning effects into genes, environment, and their interaction (how genes and environment affect each other to determine phenotypic variation) (see Chapter 4). This reveals exactly how genes, the environment, and their interaction can affect phenotypes, including behavioural phenotypes.

To use a simple morphological example to make this point very clearly while noting the principles are exactly the same for behaviour: if we could take three plant-clones (three distinct plant genotypes (Figure 1.1) and grow each of them in two highly controlled environments that only differed from each other by how much water was available and all else was exactly the same, then the differences in plant heights within each environment would be due to just the genes, and the average difference in heights between environments would be due to environmental differences alone.

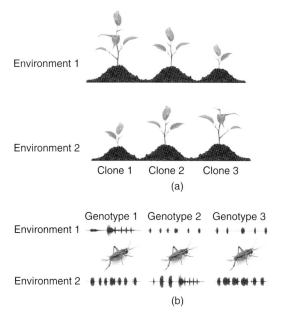

Figure 1.1 A pictorial explanation of genotype-by-environment interactions (GxE). In (a) we show a plant GxE – for simplicity's sake (see explanation below) – and in (b) cricket calling behaviour as a hypothetical behavioural example. (a) Three plant genotypes (clones) grown in two environments that only differ in how much water each plant receives, but everything else about the environments is identical. This means that each plant experiences exactly the same conditions within each environment and differences in water between environments. Therefore, plant size differences within each environment are due to just the genetic differences between plants. However, because each plant genotype is found in each environment, any difference in the average plant phenotype across environment is due to the environmental (water) differences alone. The changes in relative size across environments (i.e. Clone 1 is biggest in Environment 1, but smallest in Environment 2) represents a genotype-by-environment interaction. So plant size variation is due to genetic differences, environmental differences and an interaction between the genetic and environmental differences. The same principles apply to any phenotype, including behaviour. (b) This figure shows the same interaction-type across cricket calls where the sonograms above and below the cricket images show the hypothetical songs females of each hypothetical genotype are most attracted to across two imaginary environments. In Environment 1, call rates are slower than in Environment 2 (there is an environmental effect on preferred calls), and each genotype prefers different calls (a genetic effect), but the type of call preferred depends on the environment sampled (gene-by-environment effect).

And if the effects of the genes on the phenotype varied across the two environments (i.e. the biggest genotype in environment 1 is the smallest in environment 2), then we have a genotype-by-environment interaction (we additionally include a hypothetical behavioural example as well; see Figure 1.1). To put that into the simplest terms:

$$P = G + E + GxE \tag{1.1}$$

where P = the phenotype, G = the genotype, E = the environment, and GxE = the interaction between genotype and environment, and this is as true of behaviour as it is of morphology. And if we are talking about variation around average behaviours, then we have:

$$V_P = V_G + V_E + V_{GxE} \tag{1.2}$$

where V_P = phenotypic variation, V_G = genetic variation (averaged over environments), V_E = environmental variation (averaged over genotypes), and V_{GxE} = the variation due to the interaction between G and E. From this, we can estimate the proportion of variation in the phenotype (deviation from the mean) that is due to variation in the genes as the ratio of the genetic variation divided by the phenotypic variation as a whole (V_G/V_P), which is known as the broad-sense heritability (H^2) – a measure of the heredity of a phenotype (Falconer 1981).

Thus, by this simple variance partitioning exercise, we can attribute phenotypic variation into a genetic and an environmental component, and if H^2 = 100% then (ignoring maternal effects, for example) the variation between individuals within a population is all due to variation in genes (i.e. the phenotype equals the genotype) and if it is 0%, then all variation is due to the environment variation (the phenotype does not accurately describe the genotype). Note that these are *local* estimates – they are population, environment(s), and time specific because they depend on the genotype and environment distributions of the population sampled at that point in time. It is equally important to remember that just because a trait has zero heritability *this does not mean it has no genetic component*. Remember that heritability describes variation determinants and, for example, finger number has zero heritability despite being clearly determined by genes because there is (effectively) *no variation* in finger number due to genes – everyone (to a first approximation) has five fingers per hand at birth.

The point of the above is merely to illustrate the relative ease of hypothetically dissecting behavioural variation into genetic, environmental, and interactive effects using standard analysis of variance (ANOVA) (Anholt and Mackay 2004; and see Zar 1999; Sokal and Rohlf 1981), despite the problems ANOVA has with correctly assigning variation for some gene–environment (G/E) relationships (e.g. with some reaction norms ANOVA can fail to detect G or E effects, instead falsely assigning all variation to one effect or another: Lewontin 1974; Figure 1.2). While this is old and obvious for many, the same is

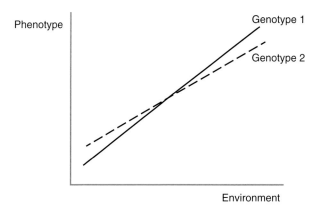

Figure 1.2 An example of how ANOVA can in principle fail to correctly assign phenotypic variation to causal factors. Here, the phenotypic reaction norms for two genotypes (1 and 2) are shown across an environmental gradient. There are clear environmental effects since both reaction norms increase across the gradient, and there are clear genotype effects on the phenotype as the reaction norms of the genotypes differ. However, if all environments were considered equally, there would be no overall effect of genotype because the two genotypes would have the same phenotypic mean across the gradient, while if environments towards the origin were sampled more heavily, an effect of genotype would be detected. Source: Redrawn from Lewontin (1974).

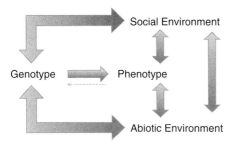

Figure 1.3 The complex interplay between nature and nurture that affects a phenotype like behaviour. Here we can see, for example, that the genes expressed in a focal animal (genotype) contribute to the social environment and the social environments (and the genes of other individuals expressed in it) can influence gene expression in the focal individual (genotype), while both also directly act on the (focal) phenotype, which itself can also affect the social environment and genotype. The latter effect is indicated by the smaller dashed arrow, which indicates epigenetic pathways whereby parental phenotypes can alter offspring gene expression, for example. These same pathways exist between the abiotic environment and genotype/phenotype and there is a link between abiotic and biotic environments – social environments can be affected by abiotic factors, for example (e.g. Simmons and Bailey 1990). Note that the social environment is effectively all non-self genes that are part of the broader environment, and this can include endosymbionts. And for simplicity's sake, we have not included gene–gene interactions, which even with small numbers of genes can be huge (e.g. with 10 genes influencing a trait, there are 180 two-way interactions, 1920 three-way interactions and 3360 four-way interactions. And, for example, the number of four-way interactions with 100 genes increases to 6.3×10^7). Source: Wade (2000).

not true for everyone. For a more thorough explanation of statistical genetics, consult a dedicated text (e.g. Falconer 1981; Roff 1997; Lynch and Walsh 1998; and see Chapter 4), but the take-home message here is that complex traits will be influenced by genes, environment, and their interaction (e.g. Boake 1994; Anholt and Mackay 2004; Hunt and Hosken 2014). This picture is further complicated by gene–gene interactions (Anholt and Mackay 2004, 2015), which rapidly increase as gene number (loci and alleles/locus) increases (Wade 2000), as well as epigenetic feedbacks and a social environment which influences, and is influenced by, genes in the focal organism (Figure 1.3). All of this makes predicting behavioural phenotypes highly probabilistic because the phenotype of each individual will be determined by the unique biochemistry of their genotype and how this interacts with the unique set of environmental and developmental conditions they experience.

So what evidence is there that behaviours are mostly polygenic and that the variation in behaviours can be partitioned as described above in Eq. 1.2? The fact that many behaviours are not discrete but tend to be continuous is indirect evidence (e.g. Sokolowski 2001), and with the advent of cheap sequencing we can now count the numbers of genes that correlate with behavioural variation – for example, more than 250 genes showed twofold expression differences in divergent *Drosophila* lines selected for behavioural differences in geotaxis (Toma et al. 2002). Statistical estimates of the relative contribution of G, E, and GxE to behavioural variation have also been undertaken in a wide range of taxa (Table 1.1), with behaviours tending to have heritabilities that are smaller than for general morphology and on a par with life history traits (reviewed in Mousseau and Roff 1987; Roff 1997). Behaviours studied include mate preference, aggression, dominance, and even personality, and clearly show G and E effects, as well

Table 1.1 A small sample of behaviours that have been explored using statistical genetics approaches to determine the heritability (here narrow-sense heritability: the proportion of variation in a behavioural phenotype explained by the additive action of genes) of behavioural phenotypes. Estimates listed here range from 15% to 71% of the variance explained by genetic effects, with the remaining 85–29% due to environmental and interactive effects.

Behaviour	Species	Estimation method	Heritability estimate
(1) Mate preference	*Drosophila simulans*	AS	0.26 (± 0.11)
(2) Mate preference	*Achroia grisella*	P-O	0.21 (± 0.13)
(3) Attractiveness[a]	*Drosophila simulans*	P-O	0.29 (± 0.15)
(4) Foraging ability	*Panorpa vulgaris*	P-O	0.15 (± 0.05)
(5) Post-mating sociality	*Nauphoeta cineria*	P-O	0.33 (± 0.28)
(6) Exploration	*Parus major*	P-O/F-S/AS	0.22/0.37/0.54
(7) Risk taking	*Parus major*	AS	0.19 (± 0.3)
(8) Dominance	*Pan troglodytes*	AM	0.71 (± 0.01)
(9) Cognitive ability	*Homo sapiens*	TS	0.41–0.66
(10) Boldness	*Ovis canadensis*	P-O[b]	0.21 (± 0.23)

AM, animal model (pedigree); AS, artificial selection; F-S, full-sib; P-O, parent-offspring regression; TS, twin studies.
a)   Includes the sum of all male courtship behaviours.
b)   Mother-offspring regression which includes maternal effects.
Source: (1) Sharma et al. (2010); (2) Jang and Greenfield (2000); (3) Taylor et al. (2007); (4) Missoweit et al. (2007); (5) Moore (1990); (6) Dingemanse et al. (2002) and van Oers et al. (2004); (7) van Oers et al. (2004); (8) Weiss et al. (2000); (9) Haworth et al. (2010); (10) Reale et al. (2000).

as GxE interactions when these have been tested (e.g. Jia et al. 2000; Miller and Brooks 2005; Narraway et al. 2010; Ingleby et al. 2013) (see also reviews in Ingleby et al. 2010; Hunt and Hosken 2014).

Human twin studies report generally similar results – there are clear genetic and environmental effects on behaviour (e.g. Bouchard 2004; Haworth et al. 2010). These studies often follow identical twins that were separated from birth that can then be compared with twins reared in the same home, thus providing essentially the experimental design discussed above (see Figure 1.1). This is because identical twins are genetic clones, and hence similarities across environments (different homes) largely represent the effects of shared genes, for example. These and related pedigree studies find that, as expected, G, E, and GxE all influence a range of human behaviours. And even in cases where genes of large effect, or gene regions that affect behaviour, have been identified, this does not imply strict determinism, for the reasons outlined above.

So for the vast number of studies that have undertaken measurements of behaviour within an appropriate genetic design, it is abundantly clear that many behaviours are determined by genes and environment (and their interaction), and this is true of human behaviour too. We hope (but with some trepidation) that is the end of the false dichotomy of genes versus environment and an end to denials of exclusive genetic underpinnings for behaviour. Does, as appears to have been the fear, the fact that genetic variation underlies behaviour mean that apparently complicated behaviours are in fact deterministic, thereby destroying the notion of human free will and responsibility ('my genes made me do it')? We hope that the above discussion puts this anxiety to rest too.

The fact that there are GxEs in one very real sense means that the effects of genes on behaviour are unpredictable and that the smallest variation in environment can fundamentally alter the effect of genes on behaviours (even ignoring gene-by-gene effects). Understanding that the social environment provided by other members of a society is also continually changing shows that the GxE interaction is also always on the move even if G remains constant (which it will not). Furthermore, development itself, where local developmental-environment and gene feedback occurs to lead from zygote to fully differentiated multicellular organisms, ensures that relationships between genes and outcomes are inherently probabilistic. And as Figure 1.3 shows, the multiple pathways that link genes and phenotype coupled with the environmental effects and all the feedbacks, including inherited epigenetic links (which are just another maternal/paternal effect), really do mean we are dealing with probabilistic rather than deterministic outcomes (for an example of complicated maternal/social-biotic-environment interactions, see Tregenza et al. 2003). Finally, it is hoped that we all now broadly accept that to be human ultimately means rising above the imperative of the genes (Dawkins 1976). So with that out of the way, we can now move on to consider some of the (more interesting) topics discussed here more fully in subsequent chapters.

Acknowledgements

We thank Sasha Dall for valuable feedback on an earlier version of this chapter.

References

Anholt, R.H.R. and Mackay, T.F.C. (2004). Quantitative genetic analyses of complex behaviours in Drosophila. *Nature Reviews Genetics* 5: 838–849.

Anholt, R.H.R. and Mackay, T.F.C. (2015). Dissecting the genetic architecture of behaviour in *Drosophila melanogaster*. *Current Opinions in Behavioral Sciences* 2: 1–7.

Boake, C.R. (ed.) (1994). *Quantitative Genetic Studies of Behavioral Evolution*. Chicago, IL: University of Chicago Press.

Bouchard, T.J. (2004). Genetic influence on human psychological traits: a survey. *Current Directions in Psychological Science* 13: 148–151.

Bucan, M. and Abel, T. (2002). The mouse: genetics meets behaviour. *Nature Reviews Genetics* 3: 114–123.

Dawkins, R. (1976). *The Selfish Gene*. Oxford: Oxford University Press.

Dingemanse, N.J., Both, C., Drent, P.J. et al. (2002). Repeatability and heritability of exploratory behaviour ion wild great tits. *Animal Behaviour* 64: 929–937.

Falconer, D.S. (1981). *Introduction to Quantitative Genetics*, 2e. London: Longman.

Haworth, C.M.A., Wright, M.J., Luciano, M. et al. (2010). The heritability of general cognitive ability increases linearly from childhood to young adulthood. *Molecular Psychiatry* 15: 1112–1120.

Hunt, J. and Hosken, D.J. (eds.) (2014). *Genotype by Environment Interactions and Sexual Selection*. Oxford: Wiley.

Ingleby, F.C., Hunt, J., and Hosken, D.J. (2010). The role of genotype-by-environment interactions in sexual selection. *Journal of Evolutionary Biology* 23: 2031–2045.

Ingleby, F.C., Hunt, J., and Hosken, D.J. (2013). Genotype-by-environment interactions for female mate choice of male cuticular hydrocarbons in *Drosophila simulans*. *PLoS One* 8: e67623.

Jang, Y. and Greenfield, M.D. (2000). Quantitative genetics of female choice in an ultrasonic pyralid moth, *Achroia grisella*: variation and evolvability of preference along multiple dimensions of the male advertisement signal. *Heredity* 84: 73–80.

Jia, F.Y., Greenfield, M.D., and Collines, R.D. (2000). Genetic variance of sexually selected traits in waxmoths: maintenance by genotype x environment interactions. *Evolution* 54: 953–967.

Lewontin, R.C. (1974). The analysis of variance and the analysis of causes. *American Journal of Human Genetics* 26: 400–411.

Lewontin, R.C., Rose, S., and Kamin, L.J. (1984). *Not in Our Genes. Biology, Ideology, and Human Nature*. New York: Pantheon Books.

Lynch, M. and Walsh, B. (1998). *Genetics and Analyses of Quantitative Traits*. Sunderland: Sinauer Associates.

Miller, L.K. and Brooks, R. (2005). The effects of genotype, age, and social environment on male ornaments, mating behaviour and attractiveness. *Evolution* 59: 2414–2425.

Missoweit, M., Engels, S., and Sauer, K.P. (2007). Foraging ability in the scorpionfly *Panorpa vulgaris*: individual differences and heritability. *Behavioural Ecology and Sociobiology* 61: 487–492.

Moore, A.J. (1990). The inheritance of social dominance, mating behaviour and attractiveness to mates in male *Nauphoeta cinerea*. *Animal Behaviour* 39: 388–397.

Mousseau, T.A. and Roff, D.A. (1987). Natural selection and the heritability of fitness components. *Heredity* 59: 181–198.

Narraway, C., Hunt, J., Wedell, N., and Hosken, D.J. (2010). Genotype by environment interactions for female preference. *Journal of Evolutionary Biology* 23: 2550–2557.

van Oers, K., Drent, P.J., de Goede, P., and van Noordwijk, A.J. (2004). Realized heritability and repeatability of risk-taking behaviour in relation to avian personalities. *Proceedings of the Royal Society of London Series B* 271: 65–73.

van Oers, K., de Jong, G., van Noordwijk, A.J. et al. (2005). Contribution of genetics to the study of animal personalities: a review of case studies. *Behaviour* 142: 1185–1206.

Reale, D., Gallant, B.Y., LeBlanc, M., and Festa-Bianchet, M. (2000). Consistency of temperament in bighorn ewes and correlates with behaviour and life history. *Animal Behaviour* 60: 589–597.

Roff, D.A. (1997). *Evolutionary Quantitative Genetics*. London: Chapman and Hall.

Rostant, W.G., Kay, C., Wedell, N., and Hosken, D.J. (2015). Sexual conflict maintains variation at an insecticide resistance locus. *BMC Biology* 13 (1): 34.

Rostant, W.G., Bowyer, J., Coupland, J. et al. (2017). Pleiotropic effects of DDT resistance on male size and behaviour. *Behavior Genetics* 47: 449–459.

Sharma, M.D., Tregenza, T., and Hosken, D.J. (2010). Female mate preference in *Drosophila simulans*: evolution and costs. *Journal of Evolutionary Biology* 23: 1672–1679.

Simmons, L.W. and Bailey, W.J. (1990). Resource influenced sex roles of zaprochiline tettigoniids (Orthhoptera: Tettigoniidae). *Evolution* 44: 1853–1868.

Smith, D.T., Hosken, D.J., Rostant, W.G. et al. (2011). DDT resistance, epistasis and male fitness in flies. *Journal of Evolutionary Biology* 24: 1351–1362.

Sokal, R.R. and Rohlf, F.J. (1981). *Biometery*, 2e. San Francisco: W.H. Freeman & Co.

Sokolowski, M.B. (2001). *Drosophila*: genetics meets behaviour. *Nature Reviews Genetics* 2: 879–890.

Taylor, M.L., Wedell, N., and Hosken, D.J. (2007). The heritability of attractiveness. *Current Biology* 17: R959–R960.

Toma, D.P., White, K.P., Hirsch, J., and Greenspan, R.J. (2002). Identification of genes involved *in Drosophila melanogaster* geotaxis, a complex behavioral trait. *Nature Genetics* 31: 349–353.

Tregenza, T., Wedell, N., Hosken, D.J., and Ward, P.I. (2003). Maternal effects on offspring depend on female mating pattern and offspring environment in yellow dung flies. *Evolution* 57: 297–304.

Wade, M.J. (2000). Epistasis as a genetic constraint within populations and an accelerant of adaptive divergence between them. In: *Epistasis and the Evolutionary Process* (ed. J.B. Wolf, E.D. Brodie and M.J. Wade). Oxford: Oxford University Press.

Weiss, A., King, J.E., and Figueredo, A.J. (2000). The heritability of personality factors in chimpanzee (Pan troglodytes). *Behavior Genetics* 30: 213–221.

Zar, J.H. (1999). *Biostatistical Analysis*, 3e. Upper Saddle River: Prentice Hall.

2

Ultimate (Re)Thinking for Behavioural Biology

Sasha R. X. Dall[1], John M. McNamara[2] and Alastair J. Wilson[1]

[1] *Centre for Ecology & Conservation, University of Exeter, Penryn Campus, Penryn, TR10 9EZ, UK*
[2] *School of Mathematics, University of Bristol, Bristol, BS8 1TW, UK*

How organisms allocate resources over their lifetimes, including the deployment of morphological and physiological investments, determines their individual contributions to the differential persistence of their lineages over evolutionary time and, thus, their Darwinian fitness (Coulson et al. 2006). Since such life history strategies are often manifest in individual patterns of action and activity, this arguably places the study of behaviour at the heart of evolutionary biology. Nevertheless, behaviour is one of the hardest classes of phenotypic traits to quantify, and therefore study empirically, precisely because it is a manifestation of how the rest of the phenotype interacts with its environment. As such, individual behaviour is a continuous stream of (often multivariate) events that must be abstracted from and quantized according to the scientific goals of any given study (hence the importance in defining behavioural endpoints – see Chapter 6). It is therefore imperative to specify the logic of behavioural hypotheses as rigorously as possible, right from the outset of any programme of research (Martin and Bateson 2007).

It has long been recognized that there are multiple, complementary levels at which it is possible to hypothesize about organismal traits, given the centrality of Darwinian processes in biology (Huxley 1942). Ernst Mayr (1961) articulated perhaps the most basic distinction between thinking about the processes that give rise to the trait in its specific, current incarnation and reasoning about the factors responsible for its evolutionary origin and maintenance; termed proximate and ultimate explanations, respectively. While the utility of this dichotomy has been debated (e.g. Laland et al. 2011), its basic heuristic value remains widely appreciated in much of the biological sciences (Nesse 2013). In essence, it is possible that any particular explanation for how or why a trait evolved can both be true *and* there be a range of potential explanations for how the trait works during an organism's lifetime, or vice versa. In other words, for any given explanation for a trait at one level, a range of explanations at other levels are feasible. Recognizing this is important to avoid placing explanations that act at different levels in opposition to each other scientifically. Furthermore, it is important to explicitly recognize that understanding the details of how a trait is controlled physiologically, for instance, does not obviate the need to explain its evolution and maintenance, or vice versa. Acknowledging that there are links between a trait's operational details and its

Genes and Behaviour: Beyond Nature-Nurture, First Edition.
Edited by David J. Hosken, John Hunt and Nina Wedell.
© 2019 John Wiley & Sons Ltd. Published 2019 by John Wiley & Sons Ltd.

evolution (even if they involve feedbacks: Laland et al. 2011) does not change such basic heuristic value of the proximate-ultimate distinction.

Nevertheless, Niko Tinbergen (1963) recognized that further refinement of the proximate-ultimate taxonomy was necessary when studying behaviour. This is because what an organism does over any given period of study will be the product of processes that can be heuristically thought of as operating over different timescales. This is most apparent for processes that operate over the lifetimes of the particular organisms under scrutiny, since behaviour emerges as an organism's morphology and physiology interact with its environment and so is relatively plastic. Hence, Tinbergen reasoned that proximate explanations could be further subdivided into reasoning about processes that stimulate the behaviour (control the specific phenotype–environment interaction) on a moment-by-moment basis and those that underpin the development of behavioural responses (the aforementioned control processes) over the organism's lifetime. These have come to be known as causal (or mechanistic) and developmental (or ontological) explanations, respectively. Equivalently, Tinbergen also reiterated (following Huxley 1942) that hypotheses about so-called ultimate factors underpinning behaviour could be usefully distinguished according to the timescale over which the putative evolutionary processes involved operate.

In this way, it has often been useful to distinguish the contemporary selection pressures operating on a behaviour in a population (i.e. those responsible for its evolutionary maintenance), generating what have become known as functional explanations, from factors that explain the macroevolutionary trajectories of behavioural traits, or so-called phylogenetic explanations (see Box 2.1 for an illustrative example of distinguishing 'Tinbergen's 4 Whys').

Box 2.1 Tinbergen's Four Whys: An Example

Here we provide an example of how to simultaneously hypothesize about a behaviour – male starling singing – at all four of the different levels proposed by Niko Tinbergen (1963). As is traditional, it is useful to posit reasoning at the different levels as answers to a question; in this case: **'Why does a male starling sing?'**

- *Proximate answers.* These will involve hypothesizing about processes operating over its lifetime, which explain why or how any given male starling (or set of males) will have come to sing rather than express any other behaviour (or remain inactive).
  - *Causal (mechanistic) answer.* A male starling might be stimulated to sing because another male sings nearby or is seen approaching its territory.
  - *Developmental (ontological) answer.* Increased day length (from short days: ~8 hours of daylight) may stimulate male starlings to grow their testes, which increases circulating levels of testosterone. High levels of circulating testosterone 'primes' the song centres in male bird brains, which may induce males to respond to perceived sexual competitors by singing.
- *Ultimate answers.* These will involve hypothesizing about the evolutionary processes responsible for the tendency of males in starling lineages to sing.
  - *Functional answer.* A male starling may have evolved to sing in the appropriate context because this deters other males from encroaching on its territory and mating with any females that are around.

- *Phylogenetic answer*. Starlings are members of the avian lineage, which evolved in forested habitats where visual signals do not propagate over long distances. Bird ancestors were arboreal and so lived in complex three-dimensional habitats where substrate-based (e.g. chemical) signalling was of limited value. Acoustic signals, on the other hand, offered an effective means of long-distance communication. This could be why male starlings sing rather than signal in other ways to deter males from approaching their territories.

However, while behavioural variation is often underpinned by genetic factors, and evolution necessarily involves genetic inheritance, patterns of behavioural inheritance do not feature explicitly in the phrasing of 'Tinbergen's 4 Whys' in either proximate or ultimate contexts. In the latter, for instance, the role of contemporary Darwinian selection in providing 'functional explanations' is central, yet a key tenet of the modern synthesis is that adaptive phenotypic evolution requires both selection and genetic variation. This was known to behavioural researchers at the time. For example, Fuller and Thompson (1960) advocated applying quantitative genetic (QG) methods from animal breeding to test behavioural heritabilities, while Tinbergen's own work emphasized the importance of putatively genetically based factors leading to innate behaviours and fixed action patterns (Tinbergen 1951). Nonetheless, details of the genetic inheritance of behaviour were not central to ethology at the time and nor, at least in an empirical sense, have they been integral to the development of behavioural ecology subsequently. In fact, the success of the latter field has stemmed largely from the ability to generate evolutionary insights about behaviour without explicit genetic knowledge. Thus, behavioural ecologists have investigated diverse behaviours expressed by myriad species in their natural habitats. In contrast, behavioural geneticists have – until recently – been more constrained, focusing attention on a few behaviours expressed in model organisms under laboratory conditions.

As this edited volume makes clear, the historical separation between behavioural ecology and genetics is no longer a boundary to researchers in either field (see Chapters 6, 7, and 8). With genetic and genomic tools and data now routinely used to study behaviour (see Chapter 11), it is time to revisit Tinbergen's questions to ensure they remain relevant and useful for a new generation of researchers. In this chapter, we will focus our attention on reasoning from the ultimate, evolutionary perspectives and not say much more regarding proximate hypothesizing about behaviour. This should in no way be taken to imply that we view the latter as somehow subordinate to ultimate approaches (we are committed Tinbergenians after all – e.g. McNamara and Houston 2009); the focus on ultimate reasoning reflects our expertise, as well as the topic of this edited volume.

2.1 Evolutionary Reasoning in Modern Behavioural Biology

There have been a number of re-evaluations of Tinbergen's explanatory framework over the years, particularly in 2013 to mark the 50th anniversary of the '4 Whys' paper (e.g. Barrett et al. 2013; Bateson and Laland 2013; Laland et al. 2013; Nesse 2013).

Most have focused on whether and/or how modern insights should refine the way that explanations at the different levels are formulated. From the perspective of ultimate hypotheses about behaviour, a number of such updates have focused on issues associated with studying contemporary selection pressures, or the functional explanations for behaviour. This is largely driven by the influence of fields such as behavioural ecology, which have specialized in studying behaviour from this perspective. For instance, Cuthill (2005) begins by reviewing some of the philosophical issues with studying how behaviour functions in contemporary contexts, given that any evolutionarily causal processes are, by definition, historical. But he then goes on to focus on discussing how modern approaches and techniques have allowed the so-called 'fitness value' of trait variation to be measured so as to increasingly approximate theoretically justified individual-level proxies for Darwinian fitness (e.g. reproductive value: Houston and McNamara 1999). Most functional hypotheses focus on specifying the links between behavioural variation and fitness (to predict the behaviour that functions best), which places a premium on measuring 'Darwinian fitness' as accurately as possible. Nevertheless, it is important not to neglect the complexities associated with deducing the trait values that maximize fitness (or whichever proxy is empirically accessible) in any given context.

Ostensibly, it seems straightforward to identify what the best thing to do is in most contexts for many organisms. For instance, surely maximally fit animals will *only* eat the most profitable food – that gains them the most nutrition at the lowest cost – they encounter? Well, not when costs are accounted as time taken to consume food items, it turns out (MacArthur and Pianka 1966; Emlen 1966). It is possible to maximize the long-term rate at which nutrition is obtained (a likely proxy for fitness while foraging: Stephens and Krebs 1986) by eating less than maximally profitable food (in terms of energy gained per unit time spent consuming it) when it is encountered if it takes longer to find more profitable food than it takes to consume the mediocre fare to hand. Generalizing such logic suggests that animals should be selected to specialize on the most profitable foods in rich environments (where such foods are abundant) but opportunistically consume less profitable food as the availability of the best nutrition declines in poorer habitats, a functional prediction that is widely corroborated (Sih and Christensen 2001). However, it is only by exposing such evolutionarily economic logic mathematically that counterintuitive yet empirically verified insights, such as the availability of some (less profitable but edible) food should not influence an animal's decision to consume it (Stephens and Krebs 1986), become apparent.

Thus, even specifying functional hypotheses for behavioural responses to some of the simplest ecological problems that animals face has benefitted substantially from formal (mathematical and computational) modelling (Kokko 2007). Indeed, exposing the logic of hypothesizing about selection on behaviour driven by the more complex (and common) ecological problems that vary over the lifetimes of animals (requiring dynamic solutions: Houston and McNamara 1999) or influencing selection on other organisms (requiring game-theoretic solution concepts: McNamara 2013) is only possible via formal modelling (Kokko 2007).

As far as modernizing broader ultimate reasoning is concerned, Bateson and Laland (2013), for instance, mainly focus their anniversary update to Tinbergen's framework on detailing how recent findings about the widespread existence of social learning and other 'extragenetic' inheritance mechanisms (e.g. parental effects, imprinting, epigenetic effects on gene expression) influence hypotheses about function and phylogeny (as well as the proximate levels of Tinbergenian enquiry). Notwithstanding both the importance of and inherent interest in uncovering the detail of transmission mechanisms, we argue that there is still much to be done to integrate genetic (let alone epigenetic!) understanding. To this end, we focus our attention on how widely used quantitative genetic approaches to formalizing the evolutionary process (over the short term) can fit into the Tinbergen framework.

Our main motivation is to highlight that using such approaches facilitates our ability to specify precise hypotheses about how behaviour is likely to evolve within populations over the short term. As well as being important in its own right, this in turn has the potential to enhance insights from comparative and phylogenetic approaches in behavioural biology by offering a formal framework for linking ideas about function (selection pressures) to short-term evolutionary trajectories (since evolution within populations is typically the driver of divergence among populations). Such bridging between micro- and macroevolutionary views of behavioural variation will be challenging but important work. Contemporary approaches to studying phylogeny largely rely on inferring how traits have evolved in the past (along with patterns of lineage branching and extinction) from piecemeal information based on molecular genetic, geological, and extant phenotypic data (Cornwell and Nakagawa 2017). However, a broader, modern rationale for understanding behavioural evolution should arguably not only include elucidation of how and why traits came to exist in their current forms, but also prediction of if, when, and how they will continue to evolve under contemporary selection.

2.2 A Quantitative Genetic View of Behavioural Evolution

The proximate-ultimate dichotomy of interest outlined above is a strong feature of research in behavioural genetics, just as it is in behavioural research more widely. From a proximate perspective, functional and developmental genetic and genomic studies have long employed a wide array of techniques to identify specific genetic loci with influence over particular aspects of behavioural expression (see Chapters 5 and 6). However, evidence from this work, coupled more recently with results from genome-wide association studies in humans (Chabris et al. 2015) and rodents (e.g. Parker et al. 2016), leads us to expect that most behaviours of ecological significance will have a complex genetic architecture (see Chapter 1). That is, they will rarely be 'Mendelian' traits (i.e. determined by one or a few genes with major effect) and are expected to depend on an unknown (but large) number of loci with small effects. In general, this means that a detailed understanding of the mapping of sequence variation at individual loci to its phenotypic consequences may offer little insight into contemporary trait evolutionary dynamics. Of course, exceptions to prove the rule are inevitable, and studies scrutinizing the genetic basis of evolved behavioural differences between populations or species provide key macroevolutionary insights (e.g. Bendesky et al. 2017). However, to the extent that the focus is on microevolution, a proximate understanding of genetic processes may not contribute much to an ultimate understanding (beyond helping to identify targets for selection in behavioural studies; see below for further discussion).

Fortunately, the field of quantitative genetics (QG) offers a theoretical framework, and a set of broadly applicable statistical methods, to study phenotypic evolution of genetically complex traits (Falconer and Mackay 1996) (see Chapter 4). In simple terms, classic quantitative genetic approaches utilize phenotypic data and knowledge of relationships or relatedness (potentially inferred from molecular or genomic data, e.g. Bérénos et al. 2014) to determine levels of genetic variance for traits in a population. If closely related individuals (which by definition are genetically similar) are phenotypically similar then, subject to controlling experimentally or statistically for possible confounding effects, we can conclude that the trait of interest has a high degree of genetic determination.

A host of statistical methods can be used to estimate genetic variance, but many are applicable to only some data structures (e.g. analysis of variance (ANOVA) is only useful for analysing data from controlled breeding designs). Linear mixed effect models and, in particular, the so-called 'animal model' have emerged as the preferred method in evolutionary ecology over recent decades due to their greater power and flexibility relative to older techniques (Wilson et al. 2010). Regardless of parameter estimation method, an appealing aspect of quantitative genetics is the availability of theoretical models such as the breeder's equation (Lush 1937; Lerner 1958) and Lande's (1979) multivariate analogue to predict how traits will evolve (in the short term) under selection. These models allow us to combine estimates of genetic variance, and in a multivariate context also the among-trait covariance (expected to be non-zero if, for instance, two traits are influenced by common genetic loci) with estimates of the strength and direction of selection.

In the current context, two key points are worth noting. First, quantitative genetic approaches are genetically naïve – just less so than traditional adaptationist approaches that (implicitly or explicitly) adopt the 'phenotypic gambit' (PG) (Grafen 1984), which can be defined as hypothesizing about the evolutionary basis of a trait as if it was inherited via the simplest conceivable genetic system (e.g. a haploid locus: Grafen 1991). Under the PG, patterns of phenotypic variation are thus assumed to be valid proxies for underlying genetic variation. Such approaches maximize broad applicability by assuming that natural selection is uniformly effective. In simplistic terms (and with some important exceptions, discussed below), PG assumes an absence of genetic constraints on evolution. Accepting this position allows genes underpinning phenotypes to be ignored; understanding adaptation becomes a matter of determining the fitness costs and benefits of particular phenotypes (i.e. the nature of selection) and deducing (from functional and physicochemical constraints) which phenotypic states have the highest fitness. Although quantitative genetics rejects the validity of this premise in general, it replaces total ignorance of genetics with an assumption that a simple model of trait genetic architecture will suffice. Specifically, it assumes the 'infinitesimal model' in which a trait is influenced additively by an infinite number of loci of infinitely small effect (and that selection is weak). This allows inheritance of any trait in any population to be modelled using population-level statistical parameters (e.g. heritability). Whether the infinitesimal model is strictly 'true' is not in question. Clearly it can't be as, for example, genomes do not contain infinite numbers of genes and non-additive processes (e.g. epistasis) are well documented. However, the pertinent question is whether it provides a sufficiently reasonable approximation that we can make valid inferences about trait evolutionary change accounting for both selection and genetics (see Box 2.2 for examples contrasting the PG and QG approaches to analysing behaviour from an evolutionary perspective).

Box 2.2 'Phenotypic Gambit' (PG) vs 'Quantitative Genetic' (QG) Approaches to the Evolution of Reproductive Behaviour

Consider two parents caring for their common young. Focus on the parental effort of each.

- From a QG perspective, the effort of a parent is regarded as a trait that can, if heritable, evolve under selection. Typically, this trait, or a proxy of it, would be directly observable, for instance as the rate of nest provisioning. However, the effort of one parent might also be influenced by the effort of its partner, and so by that partner's genotype. In other words, variation among individual parents in the observed trait will arise through both direct and indirect genetic effects (IGEs). Assuming effort is costly, it would actually be selected against through parental (direct) fitness, but selected for through offspring (direct) fitness (e.g. higher parental effort leading to higher offspring survival).
- In contrast, a behavioural ecologist adopting the PG would not really regard parental effort itself as under selection. Rather, it is the rule for adjusting own effort in response to partner's effort that is under selection (the choice of effort might also be influenced by other variables such as nutritional status or care ability: McNamara et al. 1999). The actual effort is just an outcome of following this (potentially multidimensional) rule. Of course, in hypothesizing that the response rule might be optimal (evolutionarily stable), one would have to account for why in past generations partners might have had a variety of efforts. This could be due to genetic variation or experience during development (e.g. an individual had a disease that rendered it less good at care), or a combination of both.

This example highlights one notable feature of the contrasting approaches. In QG, anything can be modelled as a trait. Although latent traits are sometimes considered, in practice it is more common to focus on directly observable behaviours. In this example, the trait would be observed parental effort, the degree of genetic determination of which can vary. In the simpler case where there are no IGEs, this variable extent of genetic determination for a trait is expressed by the heritability (which can take any value from 0 to 1). For a PG modeller, the only interesting traits are the biological mechanisms that underpin observed behaviour (e.g. physiological processes that are more directly coded for molecular-genetically). So in this example, it would be the rule for choosing parental effort that is viewed as the target of selection, and therefore the trait of interest. In the context of developing evolutionary theory to explore general phenomena, the rule would usually be treated as being (fully) genetically determined (e.g. inherited asexually).

Timescale and Predicting the Future

In some sense, we can distinguish theoretical analyses to: (i) understand past evolutionary forces that have shaped present phenotypes, (ii) explain current (immediate) phenotypic change, and (iii) predict more long-term evolutionary change. The PG is perhaps the obvious tool to deal with (i), whereas QG is more appropriate for (ii). But what about (iii)? How far into the future can either approach take us?

(Continued)

Box 2.2 (Continued)

Let us focus on a specific example. Consider how the reproductive strategy, and in particular the lay date, of a small passerine bird is affected by the advancement of spring under climate change. Before looking at the effect of climate change, consider the lay date under the reproductive strategy that evolved before climate change. Hypothetical empirical work shows that the earlier the lay date, the greater the reproductive success from the current brood. Thus, any PG model that just looked at lay date as the decision variable would conclude that there is selection for advancement in lay date. Similarly, a QG analysis that took the trait as lay date might find it was heritable and estimate a linear selection gradient, thus coming to the conclusion that the mean lay date should advance. There has, however, been such 'selection pressure' for many generations without any evidence of advancement in lay date.

Both ways of analysing the situation got it wrong because they specified the trait under selection too simplistically. It seems likely that the birds need to get themselves into good condition before they can breed. Furthermore, their decision of when to lay may also depend on the temperatures experienced in early spring. There are obvious functional reasons why we might expect this. For example, for great tits, higher than average temperatures in early spring may indicate early emergence of caterpillars that the parents use to feed young. Given the above influences, a strategy is not specified by a lay date but by a norm of reaction that specifies whether to lay now as a function of current condition, temperatures experienced, and time of year (as cued by photoperiod). A PG modeller would approach this situation by specifying the statistics of temperature variation, for example the correlation between temperatures experienced so far and those in the near future. One would then specify reproductive success as a function of body condition, early spring temperatures, and lay date. One would also need to specify the dynamics of condition – how it increased over time. It would then be possible to deduce the optimal norm of reaction.

A QG modeller would see this primarily as a missing trait problem – a univariate (lay date by itself) model won't work if genetically correlated traits (e.g. lay date and condition) are both under selection, and there is likely genetic covariance between them (assuming condition has genetic variance and is correlated with lay date, which is likely given the causal dependence of laying on condition). There is an additional layer of complexity if both genetics and selection are temperature sensitive, although this does not necessarily mess up the QG prediction unless environmental variation drives a relationship between heritability and selection. A QG model would also include genetic variation for reaction norm parameters beyond the intercept (specified as a gene-by-environment interaction, or GxE) and environment-dependent selection. Thus, a central message here is that whichever method you use, you will get it wrong unless you have good understanding of the underlying biology. QG might (more often) start from the phenomenological trait of interest and work backwards, adding more complexity underpinning that trait to see if it helps, while PG models focus a level below the observed behavioural expression to start with.

So now suppose that we have advancement of spring under climate change. The immediate phenotypic effect may be a change in lay date under the original plastic response. However, the norm of reaction is no longer liable to be adaptive and we would expect to see gradual evolutionary change in this reaction norm. The PG approach has nothing to say about the rate and immediate shape of this change, since these will depend on the underlying mechanisms that are used to control lay date, how these are coded for genetically, and the level of standing genetic variation. Nevertheless, this seems an ideal job for QG, contingent on GxE (genetic variance in plasticity). Of course, even QG will get it wrong unless the right biology is included.

In contrast, both approaches run into problems with respect to analysing long-term evolutionary change. QG runs into trouble here as it typically assumes that the genetic architecture underlying the trait (G-matrix) remains constant (i.e. current constraints/adaptive potential are invariant). Yet there is evidence that G-matrices evolve (e.g. Careau et al. 2015), and so genetic constraints on adaptive evolution are unlikely to persist in the long term (see text for further discussion). The PG could work provided that there are no major genetic constraints. However, there are things that could derail this, particularly if the ecological fundamentals of the underlying functional trade-offs change. For example, it might be that some other cue in early spring becomes informative, so that the reaction norm is selected to respond to this. The change in climate might be accompanied by an increase in population density, so that competition for nest sites and food changes, etc.

All these complications are potentially surmountable if you have the insight and imagination to have anticipated them! But of course, the world is too complex for it to be reasonable to anticipate everything.

Second, and in contrast to a number of other modelling approaches in evolutionary biology, quantitative genetics concerns itself with predicting evolutionary change over a generational time step (Kokko 2007). Thus, it does not seek to predict long-term outcomes in the sense of, for example, solving for evolutionarily stable strategies (or other evolutionary equilibria). Nor does it seek to explain current phenotypic distributions as the outcome of adaptation to Darwinian selection. In fact, the view of evolutionary quantitative genetics is that genetic (co)variances structures among traits (described in the form of a so-called G-matrix) can and do constrain trait responses to contemporary selection in important ways (Blows and Walsh 2009), despite ultimately being a product of evolution themselves. Thus, as applied to natural populations under natural selection, quantitative genetic studies assume a position of agnosticism with regard to whether trait distributions should be (approximately) optimal (e.g. evolutionarily stable) or not.

It is worth noting that while G determines the response to contemporary selection, it ultimately reflects allelic variation at (unknown) loci influencing the measured traits. As evolution involves changes in allele frequencies, it is clear that G can and must evolve. Although efforts to understand this process have met with limited success to date (Arnold et al. 2008), some qualitative expectations are clear. For instance, under strong directional or stabilizing selection on a single trait, we expect genetic variance to decline as standing variation in a population is lost (alleles are purged if deleterious or fixed if beneficial). Similarly, selection favouring particular trait combinations will change the genetic covariance structure among traits. In this way, the structure of G is both an outcome of historic selection (and drift) and the determinant of contemporary response to selection. An important assumption of quantitative genetic prediction is that G can be treated as effectively constant for the timeframe (i.e. number of generations) over which phenotypic change is being predicted.

2.3   Short-Term Ultimate Reasoning: Behavioural Genetics in a Functional Context

While QG and PG approaches can sometimes yield conclusions that appear divergent (see Box 2.2), it is important to remember that theoretical models in the two frameworks have subtly different objectives (e.g. predicting evolutionary equilibria versus short-term trait dynamics), while empirical studies make different assumptions to draw inferences. So if they are better treated as complementary, not competing, tools for tackling 'ultimate' questions, what can behavioural biologists take from QG studies to date?

To highlight just three examples: first and most fundamental is that we now know that behavioural traits are generally heritable. This is true in laboratory populations and, where estimates have been made *in situ*, in the field (Dochtermann et al. 2014). This conclusion is trivial in one sense – if behaviours had no genetic underpinnings, they could not evolve. However, it is also an important confirmation of long-held assumptions, and highlights the fact that, while responsive to the environment, behavioural traits are not 'infinitely plastic' in the sense that all behavioural variation can be explained by environmental influences. This in turn often motivates the question of what maintains 'personality' variation (Dall et al. 2004; Dingemanse and Wolf 2010; Wolf and Weissing 2010; Dall and Griffith 2014), which is typically defined as among-individual behavioural variation that is stable within-(at least some) individuals within a population. Since selection is not (necessarily) expected to erode environmentally induced variance, interest in adaptive explanations for the maintenance of personality variation often stems from an implicit assumption that there is a genetic component to differences among-individuals (but see Dall et al. 2004; Sih et al. 2015).

A second emerging pattern from QG studies is that simple bivariate trade-offs may not shape behavioural and life history evolution as ubiquitously as thought. That individual organisms are constrained by resource allocation trade-offs is a starting point for many PG-driven (and other) evolutionary models, and one that is often possible to validate experimentally. However, in postulating that such trade-offs shape evolutionary trajectories, it is necessary to invoke negative genetic correlations between traits, or trade-offs among fitness components (defined so as to be under positive selection) – for example, between vigilance and feeding, reproductive investment and growth, or, in the limiting case, survival and fecundity. Predictions are sometimes upheld, of course, but at least in some wild vertebrate populations, positive genetic correlations are much more common (Kruuk et al. 2008), suggesting an axis of genetic variation in overall 'quality' (defined as expected fitness given multivariate phenotype) that is not easy to explain from either a QG perspective (but see Wilson 2014 for one attempt) or traditional functional analyses of simple resource trade-offs.

Finally, a third area where recent developments in QG could contribute usefully to behavioural research is in social evolution. Motivated by Hamilton's theoretical work (Hamilton 1964) and the use of 'inclusive fitness' models to resolve the apparent paradox of altruistic behaviour, investigations of relatedness have assumed particular prominence in this field (Abbot et al. 2011). However, from a quantitative genetic perspective, evolution occurs because of non-zero genetic covariance between a trait and (relative) fitness. Indeed this covariance *is* the predicted evolutionary change of the trait according to the Robertson–Price identity/Price equation (Robertson 1966; Price 1970) and can be directly estimated by, for example, modelling trait and fitness as response variables in a bivariate animal model (Morrissey et al. 2012). By extension, social evolution requires genetic covariance between a trait expressed in one individual (the actor) and the (direct) fitness of another (the recipient). This covariance can occur because actor and recipient are related by descent (as captured by Hamilton's rule), but can also arise as a consequence of indirect genetic effects (IGEs) (Bijma and Wade 2008). IGEs are defined as occurring when the phenotype expressed by one individual is causally dependent on the genotype of another. By ignoring the genetic basis of phenotypic variation, behavioural biologists have to date largely missed exploring the short-term dynamics of social evolution. Indeed, recent empirical work has found evidence for IGEs on many socially expressed behaviours (and/or the traits they influence) across contexts ranging from aggression and social dominance (Wilson et al. 2009) to mating behaviours (Brommer and Rattiste 2008), and parental care (Head et al. 2012), including alloparental helping (Charmantier et al. 2007).

In principle, combining IGE models with game theoretic approaches to specifying selection pressures in social contexts should facilitate the development of testable, quantitative hypotheses about social evolution. This would go a long way to addressing a common criticism of PG-based theoretical analyses (discussed in Evans et al. 2013; Kokko 2007) as being of limited use empirically.

2.4 Concluding Remarks

It is clear that explicitly considering simplified models of trait inheritance such as the 'infinitesimal model' of QG can enrich ultimate functional approaches in behavioural biology. At the very least, doing so helps identify or corroborate constraints assumed in PG models given the right kind of datasets (e.g. evidence for simple bivariate trade-offs discussed above); optimization approaches (including game theory models) only provide insight insofar as the relevant constraints have been accurately identified since they predict strategies that are optimal (evolutionarily stable) given the constraints on the system (Stephens and Krebs 1986; McNamara 2013). Furthermore, empirical QG approaches could in principle allow selection pressures to be estimated from data (given patterns of inheritance and observed evolutionary change), which might offer another angle to test predictions from PG-style analyses. However, this will only be possible if the issue of what the behavioural 'traits' are that are actually evolving is carefully resolved (see Box 2.2 and Fawcett et al. 2013 for detailed discussions of this issue).

As far as linking micro- and macroevolutionary dynamics and thereby function and phylogeny, it is less clear how to proceed. At first glance, it seems feasible to deduce details of selection (strength and direction) from PG approaches to input into the appropriate quantitative genetics models to generate detailed evolutionary projections for behavioural traits. However, in QG problems in crossing the micro–macro divide stem largely from the fact that it is not considered 'safe' to assume stability of G-matrices over evolutionary time periods of the sort that typically separate, for example, species. The question of how stable G is, or might usefully be assumed to be, has proven difficult to answer theoretically from a QG perspective (see Turelli 1988 and wider discussion in Arnold et al. 2008). Nonetheless, matrix comparison methods and tools (e.g. Hohenlohe and Arnold 2008; Melo et al. 2016) are providing increasing traction for empiricists attempting to resolve this issue. We now know that G matrices, at least for morphological and life history traits, can sometimes be quite similar among conspecific natural populations that have experienced quite different environments (e.g. Delahaie et al. 2017). However, generalizing this and similar results may be dangerous since, for example, in a recent study of long-term selection response (31 generations), Careau et al. (2015) found that artificial selection on wheel running behaviour in mice induced appreciable changes in G. In this case, selection altered G in such a way as to exacerbate constraints and reduce adaptive potential.

Finally, an alternative approach to specifying evolutionarily stable G-matrices might come from focusing PG-style analyses on the evolution of gene–gene interactions and attempting to predict evolutionarily stable genetic architectures from functional first principles (details of the ecology and resource trade-offs). For instance, by assuming that selection acts on developmental (and other phenotypic determination) systems to generate phenotypes that track (and match) ecological (including social) conditions, it is possible to specify the conditions under which expression of developmental system loci should be influenced by other loci that can act as cues of local conditions because they are differentially selected in different environments (Dall et al. 2015; McNamara et al. 2016). Nevertheless, whether it is possible to use this kind of approach to predict genetic architecture underpinning a wide range of behavioural traits (beyond simple life history switches) remains an open question.

References

Abbot, P., Abe, J., Alcock, J. et al. (2011). Inclusive fitness theory and eusociality. *Nature* 471 (7339), E1-4-10.

Arnold, S.J., Bürger, R., Hohenlohe, P.A. et al. (2008). Understanding the evolution and stability of the G-matrix. *Evolution* 62 (10): 2451–2461.

Barrett, L., Blumstein, D.T., Clutton-Brock, T.H., and Kappeler, P.M. (2013). Taking note of Tinbergen, or: the promise of a biology of behaviour. *Philosophical Transactions of the Royal Society B: Biological Sciences* 368 (1618): 20120352.

Bateson, P. and Laland, K.N. (2013). Tinbergen's four questions: an appreciation and an update. *Trends in Ecology & Evolution* 28 (12): 712–718.

Bendesky, A., Kwon, Y.M., Lassance, J.M. et al. (2017). The genetic basis of parental care evolution in monogamous mice. *Nature* 544 (7651): 434–439.

Bérénos, C., Ellis, P.A., Pilkington, J.G., and Pemberton, J.M. (2014). Estimating quantitative genetic parameters in wild populations: a comparison of pedigree and genomic approaches. *Molecular Ecology* 23 (14): 3434–3451.

Bijma, P. and Wade, M.J. (2008). The joint effects of kin, multilevel selection and indirect genetic effects on response to genetic selection. *Journal of Evolutionary Biology* 21 (5): 1175–1188.

Blows, M. and Walsh, B. (2009). Spherical cows grazing in Flatland: constraints to selection and adaptation. In: *Adaptation and Fitness in Animal Populations* (ed. J. van der Werf, H. Graser, R. Frankham and C. Gondro), 83–101. Dordrecht: Springer Netherlands.

Brommer, J.E. and Ratisse, K. (2008). "Hidden" reproductive conflict between mates in a wild bird population. *Evolution* 62 (9): 2326–2333.

Careau, V., Wolak, M.E., Carter, P.A., and Garland, T. Jr., (2015). Evolution of the additive genetic variance – covariance matrix under continuous directional selection on a complex behavioural phenotype. *Proceedings of the Royal Society B: Biological Sciences* 282 (1819): 20151119.

Chabris, C.F., Lee, J.J., Cesarini, D. et al. (2015). The fourth law of behavior genetics. *Current Directions in Psychological Science* 24 (4): 304–312.

Charmantier, A., Keyser, A.J., and Promislow, D.E. (2007). First evidence for heritable variation in cooperative breeding behaviour. *Proceedings of the Royal Society B: Biological Sciences* 274 (1619): 1757–1761.

Cornwell, W. and Nakagawa, S. (2017). Phylogenetic comparative methods. *Current Biology* 27 (9): R333–R336.

Coulson, T., Benton, T.G., Lundberg, P. et al. (2006). Estimating individual contributions to population growth: evolutionary fitness in ecological time. *Proceedings of the Royal Society B: Biological Sciences* 273 (1586): 547–555.

Cuthill, I. (2005). The study of function in behavioural ecology. *Animal Biology* 55 (4): 399–417.

Dall, S.R.X. and Griffith, S.C. (2014). An empiricist guide to animal personality variation in ecology and evolution. *Frontiers in Ecology and Evolution* 2: 3.

Dall, S.R.X., Houston, A.I., and McNamara, J.M. (2004). The behavioural ecology of personality: consistent individual differences from an adaptive perspective. *Ecology Letters* 7 (8): 734–739.

Dall, S.R.X., McNamara, J.M., and Leimar, O. (2015). Genes as cues: phenotypic integration of genetic and epigenetic information from a Darwinian perspective. *Trends in Ecology & Evolution* 30 (6): 327–333.

Delahaie, B., Charmantier, A., Chantepie, S. et al. (2017). Conserved G-matrices of morphological and life-history traits among continental and island blue tit populations. *Heredity* 119 (2): 76–87.

Dingemanse, N.J. and Wolf, M. (2010). Recent models for adaptive personality differences: a review. *Philosophical Transactions of the Royal Society of London, Series B: Biological Sciences* 365 (1560): 3947–3958.

Dochtermann, N.A., Schwab, T., and Sih, A. (2014). The contribution of additive genetic variation to personality variation: heritability of personality. *Proceedings of the Royal Society B: Biological Sciences* 282 (1798): 20142201.

Emlen, J.M. (1966). The role of time and energy in food preference. *American Naturalist* 100 (916): 611–617.

Evans, M.R., Bithell, M., Cornell, S.J. et al. (2013). Predictive systems ecology. *Proceedings of the Royal Society B: Biological Sciences* 280 (1771): 20131452.

Falconer, D.S. and Mackay, T.F.C. (1996). *Introduction to Quantitative Genetics*, 4e. Harlow: Longman.

Fawcett, T.W., Hamblin, S., and Giraldeau, L.-A. (2013). Exposing the behavioral gambit: the evolution of learning and decision rules. *Behavioral Ecology* 24 (1): 2–11.

Fuller, J.L. and Thompson, W.R. (1960). *Behavior Genetics*. New York: Wiley.

Grafen, A. (1984). Natural selection, kin selection and group selection. In: *Behavioural Ecology: An Evolutionary Approach* (ed. J.R. Krebs and N.B. Davies), 62–84. Oxford: Blackwell.

Grafen, A. (1991). Modelling in behavioural ecology. In: *Behavioural Ecology: An Evolutionary Approach* (ed. J.R. Krebs and N.B. Davies), 5–31. Oxford: Blackwell.

Hamilton, W.D. (1964). The genetical evolution of social behaviour. I. *Journal of Theoretical Biology* 7 (1): 1–16.

Head, M.L., Berry, L.K., Royle, N.J., and Moore, A.J. (2012). Paternal care: direct and indirect genetic effects of fathers on offspring performance. *Evolution* 66 (11): 3570–3581.

Hohenlohe, P.A. and Arnold, S.J. (2008). MIPoD: a hypothesis-testing framework for microevolutionary inference from patterns of divergence. *American Naturalist* 171 (3): 366–385.

Houston, A.I. and McNamara, J.M. (1999). *Models of Adaptive Behaviour, An Approach Based on State*. Cambridge: Cambridge University Press.

Huxley, J.S. (1942). *Evolution: The Modern Synthesis*. London: Allen and Unwin.

Kokko, H. (2007). *Modelling for Field Biologists and Other Interesting People*. Cambridge: Cambridge University Press.

Kruuk, L.E.B., Slate, J., and Wilson, A.J. (2008). New answers for old questions: the evolutionary quantitative genetics of wild animal populations. *Annual Review of Ecology, Evolution, and Systematics* 39 (1): 525–548.

Laland, K.N., Sterelny, K., Odling-Smee, J. et al. (2011). Cause and effect in biology revisited: is Mayr's proximate-ultimate dichotomy still useful? *Science* 334 (6062): 1512–1516.

Laland, K.N., Odling-Smee, J., Hoppitt, W., and Uller, T. (2013). More on how and why: cause and effect in biology revisited. *Biology and Philosophy* 28 (5): 719–745.

Lande, R. (1979). Quantitative genetic analysis of multivariate evolution, applied to brain: body size allometry. *Evolution* 33 (1): 402.

Lerner, I. (1958). *The Genetic Basis of Selection*. New York: Wiley.

Lush, J. (1937). *Animal Breeding Plans*. Ames: Iowa State College Press.

MacArthur, R.H. and Pianka, E.R. (1966). On optimal use of a patchy environment. *American Naturalist* 100 (916): 603–609.

Martin, P. and Bateson, P. (2007). *Measuring Behaviour An Introductory Guide*, 3e. Cambridge: Cambridge University Press.

Mayr, E. (1961). Cause and effect in biology. *Science* 134: 1501–1506.

McNamara, J.M. (2013). Towards a richer evolutionary game theory. *Journal of the Royal Society, Interface* 10 (88): 20130544.

McNamara, J.M. and Houston, A.I. (2009). Integrating function and mechanism. *Trends in Ecology & Evolution* 24 (12): 670–675.

McNamara, J.M., Gasson, C.E., and Houston, A.I. (1999). Incorporating rules for responding into evolutionary games. *Nature* 401 (6751): 368–371.

McNamara, J.M., Dall, S.R., Hammerstein, P., and Leimar, O. (2016). Detection vs. selection: integration of genetic, epigenetic and environmental cues in fluctuating environments. *Ecology Letters* 19 (10): 1267–1276.

Melo, D., Garcia, G., Hubbe, A. et al. (2016). EvolQG – An R package for evolutionary quantitative genetics. *F1000Research* 4: 925.

Morrissey, M.B., Parker, D.J., Korsten, P. et al. (2012). The prediction of adaptive evolution: empirical application of the secondary theorem of selection and comparison to the breeder's equation. *Evolution* 66 (8): 2399–2410.

Nesse, R.M. (2013). Tinbergen's four questions, organized: a response to Bateson and Laland. *Trends in Ecology and Evolution* 28 (12): 681–682.

Parker, C.C., Gopalakrishnan, S., Carbonetto, P. et al. (2016). Genome-wide association study of behavioral, physiological and gene expression traits in outbred CFW mice. *Nature Genetics* 48 (8): 919–926.

Price, G.R. (1970). Selection and covariance. *Nature* 227 (5257): 520–521.

Robertson, A. (1966). A mathematical model of the culling process in dairy cattle. *Animal Production* 8 (1): 95–108.

Sih, A. and Christensen, B. (2001). Optimal diet theory: when does it work, and when and why does it fail? *Animal Behaviour* 61 (2): 379.

Sih, A., Mathot, K.J., Moirón, M. et al. (2015). Animal personality and state – behaviour feedbacks: a review and guide for empiricists. *Trends in Ecology and Evolution* 30 (1): 50–60.

Stephens, D.W. and Krebs, J.R. (1986). *Foraging Theory*. Princeton: Princeton University Press.

Tinbergen, N. (1951). *The Study of Instinct*. Oxford: Clarendon Press.

Tinbergen, N. (1963). On aims and methods of ethology. *Zeitschrift fur Tierpsychologie* 20: 410–433.

Turelli, M. (1988). Phenotypic evolution, constant covariances, and the maintenance of additive variance. *Evolution* 42 (6): 1342–1347.

Wilson, A.J. (2014). Competition as a source of constraint on life history evolution in natural populations. *Heredity* 112 (1): 70–78.

Wilson, A.J., Gelin, U., Perron, M.C., and Réale, D. (2009). Indirect genetic effects and the evolution of aggression in a vertebrate system. *Proceedings of the Royal Society B: Biological Sciences* 276 (1656): 533–541.

Wilson, A.J., Réale, D., Clements, M.N. et al. (2010). An ecologist's guide to the animal model. *Journal of Animal Ecology* 79 (1): 13–26.

Wolf, M. and Weissing, F.J. (2010). An explanatory framework for adaptive personality differences. *Philosophical Transactions of the Royal Society of London, Series B: Biological Sciences* 365 (1560): 3959–3968.

3

How the Dual Inheritance of Genes and Culture Shapes Behaviour: A Critical Review with a Focus on Human Culture and Behavioural Diversity

Thomas E. Currie

Human Behaviour & Cultural Evolution Group, Centre for Ecology & Conservation, University of Exeter, Penryn Campus, Penryn, TR10 9FE, UK

It is estimated that around 6000 mutually unintelligible languages are spoken in the world today (Lewis 2009), and people around the world exhibit a diverse array of beliefs, social rules, rituals, clothing, and ways of life. Despite having originally evolved in the hot savannah environments of Africa, our species has spread out to colonize and inhabit every continent on earth, adapting to a great diversity of environments on the way. This remarkable behavioural diversity is set against the fact that, despite some prominent differences in biological features such as hair, skin, eye colour, and stature, humans are extremely homogenous genetically, particularly compared to wild populations of our closest primate relatives (Jorde and Wooding 2004). Our ability to adapt to new environments and create diverse ways of life is the result of the fact that we learn much of our behaviour from other people or, in other words, we have something we generally refer to as culture.

In this chapter, I will discuss how thinking of culture as a distinct inheritance system, that itself can undergo descent with modification, adds to our understanding of behaviour and adaptation. I will illustrate the utility of this approach by discussing how the dynamics of social learning make maladaptive behaviours more likely, how understanding cultural history helps us to understand present-day behavioural diversity, and how culture has aided the evolution of co-operation on a scale not seen in other species. I will then discuss the potential for coevolution of genes and culture, including the relationship between patterns of genetic diversity and cultural diversity, and how cultural evolution can shape genetic evolution.

3.1 Culture and Behaviour

'Culture' is famously one of those terms that has many colloquial meanings. A paper by anthropologists Kroeber and Kluckholm (1952) listed 164 different definitions that they had identified in the literature. In this section, I will clarify what I mean by culture and why it is useful to draw a distinction between behaviour and culture.

We can think of behaviour as being the actions and responses of organisms that result from the information they receive from internal and/or external stimuli (Levitis et al. 2009). Some behaviours may be largely innate (see Chapters 1, 4, and 5), by which

Genes and Behaviour: Beyond Nature-Nurture, First Edition.
Edited by David J. Hosken, John Hunt and Nina Wedell.
© 2019 John Wiley & Sons Ltd. Published 2019 by John Wiley & Sons Ltd.

genes build organisms that engage in some automatic response to some internal or external stimulus (e.g. the involuntary, reflexive jerk of your leg when hit just below the kneecap). However, behaviour can also be learned, by which exposure to a stimulus causes responses to change over developmental time, either due simply to repeated exposure to a stimulus (e.g. habituation or sensitization) or by making associations between responses and outcomes (associative learning) (Heyes 1994). It is important to note that some innate behaviours can be modified or shaped by learning, and that many behaviours will have both innate and learned components (see Chapters 1, 4, and 5).

In humans and many other species, social learning is a particularly important form of learning (see Chapter 9). Social learning refers to changes in behaviour that result from some kind of interaction or observation of other individuals that can come about through a variety of mechanisms (Hoppitt and Laland 2013). Learning mechanisms themselves must of course be under some kind of genetic control. Our ability to speak a language at all is due to the evolution and development of biological features that control breathing and vocalization, as well as the construction of brains that are able to acquire and process linguistic information and enable infants to learn the particular language they are exposed to (Pinker 2003). However, the specific content of the language that individuals end up speaking (the particular words and grammatical features) is due to social learning.

Just as genes provide *information* that shapes development and affects the phenotype of organisms (including behaviour), many researchers have found it useful to think of culture as the information that individuals acquire through various forms of social learning that is capable of affecting their behaviour (Richerson and Boyd 2005; Mesoudi 2011). We can describe *cultural behaviours* as those behaviours that are at least partly the product of social learning, and we can similarly refer to these behaviours and other products of social learning such as tools, clothes, laws, and social roles as cultural traits.

Cultural traits and behaviours may cover a variety of functional roles. Some variation in traits may represent non-adaptive variation (e.g. much difference in content of languages, music, or artistic styles). However, other cultural traits and behaviours may have important fitness consequences for the individuals possessing them. Some such traits and behaviours may reflect cultural adaptations to local environmental conditions. For example, the Inuit are able to live in permanently frozen northern latitudes due to a suite of cultural behaviours and other traits relating to clothing, housing, and subsistence (Boyd et al. 2011). Some cultural behaviours may reflect existing social conditions or other cultural behaviours. For example, in Ancient Hawaiian society it was forbidden (*kapu*) for 'commoners' to look directly at a chief, and they had to prostrate themselves when in sight of him (Kirch 2010). Learning the correct deference behaviours was vitally important as violation of these norms could result in death. Many cultural traits may have less dramatic consequences but can still affect survival and reproduction, particularly when the 'correct' cultural trait depends on what others are doing or what cultural practices already exist (e.g. kinship systems and inheritance practices can affect the reproductive success of offspring) (Mace 1998; Ji et al. 2016). Finally, in humans in particular, cultural traits and behaviours are used to demarcate groups. Common group identity markers include the type of language spoken (Fought 2006) and costly social rituals (Sosis et al. 2007). As we shall see below, this structuring of humans into distinct groups has important consequences for co-operative behaviour and the transmission of culture.

The capacity for culture itself also can be considered an adaptation, to the extent that social learning enables individuals to survive and reproduce more effectively than they would through individual learning alone (Boyd and Richerson 1985a; Richerson and Boyd 2001). Models show that one of the benefits of cultural learning over individual learning is that it allows users to adapt to changing environments faster than could be accomplished by genetic evolution (Boyd and Richerson 1985a). Richerson et al. (2001) have proposed that climatic variability in the Pleistocene favoured an increased reliance on social learning and led to the evolution of our great capacity for culture.

This perspective of culture as socially learned information also helps highlight that culture is not necessarily unique to humans. Social learning has been demonstrated in a wide range of species. Early studies of Japanese macaques indicated that a behaviour of washing potatoes in sea water before eating spread through social processes (de Waal 2001). Chimpanzees (Whiten et al. 1999) and orangutans (van Schaik et al. 2003) show behavioural variations across populations. While some of these great ape behaviours could conceivably reflect individually learned solutions to different ecological conditions (e.g. different styles of termite fishing), other behaviours appear somewhat arbitrary (e.g. such as clasping hands before grooming). Both whales (Filatova et al. 2010; Cantor et al. 2015) and birds (Nelson and Marler 1994; Marler and Slabbekoorn 2004) have been shown to have dialects in their vocalizations that are at least partially socially learned. Long-term observations of humpback whales off the coast of Massachusetts, USA, show evidence for a novel feeding behaviour spreading amongst a large proportion of the population (Allen et al. 2013). In 1980, one individual added an extra element to a certain hunting technique by striking the water several times before the regular behaviour (shared by the rest of the population) of blowing bubbles around schools of fish to disorient them. Analysis of a 27-year database indicated that this water-striking practice (known as lobtail feeding) had eventually been adopted by 37% of the population. Furthermore, the majority of these lobtail feeders adopted the behaviour after being in close association with other individuals that were already practising this technique.

These observational studies of wild populations have been supplemented by experimental studies both in the lab and in the field that help rule out alternative explanations, and enable the proximate mechanisms behind social learning to be investigated (Whiten and Mesoudi 2008). A common experimental technique is to train individuals as 'demonstrators' in different groups with different solutions to some artificial foraging-related task, such as getting food from a plastic box. For example, one solution is to use a stick to lift up an obstacle, the other solution is to poke the obstacle. If individual learning is dominant in such situations then the expectation is that both behaviours should be equally common within groups. However, studies of chimpanzees on such tasks indicate that naïve individuals tend to perform the behaviour that was taught to the demonstrator, and that group differences in the persistence of the behaviours can be stable over multiple runs.

Researchers have debated whether this kind of copying behaviour reflects direct imitation or is actually the result of emulation (by which an individual is copying the goal of the behaviour rather than specific actions that lead to the goal) (Whiten et al. 2009). By adding an arbitrary behaviour in these tasks, such as placing a plastic token in one of two different holders, researchers can investigate the extent to which individuals also copy elements that are not essential to successful completion of the task or achievement of a goal. Studies on chimpanzees show that these arbitrary behaviours also spread, which strengthens the interpretation that true imitation is involved (Bonnie et al. 2007). In field experiments in South Africa (van de Waal et al. 2013), vervet monkeys were trained to eat corn that had been dyed a particular colour (e.g. blue), with the opposite colour (e.g. pink) being artificially made bitter. Four months after this training period, the dyed corn was again presented to the monkeys, but this time both colours were non-bitter. Young individuals that had been born into these groups, and were now able to eat solid food, preferentially chose the colour of corn that their group had been trained on. Interestingly, during the experimental period 10 males moved between the groups (which had been trained with the alternative bitterness/colour association). The majority of these males (7 out of 10) changed their colour preference to that of the group they moved into, which indicates that social learning could override the previous negative experiences associated with alternative behaviour, and potentially indicates a tendency to copy the most prevalent behaviour of the group (see below).

Whilst noting these continuities between human and non-human animal culture, the complexity of human culture and the diversity of behaviours across human populations seem to be vastly greater in humans compared to other species. This suggests that there are also important differences between human and animal culture. Cultural behaviours in humans appear to build on other cultural behaviours in a heavily cumulative manner (Boyd and Richerson 1996; Caldwell and Millen 2008; Kempe et al. 2014). Behaviours and material products such as tools can be modified, combined, and improved upon such that beneficial modifications and innovations can be retained and expanded upon over several generations. This leads to our ability to solve complex, adaptive problems that would be impossible for a single individual to solve within their own lifetime – we literally do not have to reinvent the wheel. This is most obvious in the evolution of technologies that show increasing abilities to hunt, cultivate food, process and transmit information, extract energy from the natural world, and, unfortunately, to wage war (Kurzweil 2000; Mesoudi et al. 2013; Morris 2013). This capacity for cumulative culture requires high-fidelity transmission, otherwise beneficial behaviours will be gradually lost from one generation to the next (Lewis and Laland 2012).

Social learning in humans appears to have a number of features that maintain accurate transmission of behaviour from one generation to the next. Language facilitates direct teaching of ideas and intentions and helps correct 'errors' in the transmission of skills, practices, etc. Humans also seem to be extreme copiers, often engaging in 'overimitation' whereby they will copy exactly what others do (Whiten et al. 2009), which may be particularly important in situations that are causally opaque or do not lead to immediate rewards. Humans also seem particularly sensitive to following social norms and conventions, including situations where these might be costly (Cialdini and Trost 1998; Sosis et al. 2007; Henrich 2015).

3.2 Cultural Evolution

As we have seen, in species that engage in social learning, individual behaviour (and phenotypes more generally) can be influenced by both information from genes and information acquired through individual and social learning. However, unlike purely individual learning, cultural information can be inherited from one generation to the next. Culture therefore represents a parallel system of inheritance to genetic inheritance (also see Chapter 10). This has implications for the evolutionary processes that shape the behaviour of humans and other cultural species.

The standard neo-Darwinian paradigm views evolution as changes in gene frequencies within a population over time (Scott-Phillips et al. 2014). Genetic changes arise through mutations in the genome, and these mutations are passed on to subsequent generations via reproduction. Changes in the distribution of alleles (varieties of the same gene) in a population can occur via a number of process but chief among them are genetic drift (random fluctuations, which can potentially lead to substantial increases or decreases in the representation of alleles) and selection (alleles that lead to increased survival and/or reproduction are better represented in subsequent generations) (Futuyma 2013). Selection is important because it leads to adaptation and the appearance of organismal design without the need to invoke a deliberate, purposeful guiding force or designer (Darwin 1859; Dawkins 1986). However, as we have seen, adaptively relevant information can be passed on through social learning, and survival and reproduction may be linked not to genetic variation but to variation in cultural traits and behaviours. Focusing too narrowly on genes, therefore, may lead to problems in adequately explaining important variations in phenotypes and their systematic changes. If we understand evolution more generally as change in the inherited characters of a population over time (Ridley 2004; Richerson and Boyd 2005; Laland et al. 2015), then we can see that systematic changes in behaviour can come about as a result of either genetic evolution or cultural evolution (or both).

3.2.1 Processes of Cultural Evolution

At a broad scale, we can recognize that there are processes of variation, inheritance, and selection that occur in cultural evolution that are analogous to those of genetic evolution. In this section, I describe what variation, inheritance, and selection look like in cultural systems. While drawing parallels between these two systems is an important exercise, it is also important to acknowledge the differences between genetic and cultural systems, and the consequences this has for how cultural evolution works.

3.2.1.1 Variation

Any evolutionary system requires the generation of variation. In genetic evolution, variation is generally thought to be generated through random genetic mutation (by which we mean that these changes occur without reference to their likely phenotypic effects). Indeed, most mutations are either silent (i.e. they don't produce a change in phenotype), selectively neutral, or actually have a deleterious effect on fitness (Eyre-Walker and Keightley 2007). Random mutation may also play a role in generating cultural variation. In cultural systems, 'mutations' can result when individuals make copying errors such as misuse of or misremembering words. Consistent with this idea, Pagel and colleagues (Pagel et al. 2007; Calude and Pagel 2011) have shown that across different languages, the rate at which words are estimated to change is related to the frequency with which those words are used. Words that are used less frequently change at a faster rate potentially because copying errors are more likely to go unnoticed, meaning new forms of words can spread and become fixated. New traits may also come about through 'happy accidents', where a beneficial innovation is discovered by chance rather than from purposeful planning. Many foodstuffs or traditional medicines may have been discovered this way. A more recent example is the discovery of the antibiotic penicillin by Alexander Fleming. Reportedly, Fleming noticed that bacterial growth in a petri dish had been inhibited by a *Penicillium* mould, due to accidental contamination as the result of a laboratory window being mistakenly left open (Diggins 1999).

Accidental, unintentional, or otherwise non-directed changes in cultural traits can and do occur. However, many changes may also be distinctly non-random, and instead variation can be guided towards a particular goal (Boyd and Richerson 1985b). For example, people undoubtedly make adjustments to technological items with the intention of improving their efficiency, power, functionality, etc. Sometimes changes to whole languages are enacted purposefully. Several instances have been recorded in New Guinea where communities have made a conscious decision to change aspects of their language in order to differentiate themselves from other groups (in one case switching certain masculine and feminine elements to be the opposite of neighbouring dialects) (Thomason 2007).

If organisms are able to design solutions to an adaptive problem themselves, this may at first appear to invalidate the need to talk about culture in evolutionary terms. Several points can be made in response to this argument. An evolutionary process only requires that variation be generated and does not require a particular type of process such a blind mutation to generate this variation. Evolutionary theory rests on population thinking, and even if an individual develops an improved trait, that trait still needs to spread through the population, and many other processes (as we shall see below) may affect whether or how this occurs. Many adaptive problems could have multiple, equally appropriate solutions (e.g. chopsticks versus cutlery). Guided variation might be part of the explanation as to why populations find these solutions at all, but population-level processes would be needed to explain why different populations might converge on different solutions. Even if variation is guided, it does not mean that complex adaptive traits will necessarily emerge fully formed. Many problems that we face are complex and causally opaque such that any improvements we make may only be incremental. The presence of guided variation certainly can affect evolutionary dynamics by enabling a population to reach an adaptive peak more quickly, but it does not automatically invalidate the utility of an evolutionary approach to understanding cultural change.

3.2.1.2 Inheritance

Inheritance in genetic systems occurs when individuals create offspring that resemble themselves through some form of reproduction. In cultural systems, the analogue of inheritance (which is more commonly referred to as transmission) occurs when individuals learn a behaviour, or otherwise acquire a trait from another individual, and then exhibit that trait (Cavalli-Sforza and Feldman 1981; Boyd and Richerson 1985b; Mesoudi et al. 2004). A common distinction made between these two systems is that whereas the predominant mode of inheritance in biology is vertical (i.e. parent to offspring, tied to reproduction), in cultural evolution traits can be transmitted from parents to offspring and can also be transmitted from other individuals too. Some modellers have referred to transmission between individuals of the same generation as horizontal transmission, while non-parent-to-offspring transmission across generations is referred to as 'oblique' transmission (Cavalli-Sforza and Feldman 1981). While this distinction is somewhat arbitrary, it could be adaptive in some cases to learn primarily from individuals of similar age (e.g. when traits are evolving quickly) or from older generations (e.g. when ability in a certain task is correlated with experience and that task is still relevant). For some processes, mode of transmission is indeed an important and relevant distinction. However, for many processes this distinction is not important, and in many cases equating biological inheritance with vertical transmission and cultural inheritance with horizontal transmission is overly simplified.

In discussing these issues, it is important to distinguish between transmission at the micro-level (i.e. within populations or species) and at the macro-level (i.e. between populations or species). For example, a trait may be spread horizontally or obliquely within a population, but if there are barriers to transmission between populations then transmission would be characterized by vertical transmission at the macro-scale, that is, individuals will exhibit traits that resemble those of previous generations from the same population, and when cultures split, the descendant, 'daughter' cultures will maintain many of the traits of the original, 'parental' culture (see section 3.4). Similarly, horizontal transmission at the macro-scale involves the transfer of traits or information between populations, and could occur even if transmission within populations tended to be predominantly from parents to offspring.

Different modes of transmission can affect the rate at which a novel trait can spread through a population. When genetic information is only transmitted vertically then the rate at which an allele will increase in frequency within a population is constrained by the rate at which individuals reproduce (and by the strength of selection). However, for cultural traits, the rate-limiting step is how often the trait can be observed and successfully copied (Cavalli-Sforza and Feldman 1981). This means that potentially a cultural trait can be adopted by a large proportion of the population extremely quickly. For example, the use of hybrid corn by farmers spread rapidly across the United States in approximately 25 years beginning in the 1930s (Griliches 1960) (see Kandler and Steele (2009) for an example of how the dynamics of such spreads have been analysed using cultural evolutionary theory). In some ways, this ability to spread horizontally makes cultural evolution somewhat epidemiological, and indeed some authors have argued that culture (or at least some aspects of it) can be thought of as like 'viruses of the mind' (Dawkins 1993; Brodie 2009). Therefore, as we shall see later, this horizontal mode of transmission potentially has consequences for genetic fitness (see section 3.3.1).

Horizontal transmission is not unique to cultural systems, however, and the messy truth is that horizontal transfers of genetic information are widespread in biology. At a microevolutionary level, horizontal transfer of genetic material between bacterial cells is extremely common and thought to be an important means of resistance to antibiotics (Barlow 2009). At the macro-level, genes (commonly transposable elements) from one species can become incorporated into the genome of another species via some kind of vector (virus, bacteria, etc.) (Schaack et al. 2010; Peccoud et al. 2017), and hybridization and introgression (Futuyma 2013; Harrison and Larson 2014) have been increasingly recognized as major processes in evolution (Maynard Smith and Szathmáry 1995; Rivera and Lake 2004; Ge et al. 2005). Indeed, recent studies have shown how early human ancestors interbred with Neanderthals and Denisovans and that modern human genomes contain genes from these species (Wills 2011; Huerta-Sanchez et al. 2014). Furthermore, cultural transmission is not necessarily dominated by non-vertical transmission, particularly at the macroevolutionary scale. As we shall see below, cultural evolutionary processes may inhibit transmission of cultural traits between populations, meaning that the historical relationships between cultures may be usefully represented using the kind of phylogenetic techniques that are used to represent macro-scale patterns of descent in biological species.

So far, we have talked in general terms about the transmission of information, but what is the mechanism by which this form of inheritance occurs? In seeking to show the general applicability of Darwinian thinking and replication, Richard Dawkins (1976) coined the term 'meme' to describe a theoretical unit of cultural inheritance that is analogous to genes. While some authors have tried to develop the meme concept (Dennett 1996; Aunger 2000; Blackmore 2000; Dennett 2006; Brodie 2009), the existence of discrete cultural units is not a necessary precondition for an evolutionary approach to culture (Laland and Brown 2002). The key point is that information is transmitted between individuals with sufficient fidelity that the same phenotypic character can be observed over a relevant time span. Models have been developed in which cultural traits are modelled as continuous rather than discrete (Richerson and Boyd 2005). In contrast to genetic systems, where we know a huge amount about the mechanisms of inheritance, the psychology and neuroscience of cultural transmission are less well understood and is an important area for future research. However, we can take comfort from the fact that, famously, Darwin knew little about the mechanisms of biological inheritance in developing his ideas. Similarly, our current ignorance about the details of cultural transmission should not prevent us from developing and testing cultural evolutionary models, including understanding how different possible inheritance mechanisms might affect the processes of cultural evolution.

3.2.1.3 Selection and Fitness

A consequence of the fact that genetic and cultural information have different channels and modes of transmission is that genetic fitness and cultural fitness can be quite distinct (El Mouden et al. 2014). Genetic fitness relates to the representation of a particular gene in subsequent generations and to production and subsequent reproduction of one's own offspring (direct fitness), or indirectly through the offspring of related individuals (weighted by the degree of relatedness). By cultural fitness, I mean the 'transmission success' of a particular cultural trait in terms of it being transmitted between individuals (i.e. the representation of a trait in the population at a later point in time due to social learning). In some cases, genetic fitness and cultural fitness will be driven by the same processes. If offspring inherit the cultural traits of their parents, and if individuals possessing certain cultural traits leave more offspring than individuals possessing other cultural traits, then the representation of cultural traits in subsequent generations can be affected by natural selection just like any other phenotypic trait. I will return to the relationship between genetic and cultural fitness later in this chapter but in this section, I discuss other selective processes acting on culture.

If individuals choose who to copy at random, then the frequency of a trait in a population will not change substantially over time (this is with the caveat that population size is important in this respect – in small populations, drift-like processes may be an important component of change over time). Models of cultural evolution have attempted to explore the effects of different psychological learning biases that may create a 'selection-like' process, in that they lead to certain cultural traits being more readily transmitted than others (i.e. they affect cultural fitness) (Boyd and Richerson

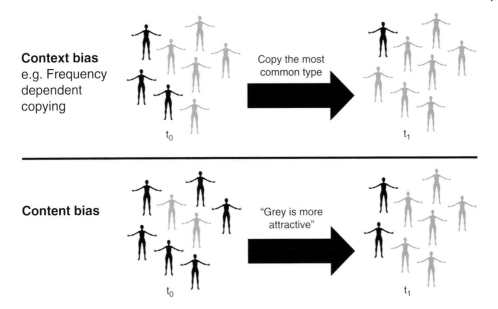

Figure 3.1 Transmission biases can affect the prevalence of cultural traits over time. Context biases involve information about the situation in which individuals find themselves. In the top example, the individuals are following a learning rule that indicates that they should preferentially copy (with some probability and some possibility for error) the most common behaviour in the population. In this case, at time t_0, the most common behaviour is for individuals to wear grey. At time t_1, two of the individuals that were wearing black have followed this rule and switched to wearing grey. Content biases (*lower panel*) relate to information about the trait itself. In this example, grey is considered a more attractive colour (which could be an innate or learnt preference) and individuals wearing grey are more likely to be copied as a result. By time t_1, four individuals have switched from wearing black to wearing grey.

1985b; Henrich 2001; Richerson and Boyd 2005; Mesoudi et al. 2006a; Henrich et al. 2008; Mesoudi 2008). These transmission biases are often categorized as relating to either content or context (Figure 3.1). Content (or direct) biases refer to properties of the cultural trait themselves that make them more or less likely to be copied. Context biases, on the other hand, refer to biases in transmission that are the result of the situation in which a trait is being transmitted. These biases may reflect properties of the individuals who are copied (model-based biases) or may relate to population-level properties. Examples of model-based biases would be preferences to copy prestigious or successful individuals (Henrich and Gil-White 2001; Mesoudi 2008). The main examples of population-level biases relate to the existing representation of a cultural trait in the population. Copying can be frequency dependent such that individuals actively assess the frequency of a trait in the population and choose to copy the most common trait (Mesoudi and Lycett 2009). This leads to conformity in the sense that rarer traits will have an even lower probability of being adopted and a population will tend towards homogeneity over time (even in the absence of ideas that a particular behaviour *must* be followed, or mechanisms for punishing those who do not exhibit such behaviours) (Morgan and Laland 2012; Aplin et al. 2015). Anticonformity biases, in which individuals would preferentially adopt rare traits, are also possible.

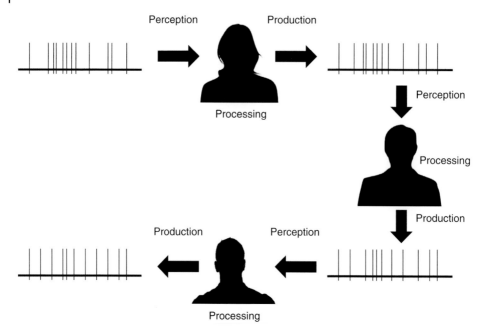

Figure 3.2 Cultural evolution may be affected by cultural attractors that shape information as it is transmitted between individuals. In this example, a piece of music (*top left*) is represented by a horizontal line with vertical lines representing the beats. A preference for (or bias towards) a regular beat in music acts to make an initially arhythmic piece of music into something that has a more regular beat as it goes through repeated stages of individuals hearing the music (perception), processing it, and then attempting to reproduce the piece of music (production). The actual psychological and/or behavioural mechanism by which such an attractor works could in theory take place at any stage in the transmission process (e.g. perceptual constraints could mean certain sounds aren't even registered, or more attention could be paid to certain features during cognitive processing, or mechanical constraints could make certain behaviours easier or more likely than others).

Some researchers have argued that these selection-like processes are not the most important factor affecting change in socially transmitted traits, but rather information received by individuals is transformed and altered by pre-existing knowledge or psychological biases. Under this view, certain ideas or concepts may represent 'cultural attractors' and channel behavioural variation towards certain outcomes (Figure 3.2). These biases may result from aspects of our innate, evolved psychology, and potentially have themselves been shaped by natural selection. For example, socially learned dietary choices may be influenced by innate preferences for sweet, salty, and fatty foods that would have produced adaptive dietary behaviours throughout much of our evolutionary history. Lab-based experimental studies are an important way of studying these issues given their ability to construct scenarios that isolate different mechanisms and alternative explanations, and reduce some of the noise present in real-world datasets (Mesoudi 2007; Whiten and Mesoudi 2008; Mesoudi 2016). For example, experimental studies have shown that colour terminology tends to converge on a relatively limited number of terms, potentially as a result of features of our perception systems that are universal across human societies (Xu et al. 2013). Studies of the transmission of music have indicated that humans impose and pass on elements of rhythm when such elements were not present in the original recording (Ravignani et al. 2016). This mirrors observed cross-cultural regularities in features of human music (Savage et al. 2015).

Cultural attractors and ideas of cultural transmission as a transformative process were originally framed as an alternative to the selective mechanisms described above, which assumes that fidelity of transmission is generally high. However, recent approaches have recognized that both processes are probably at play, and that transformative ideas are compatible with existing models of cultural evolution. Cultural attractor theory focuses attention on what actually happens psychologically when information is socially transmitted and may be particularly important in understanding widespread or universal features of human culture. However, cultural attractors have not received the same level of formal modelling as selective mechanisms and it is unclear to what extent such transformative processes affect the dynamics of cultural evolution at the population level in ways that are not captured by existing models of guided variation and content biases.

While transmission biases and cultural attractors have received large amounts of theoretical attention, the existence or importance of these biases is still an area of emerging research. Cross-cultural studies of these biases will be important as it is likely that the importance attached to different sources of information and weighting of different psychological biases will vary between cultures. For example, food taboos present in some cultures may go against innate preferences, yet may be adaptive especially if information comes from experienced or knowledgeable individuals (Henrich and Henrich 2010), or if they are used as costly markers of group membership (Sosis and Alcorta 2003). An overall picture of how various transmission biases are integrated and combined to shape human behaviour in different cultures and contexts remains an important area for future study.

3.3 Insights from Cultural Evolutionary Approaches

The sections above have shown how analogies can be drawn between cultural and genetic evolution, and how evolutionary thinking can be applied to cultural inheritance systems. However, drawing analogies between the processes of biological and cultural change is not by itself very useful. It is therefore important to articulate how this perspective provides a more complete framework for studying behaviour in humans and other cultural species. In the following sections, I provide examples from four areas of study where cultural evolutionary theory helps provide a better understanding of behavioural diversity: adaptation and maladaptive behaviour, cultural history, the evolution of co-operation, and gene–culture coevolution.

3.3.1 Adaptive and Maladaptive Behaviour

The psychological mechanisms by which we acquire cultural information from others is likely to have been shaped by natural selection to enable individuals to behave adaptively – individuals with brains that learn the wrong things and behave inappropriately end up leaving fewer offspring. However, the dual-inheritance perspective outlined above also recognizes that the dynamics of cultural transmission can lead to individuals behaving in a suboptimal or maladaptive manner. Laland and Williams (1998) used transmission chain experiments in which small groups of guppies, *Poecilia reticulata*, were initially trained to follow a more circuitous and energetically costly feeding route or a less costly shorter route. These founding individuals were gradually replaced by untrained individuals. Even several days after all the founders had been removed, the groups of untrained fish tended to follow the feeding behaviour of their founders. This indicates that these feeding routes were being socially learned, and that more costly behaviours were being passed on, even when less costly alternatives were available (see also Bates and Chappell 2002). In humans, Durham (1991) argues that the ritual practice of cannibalism amongst the Fore of New Guinea was ultimately maladaptive in that it led to the spread of a deadly, neurodegenerative disorder known as *kuru*. It is estimated that in the 1940s and 1950s, hundreds of Fore people were dying each year due to *kuru*, out of a total population of only approximately 12 000 (Liberski 2013).

Such maladaptive behaviours may be sustained due to a number of factors and features of social learning. Model-based copying biases based on status/prestige or similarity can result in the spread of maladaptive behaviours if the models being copied exhibit such behaviours (Mesoudi 2009). Even when social learning is unbiased in this way, other factors that relate to observing the trait and making links between behaviours and outcomes will affect what behaviours get adopted. Tanaka et al. (2009) wanted to understand why a large market still exists for 'complementary' medicines and 'traditional remedies', which sometimes have poor safety records and for which there is little evidence that they work as intended. They developed a mathematical model of the conditions under which efficacious, non-efficacious or even maladaptive medical treatments are able to spread by social learning. The key features of this model were that people adopt a practice based on having witnessed it, but can abandon it if it does not work. Under some circumstances, maladaptive practices could spread because they resulted in longer periods of treatment (i.e. the fact that they don't work means that people take longer to recover), meaning that more people saw the practice and adopted it. As long as the rate of conversion due to this process is greater than the rate at which individuals will abandon the practice due to its lack of effectiveness, then the maladaptive behaviour can spread. In the *kuru* example above, the decades-long gestation period of the disease meant that the link between cannibalism practices and the disease was difficult to establish, and *kuru* was instead attributed to malevolent sorcery by hostile individuals (Lindenbaum 2008).

Under a dual-inheritance perspective, we might expect to see cultural species exhibiting some maladaptive behaviour; social learning mechanisms should on average produce adaptive outcomes but occasionally will lead to stable behaviours that are at odds with genetic fitness. However, how common maladaptive behaviours actually are and the extent to which cultural inheritance leads to maladaptive outcomes are still open questions. Simple social learning mechanisms may create traditions, including arbitrary or maladaptive ones (Franz and Matthews 2010). Thornton and Malapert (2009) used field experiments in wild meerkats to show that arbitrary behavioural traditions could arise through copying the behaviour of individuals trained to visit certain landmarks in order to gain a reward. However, these traditions quickly disappeared as individual learning led some individuals to explore other landmarks. So although conformity or other biased social learning mechanisms are not required for traditions to emerge, it may be that they are necessary for enabling maladaptive outcomes to persist over substantial periods of time. Furthermore, if cultural fitness is at odds with genetic fitness then we would expect natural selection to lead to improved social learning capabilities that increasingly bias the acquisition of cultural behaviours towards more adaptive outcomes (Lumsden and Wilson 1980; El Mouden et al. 2014). It is also important to distinguish genuinely maladaptive outcomes (where there are reductions in direct or inclusive fitness from pursuing a behaviour) from behaviours that appear costly but ultimately are associated with other benefits, such that on balance they produce net increases in direct or inclusive fitness. For example, some authors (Dawkins 1993; Blackmore 2000) have argued that religious beliefs are maladaptive and that they constitute a 'virus of the mind'. However, a number of adaptive explanations for religion have also been put forward, for example that certain religious practices may be costly signals of group membership (Sosis et al. 2007) or that certain religious beliefs may help organize large-scale human societies (Norenzayan et al. 2014).

The manner in which cultural evolutionary studies analyse the possibility of maladaptive behaviour stands in contrast to other evolutionary approaches to investigating human behaviour. Behavioural ecology focuses mainly on adaptive explanations of behaviour (Smith and Winterhalder 1992; Nettle et al. 2013), while evolutionary psychology often proposes hypotheses based on maladaptive mismatches between modern and ancestral environments (Barkow et al. 1992; Buss 2015). Rather than having to commit ourselves to either end of this continuum, a cultural evolutionary approach can help integrate and bridge these other theoretical perspectives (Mesoudi et al. 2006b). Furthermore, the mismatch view of evolutionary psychology rarely explicitly considers how and why modern environments arise. Cultural evolution helps to explain how humans have shaped these environments and the fitness consequences of such niche construction, as well as the behavioural and psychological responses to such environments. Taking this approach, we can investigate under what conditions maladaptive behaviours are most likely to occur, which can serve as a motivation for testing between competing hypotheses and assessing how common maladaptive behaviours actually are.

3.4 Cultural History

Biologists recognize that evolutionary history plays an important role in explaining the diversity of life. Species may exhibit many similar traits because they share common ancestral species from which these traits have been inherited (Sober 2009). The emergence of biological novelty takes place against this background, and in some cases evolutionary history may constrain what changes are possible (Arnold 1992; McKitrick 1993; Blomberg and Garland 2002). These shared patterns of ancestry are often represented in the form of phylogenetic trees. As a description of evolutionary history, phylogenies are important both as means of structuring biological diversity (i.e. providing us with a taxonomy) and as a scaffold for testing hypotheses about the processes of evolution that have shaped this diversity over space and time (Gregory 2008; Nunn 2011). By understanding culture as a system of inheritance, we can similarly appreciate the importance of cultural evolutionary history in helping us explain both the patterns of human cultural diversity and the processes that have generated this diversity (Mace and Holden 2005; Currie 2013). An important area of cultural evolutionary research, that has enabled progress to be made in addressing long-standing questions in the social sciences, has been the application of 'tree thinking' and phylogenetic methods to study cultural macroevolution.

It has long been appreciated that languages can often be well represented by a branching tree with a nested hierarchy of historical relationships (Hock and Joseph 2009). Traditionally, linguists have relied on rather subjective means to create these family trees and to locate the origins of these languages in time and space. In the last two decades, researchers have begun to apply formal phylogenetic and phylogeographic methods from evolutionary biology to construct evolutionary trees of language and use these to more rigorously test hypotheses about the homelands and routes taken during large-scale demographic expansions (Greenhill and Gray 2005; Gray et al. 2011; Currie 2013). Various techniques have been used but such studies generally take linguistic features that are comparable across languages (e.g. similarities in lists of words with the same meaning from different languages) and analyse them using the same kind of phylogenetic inference techniques that can be applied to morphological data in biology. For example, phylogenetic studies provide support for a hypothesis that links the spread of Indo-European (IE) languages to the emergence of farming. Analyses of lexical data produce phylogenetic trees that place the origin of IE in Anatolia (in modern-day Turkey) (inferred in relation to the structure of the tree, and more recently through explicit phylogeographic modelling), and indicate that IE languages began diverging around 9000 years ago (Gray and Atkinson 2003; Bouckaert et al. 2012; Currie 2013). These kinds of dispersal hypotheses have been tested across a number of different language families in different parts of the world, including Island Southeast Asia and the Pacific (Gray et al. 2009), sub-Saharan Africa (Currie et al. 2013) (Figure 3.3), South America (Heggarty 2008), and the Middle East (Kitchen et al. 2009). Phylogenetic inference techniques have also been applied to test other cultural evolutionary hypotheses using archaeological data (Coward et al. 2008), material culture (Tehrani et al. 2010), and other aspects of culture (Eagleton and Spencer 2006; Matthews 2012; Brown et al. 2014).

Once we have trees representing how different cultures are connected, we can also apply phylogenetic comparative techniques to test other hypotheses about cultural evolution (Harvey and Pagel 1991; Mace and Pagel 1994; Currie 2013; Currie and Meade 2014). By mapping cultural traits or features of social organization onto phylogenetic trees, we can investigate the processes that lead to variation across societies in these traits. For example, Currie et al. (2010a) examined the evolution of political organization across societies in Island Southeast Asia and the Pacific, and showed that hierarchical complexity of societies evolved through sequential, incremental increases in the number of decision-making levels present in a society, with decreases in complexity also occurring. This is consistent with the idea that societies get bigger by joining together smaller groups, and that they need to develop new institutions and other social mechanisms in order to do this in a stable manner (see below). These techniques can also test adaptive hypotheses about the relationships between different traits or aspects of their ecology. Holden and Mace (2005) showed that across societies in sub-Saharan Africa, the adoption of cattle keeping led matrilineal societies (in which wealth is passed down the female line, and were typically associated with small-scale farming) to change inheritance systems and pass wealth down through the male line (patriliny) (see Figure 3.3). This is consistent with adaptive hypotheses that argue that parents should invest preferentially in one sex or the other depending on the prevailing socio-ecological conditions (Holden et al. 2003).

A number of issues remain about the application of phylogenetic methods to cultural systems. A common reservation is that horizontal transmission or other forms of

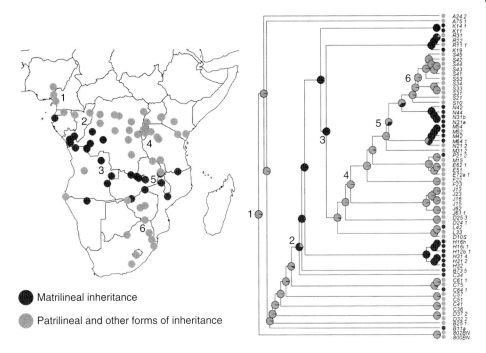

Figure 3.3 Phylogenetic techniques can be applied to study cultural evolution. In this example, the historical relationships between societies from across a large area of sub-Saharan Africa (*left*) can be represented by a phylogenetic tree (*right*). This tree was constructed using comparative linguistic data and is based on analyses in Currie et al. (2013) (codes at the tips refer to different languages). The inferred locations of numbered nodes are shown on the map and indicate the main route taken in the large-scale 'Bantu' population expansion that began from a homeland in modern-day Nigeria/Cameroon (node 1) around 3–5000 years ago. Other inferences can be made by mapping other traits onto such trees. Here, we can see variation across societies in inheritance systems in terms of whether group membership and property inheritance are traced through the female line (matrilineal inheritance), whether it traced through males (patrilineal inheritance), or whether some other system is employed (data taken from Holden et al. 2003). Based on the distribution of this trait across the tips of our tree, we can infer whether ancestral societies were more likely to be matrilineal or practise some other form of inheritance (the pie charts at the nodes represent proportional probabilities of these two possibilities) (see Currie et al. 2010a). From this analysis it can be seen that the society that was ancestral to all modern-day Bantu societies (node 1) did probably not practise matriliny, and that matrilineal inheritance probably emerged at several different points in the evolutionary history of the societies (e.g. node 2).

non-vertical inheritance (e.g. 'hybridization' or large-scale 'blending' of cultures, dialect chains, etc.) may be more common in cultural systems and thus these techniques are not appropriate (Borgerhoff Mulder et al. 2006; Nunn et al. 2006). However, transmission isolating mechanisms (TRIMs) can decrease the probability that cultural traits from another group will be copied, even in cases where individuals might physically move between and join other groups (Durham 1992). TRIMs can be active in that they identify the source of cultural traits and either favour traits that originate in the local population (e.g. active promotion of conformity, norms encouraging parochialism, or promotion of an individual's ethnic group) or discourage the adoption of traits from other populations (e.g. xenophobic norms). For example, the Academie Francaise is an official body in France that tries to prevent anglicization of the French language, and even recommends and promotes the use of new words derived from French rather than the adoption of loanwords to describe new terms (Pagel and Mace 2004). TRIMs can also be more passive and can occur if copying is frequency dependent such that individuals will tend to copy the behaviour of the majority of individuals. In such a scenario, this would mean that if a limited number of migrants enter a population, they are unlikely to transmit their behaviour to individuals such that it spreads widely in the new population. Furthermore, as previously stated above, this view often underestimates the extent to which, at the macro-scale, non-vertical forms of transmission occur in biological evolution.

Ultimately, whether horizontal transmission affects phylogenetic analyses is an empirical question. Measures of support are often used to assess the degree to which the data being used can be explained by a branching model of evolution (Douady et al. 2003; Collard et al. 2006). The use of Bayesian methods of inference, which allow us to assess and incorporate various sources of uncertainty into our analyses, is particularly important in this respect (Greenhill and Gray 2005; Holden et al. 2005). We can also assess how sensitive our inferences are to different assumptions. Previous simulation work has demonstrated that both phylogenetic inference (Greenhill et al. 2009) and comparative methods (Currie et al. 2010b) are robust to degrees of non-vertical transmission that are typically seen in the kinds of cultural datasets to which such methods have been applied. A challenge for future work in this area will be to refine these techniques so as to consider the mode of horizontal transmission and the effect it may have on analyses (Currie et al. 2010b). Just as in studies of biological evolution, the use of phylogenetic networks, cophylogenetic techniques (such as those used in analyses of host–pathogen coevolution) (Tehrani et al. 2010; Baudet et al. 2015), and comparative techniques that model different transmission processes (such as modelling the effects of both phylogeny *and* geography) (Freckleton and Jetz 2009; Towner et al. 2012) are likely to be important in this regard. There are clearly challenges involved in applying phylogenetic methods to cultural systems but these techniques have proven extremely valuable in helping us move beyond purely verbal arguments and plausible narratives. By employing these kinds of explicit quantitative approaches, we can better frame questions and provide more explicit tests of competing hypotheses about cultural macroevolution.

3.5 Culture and the Evolution of Co-operation

Humans are a highly social and co-operative species. Co-operation presents an evolutionary puzzle because individuals engage in some kind of costly behaviour that provides benefits to other individuals (West et al. 2007; Gardner and Foster 2008; Cronk 2015). Even when individuals gain some kind of net advantage from co-operating, such collective behaviours run into many potential challenges (Olson 1965). The production of public goods is vulnerable to free-riding as people can gain the benefits of the public good without paying the costs. The collective use of common-pool resources is vulnerable to the 'tragedy of the commons' as those who do not restrain themselves from using the resources will do better, thus leading to the overexploitation of the resource (Ostrom 1990). Common solutions to these problems seen in biological systems, such as reciprocity (Trivers 1971; Nowak and Sigmund 2005) and inclusive fitness (Hamilton

1964), run into difficulties when trying to explain the existence of the huge societies composed of millions of genetically unrelated individuals that we live in today.

Culture may play an important role in enabling large-scale co-operation in humans. However, social learning by itself does not necessarily promote co-operation. In fact, it may even be harmful especially if social learning is pay-off biased, as free-riders are more likely to be copied due to the fact that their pay-offs are highest (Lehmann et al. 2008). While these models provide important insights, they do not take into account the specific ways in which humans are structured into groups, which is an important feature of human social organization (Boyd and Richerson 2010). In particular, culture enables groups to construct social norms and rules (or institutions) that help stabilize co-operative behaviour (Richerson and Henrich 2012; Powers et al. 2016). Furthermore, the group-level nature of cultural variation can enable co-operative behaviours to spread through a process of cultural group selection (CGS) (Richerson et al. 2014).

Human societies are structured by what we refer to as 'institutions', which are collections of interrelated social norms and rules that proscribe roles and set expectations about social interactions, including the consequences of not following these rules (Currie et al. 2016). These rules can alter the pay-offs to different strategies and in effect change a 'game' in which the evolutionarily stable solution is to free-ride to one in which the best strategy is to co-operate (Powers et al. 2016). This can happen, for example, when the rule states that if an individual defects (i.e. doesn't uphold their end of an agreement) then they will receive some kind of punishment. For example, in irrigation systems in Nepal, communities have rules about maintenance of the irrigation system, and how much water can be taken (Anderies and Janssen 2016). If an individual is deemed to have not followed the rules then they receive a punishment. The transgressor has one of their cows placed in a pen in the middle of the village. Since the community is small, everyone in the village knows whose cow this is, and also other villagers are able to take milk from this cow. The cow is only released once a fine is paid. Therefore, someone who breaks the irrigation rules suffers directly by having to pay the fine and from losing the ability to milk the cow whilst it is in the pen. They also suffer reputational damage which, in turn, may entail opportunity costs by inhibiting future interactions with other community members.

In humans, norms and institutions are made possible through a combination of language, other forms of social learning, and aspects of our social cognition involving shared intentionality and theory of mind. Language is important in discussing, agreeing, and communicating what the rules are. Social learning is important in transmitting what the rules are to everyone in the group, and individuals can also learn what happens when rules are broken and the likely pay-offs for either co-operating or defecting by observing others. For institutions like this to work, transgressions must be punished, and be seen to be punished, in order for compliance with the rules to be maintained and co-operation to be sustained (Ostrom 1990; Boyd and Richerson 1992; Anderies and Janssen 2016).

To what extent institutions are possible in non-linguistic species or in species without our psychological capabilities is an open question (Akçay et al. 2013). These features certainly make it easier to develop and transmit rules, but conceivably mechanisms surrounding reputation (and indirect reciprocity), punishment, and expectations about behaviour and the consequences of defecting could be transmitted via social learning in other species to create behavioural patterns that were similar in some respects to institutions. Because they appear to be peculiar to humans, institutions have received relatively little attention in evolutionary models of co-operation, even when applied to humans. While institutions have been a focus of study in several social science fields, understanding how institutions interact with other culturally inherited traits and how both evolve together over time has received less formal attention and is an important topic for future research in cultural evolution (Currie et al. 2016).

Features such as institutions and social norms are group-level emergent properties of the interactions between individuals that are made possible by social learning (Smaldino 2014). This has implications for how selection can act on groups and lead to the spread of co-operative behaviour. If there is competition between groups then traits that lead to increased group survival, reproduction or recruitment can spread via a process of CGS (Richerson et al. 2014). In genetic models of group selection, the genetic variation between groups that is required for selection at this level to act upon can be destroyed by even low levels of migration of individuals between groups (Williams 1966; Okasha 2008). However, cultural evolutionary processes may create conditions under which selection between groups can become an important force. As we saw in a previous section, frequency-dependent processes and active conformism can act to maintain cultural variation between groups even in the face of physical migration of individuals (Boyd and Richerson 1985b; Bell et al. 2009). Importantly, as we are dealing with a cultural process, the extinction of groups does not have to involve the death of all group members (Soltis et al. 1995). Individuals can be incorporated within the victorious group or disperse into other groups. The key point is that the previous group and its associated norms and institutions no longer exist. For example, Lindenbaum (2008) describes how the Fore of New Guinea defeated a neighbouring group in warfare and absorbed those who were not killed; the survivors had their origins 'made invisible' and they 'became Fore' (p. 3716). Groups may also go extinct if individuals migrate based on the traits that different groups exhibit, or due to the perceived success of different groups (Boyd and Richerson 2010). Biased copying of traits, based either on the traits themselves or indirectly on the success of groups that possess them, can also lead to the spread of certain traits and has been argued to be a form of CGS.

The idea of CGS has not met with universal acceptance (see Richerson et al. (2014) and associated commentaries). This appears to be partly due to the long and controversial history of group selection in biology and a conflation of CGS models with early naïve group selection models in biology. More generally, there also appears to be a confusion around how terms are defined and used in different research traditions: What are the most salient units of analysis, and what are appropriate measures in attempting to assess these models? An important issue that is common to all group selection models is how group fitness should be measured (Okasha 2008). Is it an additive measure of individual fitness? The number of descendent groups? The stability or longevity of a group? Or simply the frequency of the group-level trait in the population? Indeed, it appears that the different 'modes' of CGS mentioned above (group extinction, migration, biased adoption) are associated with different measures of fitness. Furthermore, the relationship between cultural group fitness and individual genetic fitness in verbal descriptions of CGS is not entirely clear and is often not dealt with explicitly. This has consequences for understanding the extent to which culture may have shaped genetic evolution (see below) in relation to the evolution of genetic traits that would help canalize prosocial behaviours.

A lot of the research on CGS to date has focused on establishing whether the assumptions underpinning CGS are supported. For example, researchers have used cross-national survey data to assess the extent to which sufficient variation in traits exists between groups (Bell et al. 2009; Richerson et al. 2014). Measures of the fixation index (FST, which quantifies the degree of variation between populations) were calculated to be roughly an order of magnitude larger for cultural traits than genetic FST, indicating that the range of scenarios under which CGS might be an important force is greater than for genetic group selection. The results also indicated that for many cultural traits, the FSTs were sufficiently high that only relatively small group-level benefits would be needed to outweigh individual-level costs. In an analysis of cultural group extinction in New Guinea, Soltis et al. (1995) found that rates of extinction varied from about 2% to 32% of groups per generation, and estimated that under such conditions traits would take 500–1000 years to spread through the metapopulation. While this rate of change makes CGS an unlikely mechanism for changes in many cultural traits, it could explain changes in slowly evolving traits such as political institutions (Carneiro 1978; Spencer and Redmond 2001; Turchin et al. 2013). However, some authors have conducted empirical studies of co-operation in different populations that challenge the assumptions about the degree of variation within and between groups, and the extent to which competition between groups is associated with in-group co-operation (Lamba and Mace 2011; Mace and Silva 2016).

Future work also needs to test the predictions of CGS hypotheses, in order to assess how well they can explain real-world data (Turchin and Currie 2016). As an example of the kind of approach that is possible, Turchin et al. (2013) developed an agent-based simulation within a CGS framework and matched the outputs of these models against data on the historical distributions of large-scale societies. The results of this study suggest that between-group competition can select for increasing group size which is facilitated by the adoption of costly institutions involved in complex social organization. CGS is potentially an important process in cultural evolution, and the challenge for future work in this area is to clarify certain concepts and provide further empirical tests of both the assumptions and predictions of CGS hypotheses.

3.6 Gene–Culture Coevolution

The above discussions have largely considered cultural evolution as a separate process from genetic evolution. However, the messy truth is that the two are intertwined (Figure 3.4). We have already discussed how genes keep cultural evolution on something of a long leash, and that genes will build organisms that will on average learn from others in ways that are adaptive. In this section, I give further consideration to the possibility that the two can coevolve such that biological evolution can have important consequences for cultural evolution and vice versa, a process we can call gene–culture coevolution (Lumsden and Wilson 1980; Cavalli-Sforza and Feldman 1981; Boyd and Richerson 1985b; Durham 1991; Feldman and Laland 1996; Laland et al. 2010).

The adoption of a new cultural practice can expose genes to new selection pressures in two ways (Richerson et al. 2010). First, cultural traits can make survival and reproduction possible in novel environments and facilitate migration into environments to which genes had not previously been exposed. As noted above, anatomically modern

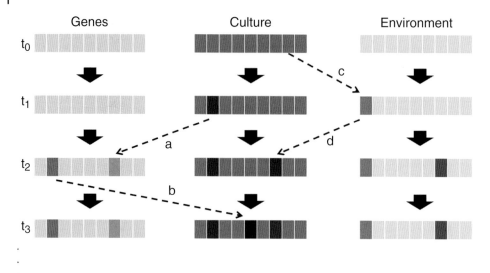

Figure 3.4 Genes and culture represent different systems of inheritance which can coevolve. Here genes, culture, and the environment are represented in an artificially abstract manner as collections of specific units at different points in time (t_0, t_1, …t_n). Changes in different aspects of these systems are represented by changes in colour. Filled black arrows represent the biological inheritance of genes, transmission of cultural information, and continuity over time in the environment (sometimes referred to as 'environmental inheritance'). Aspects of one system can permanently alter or cause evolutionary change in another system. Examples are given of (a) a cultural innovation creating a novel selection pressure on a gene, (b) a genetic mutation that facilitates a new cultural trait, (c) a cultural modification of the environment, that (d) feeds back and causes a response in a different aspect of culture.

human populations that left tropical Africa were able to move to cooler environments around the world due to the cultural evolution of suitable clothing, shelter, and the use of fire. Exposure to these new environments selected for genetic changes associated with more robust physiques and lighter skin pigmentation, that were better adapted to colder temperatures and lower light levels. Under this mechanism, the relationship between cultural evolution and biological evolution is somewhat indirect, and might be more accurately termed *culture-mediated genetic evolution*.

Second, the adoption of cultural practices can also lead to more direct selection pressures whereby the effectiveness of a cultural adaptation is supported or enhanced by subsequent genetic changes. Perhaps the most studied example of *direct gene–culture coevolution* is the adoption of dairy farming and the evolution of lactose tolerance (Holden and Mace 1997; Tishkoff et al. 2007; Itan et al. 2009; Laland et al. 2010; Richerson et al. 2010). Globally, drinking milk makes most adults ill. This is because the production of the enzyme for digesting milk (lactase) is switched off in most individuals after infancy. The ability to drink and digest milk provides certain adaptive advantages – it provides a regular source of calories and important nutrients such as vitamin D and calcium. Genetically, lactose tolerance is linked to the possession of the appropriate allele of a single gene (MCM6). The frequency of individuals possessing the lactose tolerance allele is higher in cultures that have a history of keeping sheep, goats, or cattle and commonly drink milk from these animals.

Direct gene–culture coevolution can also occur when a cultural practice or form of social organization creates a change in the social or external environment leading to a change in genetic selection pressures. A potential example is yam farming in Africa and sickle cell anaemia. Yam farming involves chopping down forests, which leads rain to collect and create standing bodies of water, which is an environment conducive to malaria-bearing mosquitoes. It is argued that this drove subsequent selection on genes that cause sickle cell anaemia, which can lead to premature death but also provides protection against malaria (Durham 1991) (but see below). The genetic changes that occur in both culturally mediated genetic evolution and direct gene–culture coevolution can of course result in further feedback processes such that the cultural change is stabilized by the genetic changes.

The theoretical possibility of gene–culture coevolution is long established, and can be more generally linked to the ideas of niche construction (Odling-Smee et al. 2003) (see Figure 3.4) and the Baldwin effect (i.e. that learned behaviour can in certain circumstances affect genetic evolution) (Sznajder et al. 2012). Whilst there has been a lot of theoretical development of the idea of gene–culture coevolution (including a large number of mathematical models), empirical studies and concrete examples of gene–culture coevolution have been fewer. The evolution of lactose tolerance remains very much the go-to example of gene–culture coevolution due to the twin facts that the genetics underlying lactose tolerance are relatively simple and it relates to a cultural innovation that occurred relatively recently in certain parts of the world.

Recent years have seen an explosion in our knowledge of human genetics and this may help to provide evidence for gene–culture coevolution. Genetic and genomic techniques enable us to identify regions of the genome that show statistical signatures of having undergone selection (Hawks et al. 2007; Cochran and Harpending 2010; Laland et al. 2010; Richerson et al. 2010). The function of many putatively selected genes can be inferred based on previous knowledge or inferences about genotype–phenotype relationships (e.g. genetic screens, knock-out studies, gene expression analysis). A pattern emerges from these analyses (Tishkoff 2015) in that they commonly relate to diet (detoxification of plant secondary compounds, alcohol metabolism, digestion of milk and dairy products, processing of high levels of dietary fat in Inuit populations), immunity or responses to pathogens (e.g. sickle cell anaemia in response to malaria, cholera resistance), or responses to physical environmental conditions (e.g. cold climate, altitude, light skin pigmentation in response to low levels of vitamin D, and short stature in rainforest-dwelling populations). While in all these cases it is clear to see how conditions could affect survival and reproduction and that biological responses would increase genetic fitness, many of these examples may reflect culture-mediated genetic evolution rather than direct gene–culture coevolution.

Another possibility is that absence of evidence really is evidence of absence. There are a number of reasons why culture may not exert enough of a selective force on genes for coevolution to be important. First, cultural innovations could reduce selective pressures on genes. If a novel environment presents some kind of adaptive problem that can be solved genetically or culturally, and a cultural solution is found, it will reduce the strength of selection acting on genes, making genetic evolution less likely. Here again is a potential reason why lactose tolerance is a convincing example; the physiological ability to digest lactose in later life is a biological process for which a cultural response is more difficult (although certain milk-processing techniques such as cheese making and ageing reduce lactose content). Another potential factor affecting the strength of selection on genes is the rate at which culture evolves. Given that cultural transmission is not linked to reproduction then cultural evolution can occur at a faster rate than genetic evolution (Perreault 2012), meaning that the effective environment is not consistent enough to incur a genetic response. This is not to say that all cultural change is extremely rapid, and many cultural traditions can persist over many generations.

We have a lot of evidence that human genes have been under selection in recent human history and we can make plausible arguments for the role of culture in either mediating or driving such adaptations. Future work will need to go beyond such plausible narratives and more rigorously test gene–culture coevolutionary hypotheses against competing explanations. For example, Itan et al. (2009) use demographic simulation models fitted to modern genetic data in combination with information from ancient DNA and archaeology in order to test more rigorously the timing, location, and strength of selection on alleles related to lactose tolerance. Ideas relating to earlier stages of human evolution may prove tricky to assess in this respect. A gene–culture coevolution account of the emergence of the human language faculty is plausible, but is hard to test as we know so little about the early stages of language evolution. However, our ability to assess such hypotheses can increase if we know more about more recent examples and the conditions under which gene–culture coevolution is most likely to occur.

So while gene–culture coevolution represents an important process in some aspects of human adaptation, the presence of culture does not automatically mean that gene–culture coevolution will be a major factor in explaining behavioural diversity. This has implications for understanding the potential importance of the effects of culture or social learning on genetic evolution in other species. There are currently few convincing examples of gene–culture coevolution in non-human animals. The socially learned behaviour of using marine sponges whilst foraging has been shown to be linked with geographical patterns of mtDNA genetic structure due to the fact that this sponging behaviour is transmitted from mothers to offspring (Kopps et al. 2014). This, however, represents correlated evolution between genes and culture rather than evidence that any particular alleles have been selected due to this sponging behaviour. Situations where socially learned behaviours have enabled species to move into novel environments or be exposed to new foodstuffs, or where social learning has led to environmental modifications may represent potential candidates for assessing the existence of culturally influenced genetic selection in other species. The human reliance on culture as a means of adaptation and our capacity for cumulative culture, with high-fidelity transmission and long-lasting behavioural traditions, may mean that gene–culture coevolution is more likely in humans than other species.

3.7 Conclusion

From the above discussion, it can be seen that there are important analogies to be drawn between the processes of genetic and cultural evolution. There are also important differences in the mechanisms of these two inheritance systems and we should not seek to constrain our understanding of evolutionary processes by insisting that genes are the only important system of inheritance. As Mesoudi (2011) has argued, cultural evolution can be characterized as Darwinian although it is not necessarily neo-Darwinian.

Variation in cultural traits exists and can be generated by a number of processes, some of which may be essentially random, some of which are more guided. These traits can be inherited or transmitted between individuals and populations, but are not restricted to the parent-to-offspring route. The frequency of traits within a population can be shaped by different transmission biases leading to selection of different traits. Furthermore, if there is competition between groups, then cultural variation between groups may explain which groups win out. Also, while the possibility that selection can occur in cultural systems is important to recognize (and underlines the relevance of taking an evolutionary approach), cultural change may also be shaped by non-selective drift-like mechanisms or transformative psychological biases. Overly rigid attempts to apply genetic mechanisms to cultural evolution have not proven to be productive frameworks for investigating human behaviour. For example, 'selectionist' approaches in archaeology (Dunnell 1980) attempted to argue that all cultural mutations were random and that changes in the frequency of artefacts found in the archaeological record could be explained by natural selection affecting the biological fitness of populations (Boone and Smith 1998). Similarly, advocates of 'memetic' approaches to cultural evolution focused too heavily on trying to draw parallels between genes as units of biological evolution and memes as units of cultural evolution, when such a step is not necessary (Laland and Brown 2002).

The modern cultural evolutionary perspective helps illustrate the importance of not restricting evolutionary explanations to genetic changes. If we focus too narrowly on genetic evolution, we risk 'missing the phenotype'. Emphasizing phenotypes is particularly pertinent for at least two reasons. First, natural selection generally acts on phenotypes rather than directly on genes. Second, for behavioural biologists the phenotype represents the features of organisms we directly observe and is the thing we want to explain. For example, in discussing the potential for maladaptive behaviour caused by cultural inheritance, El Mouden et al. (2014) recast this in genetic adaptationist terms as being simply the genetic system not yet at equilibrium, or constraints on the genetically determined mechanisms that shape behaviour. However, by taking this view, there is a risk of missing what it is about the cultural trait or the context of the situation that leads traits that are genetically deleterious to spread. While it is possible, and perfectly valid, to frame things around the causes and consequences of genetic evolution, such an approach leaves unexplained important aspects of the processes that ultimately shape phenotypes, including behaviour. Taking a dual-inheritance perspective provides a broader framework in which to assess competing hypotheses about the function of behaviours.

In this chapter, I have attempted to demonstrate the ways in which this approach is able to integrate existing findings and provide novel insights, but also stressed some of the important goals and challenges of future research in this area. I have argued that explicitly incorporating culture as a system of inheritance, which can change via evolutionary processes of descent with modification, is essential for understanding behavioural diversity in species that rely to a great extent on social learning. Rather than culture being something that sets us apart from the rest of the animal kingdom, this scientific approach to culture helps illustrate the continuities between ourselves and other species. This perspective helps demonstrate that both genetic and cultural information are important for shaping the behaviour of animals, but equally that genetic and cultural processes are not independent: culture and cultural evolution relies on genetically shaped biological capacities, and genetic evolution can in turn be shaped by culture. This is true even in our own species where culture appears an essential means by which we adapt to, and structure, the world around us.

Acknowledgements

The author is supported by funding from the European Research Council (ERC) under the European Union's Horizon 2020 research and innovation programme (Project title: '*The Cultural Evolution and Ecology of Institutions: An integrated evolutionary approach to the interrelated rules that regulate human social organization and cooperation*', Grant Agreement 716212).

References

Akçay, E., J. Roughgarden, J. D. Fearon, J. A. Ferejohn, and B. R. Weingast. 2013. Biological Institutions: The Political Science of Animal Cooperation. Available at: https://pdfs .semanticscholar.org/f869/70b055bc6a56997825ccf9319e78fcfed668.pdf

Allen, J., Weinrich, M., Hoppitt, W., and Rendell, L. (2013). Network-based diffusion analysis reveals cultural transmission of Lobtail feeding in humpback whales. *Science* 340: 485–488.

Anderies, J. and Janssen, M. (2016). *Sustaining the Commons*. Arizona State University.

Aplin, L.M., Farine, D.R., Morand-Ferron, J. et al. (2015). Experimentally induced innovations lead to persistent culture via conformity in wild birds. *Nature* 518: 538–541.

Arnold, S.J. (1992). Constraints on phenotypic evolution. *American Naturalist* 20: S85–S107.

Aunger, R. (2000). *Darwinizing Culture: The Status of Memetics as a Science*. Oxford: Oxford University Press.

Barkow, J.H., Cosmides, L., and Tooby, J. (eds.) (1992). *The Adapted Mind: Evolutionary Psychology and the Generation of Culture*. Oxford: Oxford University Press.

Barlow, M. (2009). What antimicrobial resistance has taught us about horizontal gene transfer. In: *Horizontal Gene Transfer: Genomes in Flux* (ed. M.B. Gogarten, J.P. Gogarten and L.C. Olendzenski), 397–411. Totowa: Humana Press.

Bates, L. and Chappell, J. (2002). Inhibition of optimal behavior by social transmission in the guppy depends on shoaling. *Behavioral Ecology* 13: 827–831.

Baudet, C., Donati, B., Sinaimeri, B. et al. (2015). Cophylogeny reconstruction via an approximate Bayesian computation. *Systematic Biology* 64: 416–431.

Bell, A.V., Richerson, P.J., and McElreath, R. (2009). Culture rather than genes provides greater scope for the evolution of large-scale human prosociality. *Proceedings of the National Academy of Sciences* 106: 17671–17674.

Blackmore, S.J. (2000). *The Meme Machine*. Oxford: Oxford University Press.

Blomberg, S.P. and Garland, T. (2002). Tempo and mode in evolution: phylogenetic inertia, adaptation and comparative methods. *Journal of Evolutionary Biology* 15: 899–910.

Bonnie, K.E., Horner, V., Whiten, A., and de Waal, F.B.M. (2007). Spread of arbitrary conventions among chimpanzees: a controlled experiment. *Proceedings of the Royal Society B: Biological Sciences* 274: 367–372.

Boone, J.L. and Smith, E.A. (1998). Is it evolution yet? A critique of evolutionary archaeology. *Current Anthropology* 39: S141–S173.

Borgerhoff Mulder, M., Nunn, C.L., and Towner, M.C. (2006). Cultural macroevolution and the transmission of traits. *Evolutionary Anthropology: Issues, News, and Reviews* 15: 52–64.

Bouckaert, R., Lemey, P., Dunn, M. et al. (2012). Mapping the origins and expansion of the Indo-European language family. *Science* 337: 957–960.

Boyd, R. and Richerson, P. (1985a). *Culture and the Evolutionary Process*. Chicago: Chicago University Press.

Boyd, R. and Richerson, P.J. (1985b). *Culture and the Evolutionary Process*. Chicago University Press, Chicago.

Boyd, R. and Richerson, P.J. (1992). Punishment allows the evolution of cooperation (or anything else) in sizable groups. *Ethology and Sociobiology* 13: 171–195.

Boyd, R. and Richerson, P.J. (1996). Why culture is common, but cultural evolution is rare. In: *Proceedings of the British Academy Vol 88 Evolution of Social Behaviour Patterns in Primates and Man* (ed. W.G. Runciman, J.M. Smith and R.I.M. Dunbar), 77–94. Oxford: Oxford University Press.

Boyd, R. and Richerson, P.J. (2010). Transmission coupling mechanisms: cultural group selection. *Philosophical Transactions of the Royal Society B: Biological Sciences* 365: 3787–3795.

Boyd, R., Richerson, P.J., and Henrich, J. (2011). The cultural niche: why social learning is essential for human adaptation. *Proceedings of the National Academy of Sciences* 108: 10918–10925.

Brodie, R. (2009). *Virus of the Mind: The New Science of the Meme*. Hay House, Carlsbad.

Brown, S., Savage, P.E., Ko, A.M.-S. et al. (2014). Correlations in the population structure of music, genes and language. *Proceedings of the Royal Society B: Biological Sciences* 281.

Buss, D. (2015). *Evolutionary Psychology: The New Science of the Mind*, 5e. Abingdon: Taylor & Francis.

Caldwell, C.A. and Millen, A.E. (2008). Experimental models for testing hypotheses about cumulative cultural evolution. *Evolution and Human Behavior* 29: 165–171.

Calude, A.S. and Pagel, M. (2011). How do we use language? Shared patterns in the frequency of word use across 17 world languages. *Philosophical Transactions of the Royal Society B: Biological Sciences* 366: 1101–1107.

Cantor, M., Shoemaker, L.G., Cabral, R.B. et al. (2015). Multilevel animal societies can emerge from cultural transmission. *Nature Communications* 6: 8091.

Carneiro, R.L. (1978). Political expansion as an expression of the principle of competitive exclusion. In: *Origins of the State* (ed. R. Cohen and E.R. Service), 205–223. Philadelphia: Institute for the Study of Human Issues.

Cavalli-Sforza, L.L. and Feldman, M.W. (1981). *Cultural Transmission and Evolution: A Quantitative Approach*. Princeton: Princeton University Press.

Cialdini, R.B. and Trost, M.R. (1998). Social influence: social norms, conformity and compliance. In: *The Handbook of Social Psychology* (ed. D.T. Gilbert, S.T. Fiske and G. Lindzey), 151–192. New York: McGraw-Hill.

Cochran, G. and Harpending, H. (2010). *The 10,000 Year Explosion: How Civilization Accelerated Human Evolution*. New York: Basic Books.

Collard, M., Shennan, S.J., and Tehrani, J.J. (2006). Branching, blending, and the evolution of cultural similarities and differences among human populations. *Evolution and Human Behavior* 27: 169–184.

Coward, F., Shennan, S., Colledge, S. et al. (2008). The spread of Neolithic plant economies from the near east to Northwest Europe: a phylogenetic analysis. *Journal of Archaeological Science* 35: 42–56.

Cronk, L. (2015). Human cooperation: evolutionary approaches to a complex phenomenon. In: *Handbook on Evolution and Society: Toward an Evolutionary Social Science* (ed. J. Turner, R. Machalek and A. Maryanski), 441–459. St Paul: Paradigm Publishing.

Currie, T.E. (2013). Cultural evolution branches out: the phylogenetic approach in cross-cultural research. *Cross-Cultural Research* 47: 102–130.

Currie, T.E., Greenhill, S.J., Gray, R.D. et al. (2010a). Rise and fall of political complexity in island South-East Asia and the Pacific. *Nature* 467: 801–804.

Currie, T.E., Greenhill, S.J., and Mace, R. (2010b). Is horizontal transmission really a problem for phylogenetic comparative methods? A simulation study using continuous cultural traits. *Philosophical Transactions of the Royal Society B: Biological Sciences* 365: 3903–3912.

Currie, T.E. and Meade, A. (2014). Keeping yourself updated: Bayesian approaches in phylogenetic comparative methods with a focus on Markov chain models of discrete character evolution. In: *Modern Phylogenetic Comparative Methods and their Application in Evolutionary Biology.* (ed. L. Garamszegi). Berlin: Springer-Verlag.

Currie, T.E., Meade, A., Guillon, M., and Mace, R. (2013). Cultural phylogeography of the Bantu languages of sub-Saharan Africa. *Proceedings of the Royal Society B: Biological Sciences* 280.

Currie, T.E., Turchin, P., Bednar, J. et al. (2016). The evolution of institutions and organizations. In: *Complexity and Evolution: A New Synthesis for Economics* (ed. D.S. Wilson, A. Kirkman and E. Beinhoffer). Cambridge: MIT Press.

Darwin, C. (1859). *On the Origin of Species.* Oxford: Oxford University Press.

Dawkins, R. (1976). *The Selfish Gene.* Oxford: Oxford University Press.

Dawkins, R. (1986). *The Blind Watchmaker: Why the Evidence of Evolution Reveals a Universe without Design.* New York: Norton.

Dawkins, R. (1993). Viruses of the mind. In: Dennett and His Critics: Demystifying Mind. (ed. B. Dahlbom). Chichester: Wiley.

De Waal, F.B.M. (2001). *The Ape and the Sushi Master: Cultural Reflections of a Primatologist.* New York: Basic Books.

Dennett, D.C. (1996). *Darwin's Dangerous Idea: Evolution and the Meanings of Life.* New York: Simon & Schuster.

Dennett, D.C. (2006). *Breaking the Spell: Religion as a Natural Phenomenon.* New York: Viking.

Diggins, F.W. (1999). The true history of the discovery of penicillin, with refutation of the misinformation in the literature. *British Journal of Biomedical Science* 56: 83.

Douady, C.J., Delsuc, F., Boucher, Y. et al. (2003). Comparison of Bayesian and maximum likelihood bootstrap measures of phylogenetic reliability. *Molecular Biology and Evolution* 20: 248–254.

Dunnell, R.C. (1980). Evolutionary theory and archaeology. *Advances in Archaeological Method and Theory* 3: 35–99.

Durham, W.H. (1991). *Coevolution: Genes, Culture, and Human Diversity*. Palo Alto: Stanford University Press.

Durham, W.H. (1992). Applications of evolutionary culture theory. *Annual Review of Anthropology* 21: 331–355.

Eagleton, C. and Spencer, M. (2006). Copying and conflation in Geoffrey Chaucer's treatise on the astrolabe: a stemmatic analysis using phylogenetic software. *Studies in History and Philosophy of Science* 37: 237–268.

El Mouden, C., André, J.B., Morin, O., and Nettle, D. (2014). Cultural transmission and the evolution of human behaviour: a general approach based on the Price equation. *Journal of Evolutionary Biology* 27: 231–241.

Eyre-Walker, A. and Keightley, P.D. (2007). The distribution of fitness effects of new mutations. *Nature Reviews Genetics* 8: 610–618.

Feldman, M.W. and Laland, K.N. (1996). Gene-culture co-evolutionary theory. *Trends in Ecology & Evolution* 11: 453–457.

Filatova, O., Burdin, A., and Hoyt, E. (2010). Horizontal transmission of vocal traditions in killer whale (Orcinus orca) dialects. *Biology Bulletin* 37: 965–971.

Fought, C. (2006). *Language and Ethnicity*. Cambridge: Cambridge University Press.

Franz, M. and Matthews, L.J. (2010). Social enhancement can create adaptive, arbitrary and maladaptive cultural traditions. *Proceedings of the Royal Society B: Biological Sciences* 277: 3363–3372.

Freckleton, R.P. and Jetz, W. (2009). Space versus phylogeny: disentangling phylogenetic and spatial signals in comparative data. *Proceedings of the Royal Society B: Biological Sciences* 276: 21–30.

Futuyma, D.J. (2013). *Evolution*. Sunderland: Sinauer Associates.

Gardner, A. and Foster, K. (2008). The evolution and ecology of cooperation – history and concepts. In: *Ecology of Social Evolution* (ed. J. Korb and J. Heinze), 1–36. Berlin: Springer.

Ge, F., Wang, L.-S., and Kim, J. (2005). The cobweb of life revealed by genome-scale estimates of horizontal gene transfer. *Plos Biology* 3: e316.

Gray, R.D. and Atkinson, Q.D. (2003). Language-tree divergence times support the Anatolian theory of Indo-European origin. *Nature* 426: 435–439.

Gray, R.D., Atkinson, Q.D., and Greenhill, S.J. (2011). Language evolution and human history: what a difference a date makes. *Philosophical Transactions of the Royal Society of London B: Biological Sciences* 366: 1090–1100.

Gray, R.D., Drummond, A.J., and Greenhill, S.J. (2009). Language phylogenies reveal expansion pulses and pauses in Pacific settlement. *Science* 323: 479–483.

Greenhill, S.J., Currie, T.E., and Gray, R.D. (2009). Does horizontal transmission invalidate cultural phylogenies? *Proceedings of the Royal Society B: Biological Sciences* 276: 2299–2306.

Greenhill, S.J. and Gray, R.D. (2005). Testing population dispersal hypotheses: Pacific settlement, phylogenetic trees and Austronesian languages. In: *The Evolution of Cultural Diversity: A Phylogenetic Approach* (ed. R. Mace, C. Holden and S. Shennan), 31–52. London: UCL Press.

Gregory, T.R. (2008). Understanding evolutionary trees. *Evolution: Education and Outreach* 1: 121–137.

Griliches, Z. (1960). Hybrid corn and the economics of innovation. *Science* 132: 275–280.

Hamilton, W.D. (1964). The genetical evolution of social behaviour I. *Journal of Theoretical Biology* 7: 1–16.

Harrison, R.G. and Larson, E.L. (2014). Hybridization, introgression, and the nature of species boundaries. *Journal of Heredity* 105: 795–809.

Harvey, P.H. and Pagel, M.D. (1991). *The Comparative Method in Evolutionary Biology.* Oxford: Oxford University Press.

Hawks, J., Wang, E.T., Cochran, G.M. et al. (2007). Recent acceleration of human adaptive evolution. *Proceedings of the National Academy of Sciences* 104: 20753–20758.

Heggarty, P. (2008). Linguistics for archaeologists: a case-study in the Andes. *Cambridge Archaeological Journal* 18: 35–56.

Henrich, J. (2001). Cultural transmission and the diffusion of innovations: adoption dynamics indicate that biased cultural transmission is the predominate force in behioral change. *American Anthropologist* 103: 992–1013.

Henrich, J. (2015). *The Secret of our Success: How Culture is Driving Human Evolution, Domesticating Our Species, and Making Us Smarter.* Princeton: Princeton University Press.

Henrich, J., Boyd, R., and Richerson, P.J. (2008). Five misunderstandings about cultural evolution. *Human Nature – An Interdisciplinary Biosocial Perspective* 19: 119–137.

Henrich, J. and Gil-White, F.J. (2001). The evolution of prestige: freely-conferred deference as a mechanism for enhancing the benefits of cultural transmission. *Evolution and Human Behavior* 22: 165–196.

Henrich, J. and Henrich, N. (2010). The evolution of cultural adaptations: Fijian food taboos protect against dangerous marine toxins. *Proceedings of the Royal Society B: Biological Sciences* 277: 3715–3724.

Heyes, C.M. (1994). Social learning in animals: categories and mechanisms. *Biological Reviews* 69: 207–231.

Hock, H.H. and Joseph, B.D. (2009). *Language History, Language Change and Language Relationship an Introduction to Historical and Comparative Linguistics.* Berlin: Mouton de Gruyter.

Holden, C. and Mace, R. (1997). Phylogenetic analysis of the evolution of lactose digestion in adults. *Human Biology* 69: 605–628.

Holden, C.J. and Mace, R. (2005). "The cow is the enemy of matriliny": using phylogenetic methods to investigate cultural diversity in Africa. In: *The Evolution of Cultural Diversity: A Phylogenetic Approach* (ed. R. Mace, C.J. Holden and S. Shennan), 217–234. London: UCL Press.

Holden, C.J., Meade, A., and Pagel, M. (2005). Comparison of maximum parsimony and Bayesian Bantu language trees. In: *The Evolution of Cultural Diversity: A Phylogenetic Approach* (ed. R. Mace, C.J. Holden and S. Shennan). London: UCL Press.

Holden, C.J., Sear, R., and Mace, R. (2003). Matriliny as daughter-biased investment. *Evolution and Human Behavior* 24: 99–112.

Hoppitt, W. and Laland, K.N. (2013). *Social Learning: An Introduction to Mechanisms, Methods, and Models.* Princeton: Princeton University Press.

Huerta-Sanchez, E., Jin, X., Asan, Z. et al. (2014). Altitude adaptation in Tibetans caused by introgression of Denisovan-like DNA. *Nature* 512: 194–197.

Itan, Y., Powell, A., Beaumont, M.A. et al. (2009). The origins of lactase persistence in Europe. *PLoS Computational Biology* 5: e1000491.

Ji, T., Zheng, X.-D., He, Q.-Q. et al. (2016). Kinship as a frequency dependent strategy. *Royal Society Open Science* 3.

Jorde, L.B. and Wooding, S.P. (2004). Genetic variation, classification and 'race'. *Nature Genetics* 36: S28–S33.

Kandler, A. and Steele, J. (2009). Innovation diffusion in time and space: effects of social information and of income inequality. *Diffusion Fundamentals* 11: 1–17.

Kempe, M., Lycett, S.J., and Mesoudi, A. (2014). From cultural traditions to cumulative culture: parameterizing the differences between human and nonhuman culture. *Journal of Theoretical Biology* 359: 29–36.

Kirch, P.V. (2010). *How Chiefs Became Kings: Divine Kingship and the Rise of Archaic States in Ancient Hawai'i*. Berkeley: University of California Press.

Kitchen, A., Ehret, C., Assefa, S., and Mulligan, C.J. (2009). Bayesian phylogenetic analysis of Semitic languages identifies an early bronze age origin of Semitic in the near east. *Proceedings of the Royal Society B: Biological Sciences* 276: 2703–2710.

Kopps, A.M., Ackermann, C.Y., Sherwin, W.B. et al. (2014). Cultural transmission of tool use combined with habitat specializations leads to fine-scale genetic structure in bottlenose dolphins. *Proceedings of the Royal Society B: Biological Sciences* 281: 20133245.

Kroeber, A.L. and Kluckholm, C. (1952). *Culture: A Critical Review of the Concepts and Definitions*. Papers of the Peabody Museum of Archaeology and Ethnology, vol. 47, 1–223.

Kurzweil, R. (2000). *The Age of Spiritual Machines: When Computers Exceed Human Intelligence*. London: Penguin.

Laland, K.N. and Brown, G.R. (2002). *Sense and Nonsense*. Oxford: Oxford University Press.

Laland, K.N., Odling-Smee, J., and Myles, S. (2010). How culture shaped the human genome: bringing genetics and the human sciences together. *Nature Reviews Genetics* 11: 137–148.

Laland, K.N., Uller, T., Feldman, M.W. et al. (2015). The extended evolutionary synthesis: its structure, assumptions and predictions. *Proceedings of the Royal Society of London B: Biological Sciences* 282: 20151019.

Laland, K.N. and Williams, K. (1998). Social transmission of maladaptive information in the guppy. *Behavioral Ecology* 9: 493–499.

Lamba, S. and Mace, R. (2011). Demography and ecology drive variation in cooperation across human populations. *Proceedings of the National Academy of Sciences* 108: 14426–14430.

Lehmann, L., Feldman, M.W., and Foster, K.R. (2008). Cultural transmission can inhibit the evolution of altruistic helping. *American Naturalist* 172: 12–24.

Levitis, D.A., Lidicker, W.Z., and Freund, G. (2009). Behavioural biologists do not agree on what constitutes behaviour. *Animal Behaviour* 78: 103–110.

Lewis, H.M. and Laland, K.N. (2012). Transmission fidelity is the key to the build-up of cumulative culture. *Philosophical Transactions of the Royal Society B: Biological Sciences* 367: 2171–2180.

Lewis, M.P. (ed.) (2009). *Ethnologue: Languages of the World*, 6e. Dallas: SIL International.

Liberski, P.P. (2013). Kuru: a journey back in time from Papua New Guinea to the Neanderthals' extinction. *Pathogens* 2: 472–505.

Lindenbaum, S. (2008). Understanding kuru: the contribution of anthropology and medicine. *Philosophical Transactions of the Royal Society B: Biological Sciences* 363: 3715–3720.

Lumsden, C.J. and Wilson, E.O. (1980). Translation of epigenetic rules of individual behavior into ethnographic patterns. *Proceedings of the National Academy of Sciences of the United States of America-Biological Sciences* 77: 4382–4386.

Mace, R. (1998). The coevolution of human fertility and wealth inheritance strategies. *Philosophical Transactions of the Royal Society B: Biological Sciences* 353: 389–397.

Mace, R. and Holden, C.J. (2005). A phylogenetic approach to cultural evolution. *Trends in Ecology & Evolution* 20: 116–121.

Mace, R. and Pagel, M. (1994). The comparative method in anthropology. *Current Anthropology* 35: 549–564.

Mace, R. and Silva, A.S. (2016). The role of cultural group selection in explaining human cooperation is a hard case to prove. *Behavioral and Brain Sciences* 39: e45.

Marler, P.R. and Slabbekoorn, H. (2004). *Nature's Music: The Science of Birdsong*. Elsevier Science, Philadelphia.

Matthews, L. (2012). The recognition signal hypothesis for the adaptive evolution of religion. *Human Nature* 23: 218–249.

Maynard Smith, J. and Szathmáry, E. (1995). *The Major Transitions in Evolution*. New York: Wiley.

McKitrick, M.C. (1993). Phylogenetic constraint in evolutionary theory: has it any explanatory power? *Annual Review of Ecology and Systematics* 24: 307–330.

Mesoudi, A. (2007). Using the methods of experimental social psychology to study cultural evolution. *Journal of Social, Evolutionary, and Cultural Psychology* 1: 35.

Mesoudi, A. (2008). An experimental simulation of the "copy-successful-individuals" cultural learning strategy: adaptive landscapes, producer-scrounger dynamics, and informational access costs. *Evolution and Human Behavior* 29: 350–363.

Mesoudi, A. (2009). The cultural dynamics of copycat suicide. *PLoS One* 4: e7252.

Mesoudi, A. (2011). *Cultural Evolution: How Darwinian Theory Can Explain Human Culture & Synthesize the Social Sciences*. London: University of Chicago Press.

Mesoudi, A. (2016). Cultural evolution: a review of theory, findings and controversies. *Evolutionary Biology* 43: 481–497.

Mesoudi, A., Laland, K.N., Boyd, R. et al. (2013). The cultural evolution of technology and science. In: *Cultural Evolution: Society, Technology, Language, and Religion* (ed. P.J. Richerson and M.H. Christiansen), 193–216. Cambridge: MIT Press.

Mesoudi, A. and Lycett, S.J. (2009). Random copying, frequency-dependent copying and culture change. *Evolution and Human Behavior* 30: 41–48.

Mesoudi, A., Whiten, A., and Dunbar, R. (2006a). A bias for social information in human cultural transmission. *British Journal of Psychology* 97: 405–423.

Mesoudi, A., Whiten, A., and Laland, K.N. (2004). Perspective: is human cultural evolution Darwinian? Evidence reviewed from the perspective of the origin of species. *Evolution* 58: 1–11.

Mesoudi, A., Whiten, A., and Laland, K.N. (2006b). Towards a unified science of cultural evolution. *Behavioral and Brain Sciences* 29: 329–347.

Morgan, T.J.H. and Laland, K.N. (2012). The biological bases of conformity. *Frontiers in Neuroscience* 6: 87.

Morris, I. (2013). *The Measure of Civilization: How Social Development Decides the Fate of Nations*. Princeton: Princeton University Press.

Nelson, D.A. and Marler, P. (1994). Selection-based learning in bird song development. *Proceedings of the National Academy of Sciences USA* 91: 10498–10501.

Nettle, D., Gibson, M.A., Lawson, D.W., and Sear, R. (2013). Human behavioral ecology: current research and future prospects. *Behavioral Ecology* 24: 1031–1040.

Norenzayan, A., Shariff, A.F., Gervais, W.M. et al. (2014). The cultural evolution of prosocial religions. *Behavioral and Brain Sciences* FirstView 1–86.

Nowak, M.A. and Sigmund, K. (2005). Evolution of indirect reciprocity. *Nature* 437: 1291–1298.

Nunn, C.L. (2011). *The Comparative Approach in Evolutionary Anthropology and Biology*. Chicago: University of Chicago Press.

Nunn, C.L., Mulder, M.B., and Langley, S. (2006). Comparative methods for studying cultural trait evolution: a simulation study. *Cross-Cultural Research* 40: 177–209.

Odling-Smee, F.J., Laland, K.N., and Feldman, M.W. (2003). *Niche Construction : The Neglected Process in Evolution*. Princeton: Princeton University Press.

Okasha, S. (2008). *Evolution and the Levels of Selection*. Clarendon Press, Wotton under Edge.

Olson, M. (1965). *The Logic of Collective Action: Public Goods and the Theory of Groups*. Cambridge: Harvard University Press.

Ostrom, E. (1990). *Governing the Commons: The Evolution of Institutions for Collective Action*. Cambridge University Press, Cambridge.

Pagel, M., Atkinson, Q.D., and Meade, A. (2007). Frequency of word-use predicts rates of lexical evolution throughout Indo-European history. *Nature* 449: 717–720.

Pagel, M. and Mace, R. (2004). The cultural wealth of nations. *Nature* 428: 275–278.

Peccoud, J., Loiseau, V., Cordaux, R., and Gilbert, C. (2017). Massive horizontal transfer of transposable elements in insects. *Proceedings of the National Academy of Sciences USA* 114: 4721–4726.

Perreault, C. (2012). The pace of cultural evolution. *PLoS One* 7: e45150.

Pinker, S. (2003). *The Language Instinct: How the Mind Creates Language*. Penguin, London.

Powers, S.T., van Schaik, C.P., and Lehmann, L. (2016). How institutions shaped the last major evolutionary transition to large-scale human societies. *Philosophical Transactions of the Royal Society B: Biological Sciences* 371.

Ravignani, A., Delgado, T., and Kirby, S. (2016). Musical evolution in the lab exhibits rhythmic universals. *Nature Human Behaviour* 1: 0007.

Richerson, P., Baldini, R., Bell, A.V. et al. (2014). Cultural group selection plays an essential role in explaining human cooperation: a sketch of the evidence. *Behavioral and Brain Sciences* 39: e30.

Richerson, P. and Henrich, J. (2012). Tribal social instincts and the cultural evolution of institutions to solve collective action problems. *Cliodynamics* 3: 38–80.

Richerson, P.J. and Boyd, R. (2001). Built for speed, not for comfort – Darwinian theory and human culture. *History and Philosophy of the Life Sciences* 23: 425–465.

Richerson, P.J. and Boyd, R. (2005). *Not by Genes Alone: How Culture Transformed Human Evolution*. Chicago: University of Chicago Press.

Richerson, P.J., Boyd, R., and Bettinger, R.L. (2001). Was agriculture impossible during the Pleistocene but mandatory during the Holocene? A climate change hypothesis. *American Antiquity* 66: 387–411.

Richerson, P.J., Boyd, R., and Henrich, J. (2010). Gene-culture coevolution in the age of genomics. *Proceedings of the National Academy of Sciences USA* 107: 8985–8992.

Ridley, M. (2004). *Evolution*. London: Blackwell.

Rivera, M.C. and Lake, J.A. (2004). The ring of life provides evidence for a genome fusion origin of eukaryotes. *Nature* 431: 152–155.

Savage, P.E., Brown, S., Sakai, E., and Currie, T.E. (2015). Statistical universals reveal the structures and functions of human music. *Proceedings of the National Academy of Sciences USA* 112: 8987–8992.

Schaack, S., Gilbert, C., and Feschotte, C. (2010). Promiscuous DNA: horizontal transfer of transposable elements and why it matters for eukaryotic evolution. *Trends in Ecology & Evolution* 25: 537–546.

Scott-Phillips, T.C., Laland, K.N., Shuker, D.M. et al. (2014). The niche construction perspective: a critical appraisal. *Evolution* 68: 1231–1243.

Smaldino, P.E. (2014). The cultural evolution of emergent group-level traits. *Behavioral and Brain Sciences* 37: 243–254.

Smith, E.A. and Winterhalder, B. (eds.) (1992). *Evolutionary Ecology and Human Behavior*. New York: Aldine de Gruyter.

Sober, E. (2009). Did Darwin write the origin backwards? *Proceedings of the National Academy of Sciences* 106: 10048–10055.

Soltis, J., Boyd, R., and Richerson, P.J. (1995). Can group-functional behaviors evolve by cultural-group selection - an empirical-test. *Current Anthropology* 36: 473–494.

Sosis, R. and Alcorta, C. (2003). Signaling, solidarity, and the sacred: the evolution of religious behavior. *Evolutionary Anthropology* 12: 264–274.

Sosis, R., Kress, H.R., and Boster, J.S. (2007). Scars for war: evaluating alternative signaling explanations for cross-cultural variance in ritual costs. *Evolution and Human Behavior* 28: 234–247.

Spencer, C.S. and Redmond, E.M. (2001). Multilevel selection and political evolution in the valley of Oaxaca, 500–100 BC. *Journal of Anthropological Archaeology* 20: 195–229.

Sznajder, B., Sabelis, M.W., and Egas, M. (2012). How adaptive learning affects evolution: reviewing theory on the Baldwin effect. *Evolutionary Biology* 39: 301–310.

Tanaka, M.M., Kendal, J.R., and Laland, K.N. (2009). From traditional medicine to witchcraft: why medical treatments are not always efficacious. *PLoS One* 4: e5192.

Tehrani, J.J., Collard, M., and Shennan, S.J. (2010). The cophylogeny of populations and cultures: reconstructing the evolution of Iranian tribal craft traditions using trees and jungles. *Philosophical Transactions of the Royal Society B: Biological Sciences* 365: 3865–3874.

Thomason, S.G. (2007). Language contact and deliberate change. *Journal of Language Contact* 1: 41–62.

Thornton, A. and Malapert, A. (2009). The rise and fall of an arbitrary tradition: an experiment with wild meerkats. *Proceedings of the Royal Society B: Biological Sciences* 276: 1269–1276.

Tishkoff, S. (2015). Strength in small numbers. *Science* 349: 1282–1283.

Tishkoff, S.A., Reed, F.A., Ranciaro, A. et al. (2007). Convergent adaptation of human lactase persistence in Africa and Europe. *Nature Genetics* 39: 31–40.

Towner, M.C., Grote, M.N., Venti, J., and Borgerhoff Mulder, M. (2012). Cultural macroevolution on neighbor graphs. *Human Nature* 23: 283–305.

Trivers, R.L. (1971). Evolution of reciprocal altruism. *Quarterly Review of Biology* 46: 35–57.

Turchin, P. and Currie, T.E. (2016). Cultural group selection is plausible, but the predictions of its hypotheses should be tested with real-world data. *Behavioral and Brain Sciences* 39: e55.

Turchin, P., Currie, T.E., Turner, E.A.L., and Gavrilets, S. (2013). War, space, and the evolution of Old World complex societies. *Proceedings of the National Academy of Sciences USA* 110: 16384–16389.

van de Waal, E., Borgeaud, C., and Whiten, A. (2013). Potent social learning and conformity shape a wild primate's foraging decisions. *Science* 340: 483–485.

van Schaik, C.P., Ancrenaz, M., Borgen, G. et al. (2003). Orangutan cultures and the evolution of material culture. *Science* 299: 102–105.

West, S.A., Griffin, A.S., and Gardner, A. (2007). Evolutionary explanations for cooperation. *Current Biology* 17: R661–R672.

Whiten, A., Goodall, J., McGrew, W.C. et al. (1999). Cultures in chimpanzees. *Nature* 399: 682–685.

Whiten, A., McGuigan, N., Marshall-Pescini, S., and Hopper, L.M. (2009). Emulation, imitation, over-imitation and the scope of culture for child and chimpanzee. *Philosophical Transactions of the Royal Society B: Biological Sciences* 364: 2417–2428.

Whiten, A. and Mesoudi, A. (2008). Establishing an experimental science of culture: animal social diffusion experiments. *Philosophical Transactions of the Royal Society B: Biological Sciences* 363: 3477–3488.

Williams, G.C. (1966). *Adaptation and Natural Selection*. Princeton: Princeton University Press.

Wills, C. (2011). Genetic and phenotypic consequences of introgression between humans and Neanderthals. *Advances in Genetics* 76: 27–54.

Xu, J., Dowman, M., and Griffiths, T.L. (2013). Cultural transmission results in convergence towards colour term universals. *Proceedings of the Royal Society B: Biological Sciences* 280: 20123073.

4

Beyond Genes and Environments: Indirect Genetic Effects and the Evolution of Behaviour

John Hunt[1,2], James Rapkin[1], Clarissa M. House[2] and Alastair J. Wilson[1]

[1] *Centre for Ecology and Conservation, University of Exeter, Penryn Campus, Penryn, TR10 9FE, UK*
[2] *School of Science and Health and The Hawkesbury Institute for the Environment, Western Sydney University, Penrith NSW 2751, Australia*

Like most undergraduates learning evolutionary genetics now, we (the authors) were taught the importance of recognizing that the expression of many phenotypic traits depends on both genetic and environmental effects. From this, it follows that some of the trait variation we see in a population can be due to genetic differences among individuals, but some will be driven by environments. Unfortunately, this has sometimes formed the crux of a misguided nature–nurture debate which influences public perception of behaviour to this day. The point that both genes and environments influence phenotype does not mean they offer mutually exclusive explanations for behavioural variation.

This misinterpretation was perhaps understandable at a time when the (re)discovered Mendelian laws of inheritance were being combined with Darwin's insights on natural selection in the 'modern synthesis'. Mendel's work famously showed that the expression of a number of phenotypic traits in peas depends almost entirely on alleles inherited with very little environmental contribution. However, he focused on traits that are each determined by a single locus, with a one-to-one mapping of genotype to phenotype. Most traits, and almost all behaviours, are genetically complex. While expression does depend on genes, it is not fully determined by a single genetic locus (or even by a few loci). Furthermore, most traits that are of interest to behavioural biologists are actually very sensitive to diverse types of environmental effects; for instance 'calling' effort by male crickets may well depend on genes, but we also know it varies with temperature (e.g. Hedrick et al. 2002), diet (e.g. Rapkin et al. 2017), time of day (e.g. Bertram et al. 2013), and presence of male competitors (e.g. Kasumovic et al. 2012; Callander et al. 2013).

Quantitative genetics is a body of theoretical and empirical research that makes clear why, for genetically complex traits, it is not sensible to think in absolute terms of 'nature versus nurture'. Almost inevitably, both are important. Crucially, as laid out in the seminal work by Ronald A. Fisher in the 1930s, the evolutionary response of any trait to natural selection depends on the presence (and amount) of trait variation that is due to genes. Consequently, it is both sensible and useful to ask what the relative contribution of genes and environments is to behavioural variation in a population. The

Genes and Behaviour: Beyond Nature-Nurture, First Edition.
Edited by David J. Hosken, John Hunt and Nina Wedell.
© 2019 John Wiley & Sons Ltd. Published 2019 by John Wiley & Sons Ltd.

fact that genetic variance is a prerequisite for adaptive evolution does not, however, imply that environmental effects are unimportant. Indeed, this cannot be true because natural selection itself is an outcome of phenotypes interacting with environments. Furthermore, we now know that even the apparently simple question of how much of the variation in a given behaviour is genetic can have a rather complex answer. For instance, in the presence of genotype–environment interactions (GxE), the effect of a gene on an individual's phenotype depends on the environment in which that individual finds itself. When this happens, the clear separation of trait variation into genetic and environmental components breaks down. In this chapter, we focus on a related but less widely recognized phenomenon that also blurs the traditional distinction between genetic and environmental effects – indirect genetic effects (IGEs). IGEs occur when traits depend not only on genes *and* environments, but on genes *in* environments.

The social environment – defined broadly as the environment provided by con-specifics (Moore et al. 1997) – represents one of the most dynamic and influential forces an individual experiences during its lifetime. This is particularly true when considering the evolution of behavioural traits expressed during interactions with conspecifics. For such traits, a 'focal' individual's own behaviour, together with behaviours expressed by 'partner' conspecifics, jointly determine the frequency and intensity of any social interactions. Consequently, social behaviours expressed by one individual can simultaneously be thought of as focal phenotypes we want to understand and as 'environmental effects' on traits expressed by others. Indeed, some behaviours (e.g. dominance, territoriality, courtship) are not observable outside the context of a social interaction. These are sometimes referred to as 'interacting phenotypes' since the inevitable influence of individuals on each other can make deciding whose phenotype you are observing arbitrary. More generally, the important point to recognize is that 'environmental effects' on (focal) phenotypes caused by partner traits are themselves likely to depend on genes being expressed by those partners. Whenever the phenotype of a focal individual causally depends on the genotype of one or more partner individual there is an IGE. The word 'indirect' here is used in contrast to the 'direct' genetic effect (DGE) that a focal individual's own genotype has on its phenotype.

The consequences of IGEs for the evolution of focal traits can sometimes be both dramatic and non-intuitive (Wolf et al. 1998). Ultimately, this is because in the presence of IGEs, the (social) environment itself is heritable and can evolve under selection. Clearly, this further blurs the convenient – but already indistinct – line between genes and environment (or nature and nurture). In this chapter, we examine the importance of IGEs for the evolution of behaviour. We also consider the role of the 'interaction coefficient' (Ψ), a parameter that describes the effect of a partner's trait on a focal behaviour of interest (Moore et al. 1997). We provide an overview of theoretical models to illustrate the diverse and important consequences social interactions might have for behavioural evolution. We then highlight how empiricists might go about more fully assessing the role IGEs really do play. This is accompanied by an overview of empirical studies that have to date quantified Ψ and/or estimated IGEs for behaviour. Finally, we conclude by providing some useful suggestions for future directions of research on this topic that we believe will be useful and productive.

Before progressing we draw the reader's attention to two points that define the intended scope of this chapter. First, in line with the goals of this edited volume, we focus primarily on IGEs affecting the expression of behavioural traits. However, it should be recognized that IGEs arising from behavioural interactions will often have pronounced consequences for morphological (e.g. Hunt and Simmons 2002; Wolf 2003), life history (e.g. Andersson 2014; Ellen et al. 2015) and physiological (e.g. Reimert et al. 2014) traits in both plants and animals. This is to be expected as social behaviours like aggression mediate the outcome of social competition, meaning that IGEs will influence resource acquisition and downstream traits including growth and life history (Wilson 2014).

Second, we deliberately focus our discussion on IGEs arising from interactions among unrelated individuals. Of course, many fascinating and important behavioural interactions occur among close relatives. In fact, maternal genetic effects, which arise most commonly from genetically determined parental care traits expressed in mothers, are the best studied instance of an IGE. Sibling competition is another familiar scenario in which IGEs are expected, as is co-operation among kin. We limit our discussion of these contexts since they are already extensively covered elsewhere (e.g. Kirkpatrick and Lande 1989; Lande and Kirkpatrick 1990; Wolf and Wade 2009, 2016). However, in doing so we recognize that readers more familiar with a behavioural ecology view of social evolution (especially co-operation) may find our 'glossing over' of relatedness here surprising. We therefore stress that while relatedness is in fact central to both behavioural ecology and quantitative genetic models of social evolution, models rooted in the former field assume an absence of IGEs (and typically a very simplistic model of trait inheritance through DGE). For instance, Hamilton's rule allows us to explore the consequences of relatedness between an 'actor' and a 'recipient' but assumes that the behaviour expressed depends only on the former's genotype. The key point is that IGEs should be recognized as a distinct phenomenon in social evolution, not an alternative way of conceptualizing the consequences of relatedness (see Chapter 2 for more on this). In fact, a secondary reason for focusing primarily on interactions among unrelated individuals is that it makes this exact point easier to see.

4.1 A Quantitative Genetic View of Behavioural Evolution without IGEs

To understand the evolutionary consequences of IGEs, it is actually easiest to begin with a brief introduction to traditional quantitative genetic theory. We can then show how this can be extended to include social interactions and illustrate what consequences this has for the evolution of behaviour.

Readers already familiar with the classic quantitative genetic theory utilized in the absence of IGEs could safely jump ahead to section 4.3 at this point. For those who need a refresher, let's assume as our starting point that the phenotype (z) of any individual (i) in a population is determined by genetic and environmental effects (Figure 4.1). In the simplest case, we can therefore describe i's phenotype as:

$$z_i = \mu + a_i + e_i \tag{4.1}$$

where z_i is determined by a DGE of the individual's own genotype (a_i) and the environment it experiences (e_i). Note that by convention, we will use a_i rather than g_i to denote the DGE as we focus on the additive genetic component of inheritance that leads to predictable transmission of phenotype across generations. The DGE is often referred

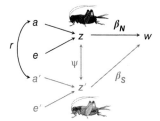

Figure 4.1 Path diagram showing the direct (*black text and arrows*) and indirect (*grey text and arrows*) effects on the phenotype of two interacting male crickets (z_i and z'_j). Variables associated with the focal male are shown without superscript, whereas variables associated with its social partner are given a prime. Each phenotype is influenced by additive genetic (a), environmental (e) and indirect genetic effects. r is the relatedness between the two males and this describes the translation between the additive genetic values of the focal individual and his social partner. Although the social environment can be provided by relatives, in this chapter we focus on those instances where interacting individuals are unrelated (i.e. $r = 0$). Ψ is the interaction coefficient that translates the phenotype of the social partner into an indirect genetic effect on the focal male. The double arrow indicates that the effect of Ψ is reciprocal. The fitness (w) of the focal male is affected by both its own phenotype (non-social selection, β_N) and that of its social partner (social selection, β_S). Source: Modified from McGlothlin et al. (2010).

to as an 'additive genetic merit' or 'breeding value' and both it and the environmental effect are usually expressed relative the overall trait mean (μ). Eq. 4.1 can of course be expanded to include more terms (and hence biological realism), but this simple idea forms the backbone of all quantitative genetic theory, irrespective of the particular trait being modelled (Falconer and Mackay 1996). If variation in trait z is associated with variation in relative fitness, then the trait is under selection. In other words, individuals in the population with particular phenotypic values will have higher fitness than average. As selection acts on phenotypes, we can measure its strength and direction without any knowledge of the trait's genetic basis of variation (Lande and Arnold 1983). The directional selection differential (S) represents the total selection acting on a given phenotypic trait and can be measured as the difference in the mean value of the trait in the population before ($\bar{z}$) and after ($\bar{z}^*$) selection (within a generation):

$$S = \bar{z}^* - \bar{z} \tag{4.2}$$

or as the covariance between relative fitness (w) and the phenotypic trait (Lande and Arnold 1983):

$$S = Cov(w, z) \tag{4.3}$$

where relative fitness is simply an individual's absolute fitness (W) scaled by the population mean fitness. Whereas Eq. 4.2 can readily be applied when (our proxy of) relative fitness is discrete (e.g. under artificial truncating selection or natural viability selection), Eq. 4.3 is more useful for studies of natural selection based on more continuous measures of individual fitness (e.g. the number of offspring produced).

Regardless of whether we are looking at social behaviours or any other type of trait, it is worth remembering that natural selection is properly thought of as the process by which fitness variation is *caused* by trait variation. Consequently, a non-zero estimate of S needs to be interpreted rather carefully; a phenotypic trait (z) may covary with relative fitness either because it affects fitness itself (direct selection) or because it is correlated with some other trait(s) that does (indirect selection) (Lande and Arnold 1983). With some important caveats (see Morrissey et al. 2010), multiple regression allows separation of direct from indirect targets of selection. For a set of traits (e.g. z_1, z_2, z_3) we can estimate the direct selection on the nth as the partial regression coefficients of w on z_n. This partial regression is referred to as a linear selection gradient and typically denoted β.

Selection can be viewed as a purely phenotypic process, but evolutionary responses to it depend on genetics. If we want to predict responses, we therefore need to know whether, and to what extent, individuals having trait values that confer higher fitness will actually transmit their phenotypes to the next generation. In simple terms, this means knowing how much of the variation in z that we observe is explained by genetic differences among individuals. From Eq. 4.1 it follows that the variance (Var) for trait z is:

$$Var(z) = Var(a) + Var(e) + 2Cov(a, e) \qquad (4.4a)$$

However, quantitative genetic studies are normally designed such that it is reasonable to assume environmental (e) and genetic (a) effects are uncorrelated, allowing simplification to:

$$Var(z) = Var(a) + Var(e) \qquad (4.4b)$$

where $Var(z)$ is the phenotypic variance and $Var(a)$ is the additive genetic variance. Commonly, these are denoted V_P and V_A respectively, so we will adopt this notation from here on. V_A thus gives us a measure of the amount of trait variation that is determined by genes which we often express as a proportion of V_P. This ratio is termed the narrow sense heritability (h^2).

The heritability is therefore the proportion of variation explained by (additive) genetic differences among individuals. It is also the proportion of the change in the phenotypic distribution that is caused by selection in one generation that we expect to be transmitted to the next. We can predict the trait's response (R) to directional selection as the change in trait mean over a single generation time step. This is done using the univariate breeder's equation, which can be parameterized in terms of either h^2 and S, or V_A and β:

$$R = h^2 S = \frac{Var(a)}{Var(z)} Cov(w, z) = V(a) \frac{Cov(w, z)}{Var(z)} = V_A \beta \qquad (4.5)$$

When we have multiple traits under selection (direct and/or indirect) that are likely to be correlated, perhaps in part due to genetic effects, the multivariate breeders equation is used instead (Lande 1979). We now need to predict not a single value of R but rather a vector of predicted changes in trait means for n traits. This model is usually written as:

$$\Delta \bar{z} = G\beta \qquad (4.6)$$

Note that the bold font is used to differentiate vectors and matrices from scalars. In Eq. 4.6 the left-hand side is the vector of predicted selection responses (i.e. a multitrait equivalent of R in Eq. 4.5) while β contains the selection gradients for each trait. Both vectors will have length n (i.e. the number of traits). Here, $\mathbf{G}$ is the additive genetic variance-covariance matrix. This is a square, symmetrical matrix with n rows and n columns that contains the required genetic data in the form of additive genetic variances for each trait (V_A) on the diagonal and additive genetic covariances (Cov_A) for each pair of traits off the diagonal. So for the simplest multivariate case (i.e. 2 traits, 1 and 2), **G** can be written out in full as:

$$\mathbf{G} = \begin{bmatrix} V_{A1} & Cov_{A1,2} \\ Cov_{A1,2} & V_{A2} \end{bmatrix} \tag{4.7}$$

Thus far, we have outlined a standard quantitative genetic approach to understanding, and predicting, how genetically variable traits (including but not limited to behaviours) will respond to selection. The validity of these models is of course dependent on many assumptions. From a behavioural genetics perspective, an important one to note is that classic quantitative genetic models are formulated to help us when we don't necessarily know the detailed genetic architecture underpinning genetic variation, but can assume that a trait is polygenic (i.e. influenced by many genes). More formally, the predictive models of change shown above assume an infinitesimal model of gene action, which means that the (additive) genetic effect (a_i in Eq. 4.1) is assumed to arise from an infinite number of genes each with infinitely small effect. In this case we can – for the purposes of understanding phenotypic dynamics – reasonably hope to replace mechanistic understanding of the genotype to phenotype map (i.e. knowledge of which genes affect the trait, where they are located, and what products they encode) with a statistical description. Of course, for researchers interested in the finer details of genetic architecture, demonstrating that a behavioural trait is heritable gives little insight. It is, however, a rather useful first step, as molecular genetic investigations of behaviours in the absence of any evidence for an additive genetic component are most probably doomed to failure from the outset.

4.2 Adding IGEs to the Traditional Theory

Having laid out the relevant theory in the absence of IGEs, we now come to the question of what changes when traits depend on the phenotypes of interacting conspecifics. Let's start by expanding Eq. 4.1 so that z depends on some specific and measurable environmental variable x (e.g. temperature, population density):

$$z_i = \mu + a_i + bx_i + e_i \tag{4.8}$$

where we subscript x (x_i) to make it clear it is the value experienced by an individual that determines its phenotype. The effect of x on z is described by the slope b, which is the average (over all individuals) plastic response in z to a unit change in x. We retain the general environmental effect e_i which represents non-genetic effects on z over and above the specific impact of variable x. So far, this doesn't really change much: population-level genetic influences on z are still fully captured by estimating the direct additive genetic variance V_A.

However, what if z is a social behaviour and x is now a trait expressed by an interacting partner j? If this is the case, we can rewrite Eq. 4.8, adding some additional subscripting to avoid confusion, and using Ψ in place of b to specifically denote a social plastic effect as:

$$z_i = \mu_z + a_{zi} + \Psi x_j + e_{zi}, \tag{4.9a}$$

Now let the partner trait (x) be genetically variable so that x_j depends on the genotype of j, Eq. 4.9a can expanded as:

$$z_i = \mu_z + a_{zi} + \Psi(\mu_x + a_{xj} + e_{xj}) + e_{zi} \tag{4.9b}$$

From Eq. 4.9b, it follows directly that i's phenotype at trait z is now subject to both the DGE (a_{zi}) and an IGE from the social partner's 'interactor' trait x, Ψa_{xj}. A consequence of this is that the total genetic variance for trait z in the population is no longer described by the direct additive genetic variance V_A. Rather, it will now also depend on genetic variance for trait x, and on the magnitude of the Ψ (see Figure 4.1).

Although not immediately obvious from Eqs. 4.9a and 4.9b, an important consideration is that there can be a correlation between an individual's DGE on its own trait, and that same individual's IGE on the phenotypes of others. In other words, and paying close attention to the subscripts, there is a correlation between a_{zi} and a_{xi} (as opposed to between a_{zi} and a_{xj}). Biologically, correlations between IGE and DGE arise from the same underlying processes as any other genetic correlation – namely pleiotropy and/or linkage disequilibrium. In the limiting case that z and x are actually the same trait, the genetic correlation will of course be +1. This would be the case if, for example, aggression is a heritable focal trait (z) but also varies plastically ($\Psi \neq 0$) with aggression as displayed by a rival conspecific (x). This type of correlation structure is really important because, mathematically $Var(a + b) = Var(a) + Var(b) + 2Cov(a,b)$ which necessarily means that the total genetic variance for z under Eq. 4.9b must depend not only on V_{Az} and V_{Ax} but also on the value of $Cov_{A(z, x)}$. As we will highlight later, the sign and magnitude of any covariance between DGE and IGE can be really important. In fact, the existence of this covariance is the reason social environments can evolve in response to direct selection on focal traits (and vice versa).

4.2.1 'Trait-Based' and 'Variance Partitioning' Models of IGEs

So far, we have assumed that we know both the focal trait (z) and the partner's interactor trait (x) on which z depends. Since both traits are known and could be measured if we wanted to parameterize the model, Eqs. 4.9a and 4.9b are considered an example of the 'trait-based approach' to modelling IGEs. However, what if we don't know exactly which aspect(s) of the partner's phenotype is relevant? Staying with the case of a dyadic interaction between focal i and partner j, we could rewrite Eqs. 4.9a and 4.9b as:

$$z_i = \mu_z + a_{zi} + a_{z'j} + e_{zij} \tag{4.10}$$

where $a_{z'j}$ is the indirect (additive) genetic effect of j on z_i (we use the prime ' to denote a trait expressed by the social partner). Assuming there is only a single interactor trait, $a_{z'j} = \Psi a_{xj}$, we can reconcile Eqs. 4.9a, 4.9b, and 4.10 by noting that while the non-genetic component of the indirect effect from Eqs. 4.9a and 4.9b (e_{xj}) is now included in the general environmental effect on z (but subscripted ij as this environmental effect is also a function of partner j) (see Figure 4.1).

Interestingly, provided appropriate data structures are available (discussed below), it is possible to parameterize the model shown in Eq. 4.10, and partition variance in z into contributions from DGE and IGE without knowledge of the interactor trait x. A nice feature of this 'variance partitioning' approach is that it is readily generalized to groups of size n interacting individuals, which should be useful as, in reality, social interactions can certainly involve more than one partner. Following Bijma et al. (2007a):

$$z_i = \mu_z + a_{zi} + \Sigma_{j \neq i}^{n} a_{z'j} + g_k + e_{zij} \tag{4.11}$$

where g is an effect of the social group (expressed relative to the overall mean), and the IGEs experienced by i are summed over all partners j in group k. Following from Eq. 4.11, Bijma et al. (2007a) also introduced the concept of a 'total breeding value' (TBV) which for any individual describes their impact (through both DGEs and IGEs) on the mean of z. The variance in TBVs is then a measure of genetic variance for z that includes both direct and indirect contributions and is specified for a trait z as:

$$V_{TBVz} = V_{Az} + 2(n-1)Cov_{Az,z'} + (n-1)^2 V_{Az'} \tag{4.12}$$

where V_{Az} and $V_{Az'}$ are the direct and IGE variances, respectively, and $Cov_{Az,z'}$ is the DGE-IGE covariance. Since variances are positive (unless zero), the first and third terms on the right-hand side of Eq. 4.12 must be positive. However, the second term – the h covariance – can be either positive or negative. As a result, and dependent on the sign of this covariance, V_{TBV} may be either greater than or less than the conventional estimate of V_A. V_{TBV} is not quite a direct replacement for V_A in, for example, the univariate breeder's equation. This is because once you have group structure, the possibility exists for multi-level selection (i.e. selection among groups as well as among individuals within groups). However, to a first approximation IGEs can be thought of as changing the total amount of genetic variance that is available to facilitate a selection response in z. If present, they can lead to either faster or slower selection responses (dependent largely on the sign of any the DGE-IGE covariance) than predicted by traditional models.

An alternative view of this same process, which is most readily understood by reverting to the trait-based (TB) perspective, is that a covariance between DGEs and IGEs leads to coevolution of the focal trait under selection and the social environment (i.e. the interactor trait). Since any evolved change in mean expression of the interactor trait (x) in the social environment will affect the mean of focal trait (z), changes in z across generation will differ from predictions made using the breeder's equation (which assumes no change in the environment). Will the response be greater or less than predicted in the absence of IGE (and by how much)? Well, this depends on the sign and magnitude of both Ψ and $Cov_{A(z, x)}$.

Thus, under a variance partitioning view, it is easiest to think of IGEs as changing the total genetic variance available to facilitate a selection response in z, whereas under the trait-based view they are perhaps more readily understood as causing coevolution between z and x. Although it is not easy to see this without recourse to an intimidating algebraic treatment, both perspectives can be shown to be equivalent and are thus equally valid ways of conceptualizing the role of IGEs (McGlothlin and Brodie 2009).

4.3 From Theory Towards Empiricism

As described above, theoretical models of IGEs have been formulated in two main ways which have become known as 'trait-based' and 'variance partitioning' approaches (see McGlothlin and Brodie 2009). These are represented for the case of a dyadic interaction by Eqs. 4.9b and 4.10, respectively. As trite as this sounds, an important starting point for any empiricist is to make sure they understand these frameworks sufficiently well to decide which type of model can be most readily (or usefully) parameterized for the question at hand.

Although the two approaches are equally valid and can be mathematically 'translated' across a range of scenarios, they do not generally: (i) have identical data requirements, (ii) make the same assumptions, or (iii) use the same parameters to make evolutionary inferences. This creates obvious potential for confusion. For instance, under the trait-based approach a non-zero coefficient Ψ may be greater than or less than zero. This has sometimes led to a distinction being made between 'positive' and 'negative' IGEs, with the sign describing the direction of the change in expected focal phenotype with increase in the partner's trait value. However, discussing positive and negative IGEs in a variance partitioning approach makes no sense since Ψ is undefined. Conversely, variance partitioning (VP) studies emphasize the sign of the covariance between DGE and IGE ($Cov_{Az,z'}$ in Eq. 4.12). This is because a positive value will lead to accelerated selection responses, while a negative covariance will generally constrain expected rates of evolution. Under a trait-based model the same inference can be drawn not from the sign Ψ but from the sign of $\Psi.Cov_{Az,x}$.

If the above example seems a bit arcane, that is because it is. However, the take-home point is simply that the two frameworks can certainly be used to obtain the same insights, but typically the models and results are presented in a way that does not make this obvious. Thus, for any particular study a pragmatic decision needs to be made about which approach – trait based or variance partitioning – should be employed. A blanket recommendation is not really sensible since they both have different strengths and limitations.

4.3.1 Using Trait-Based Models

In our experience, most behavioural biologists find the trait-based approach to IGEs more intuitive. This is probably because it places a stronger emphasis on the actual behaviours that mediate social interactions. Under the trait-based approach, the contribution of IGEs to the variation in a focal trait is a consequence of the magnitude of Ψ and the amount of genetic variation in the 'interactor trait' expressed by conspecific partners. Estimating Ψ appears straightforward, at least initially. For example, as formulated in Eq. 4.9a, Ψ could simply be estimated from a linear regression of focal z on partner x (obviously this does mean that both traits must be observed and measured in the experiment). In the case where there are multiple predictors (e.g. additional interactor traits), then Ψ becomes the vector of partial regression coefficients (Bailey and Hoskins 2014). However, strictly speaking, using simple (multiple) regression assumes a unidirectional causality that may not always be reasonable if feedback occurs within the timeframe of the observed interaction. Imagine the simple scenario alluded to earlier in which z and x are the same trait – aggressiveness – expressed in focal and partner respectively. It is certainly reasonable to expect z_i will respond plastically to x_j but perhaps less reasonable to expect that the focal individual's behaviour will not then lead to modification of partner phenotype (and so on *ad infinitum*). In such scenarios the regression coefficient estimated from observed data may not really be measuring Ψ as defined in a theoretical model (Bijma 2014).

Taking the trait-based approach, we must also estimate V_A for the interactor trait if we want to infer the presence of IGEs in a particular population. The latter can be done using any number of standard statistical methods that might also be used to estimate the DGEs (i.e. V_A for the focal behaviour). Applicable statistical methods include parent-offspring regression, ANOVA, and linear mixed effect models. Although specific analyses can certainly be complex, the premise underlying all is a simple one. If closely related individuals (which by definition share genes) are also phenotypically similar, then we can conclude there is a genetic basis of trait variation. The strength of this pattern can thus be used to estimate V_A for a trait and in the multivariate context Cov_A between traits. Thus, the extra data required here are knowledge of the pedigree structure (i.e. relationships) or genetic relatedness among measured individuals in the population. This knowledge may come from an experimental breeding design in a laboratory study or from molecular genetic data collected in a field study. We will not review these further here but the unfamiliar reader might refer to Falconer and McKay (1996), Lynch and Walsh (1998) and Wilson (2014).

Despite our view that trait-based models are more intuitive for many people, they also come with some drawbacks. The first, which relates to interpretation rather than implementation, is that there has sometimes been a tendency in the literature to erroneously conflate the presence of social plasticity (i.e. a non-zero Ψ) with the presence of IGEs. Thus to be clear, single population studies estimating Ψ do not, in themselves, tell us about the actual importance of IGEs. This is because it is perfectly possible for a large Ψ to be associated with a non-heritable interactor trait such that there are no IGEs. In our view, studies that estimate Ψ but not genetic (co)variance are nonetheless valuable, in part because they do tell us something about whether IGEs are possible, but probably more so because social plasticity is an interesting phenomenon in its own right. Second, at least relative to variance partitioning models, it is less clear how empirical analyses should be extended to cope with multiple partners interacting simultaneously. Third, perhaps the biggest limitation is that they require knowledge of exactly which trait(s) expressed by social partners causally influence focal phenotypes of interest. In a theoretical model, this is trivial but for an empiricist, causality can be hard to determine. If (heritable) interactor and focal traits are correlated but only through a shared dependence on, for instance, some uncontrolled environmental effect, a non-zero estimate of Ψ may easily lead to erroneous inference of IGEs. Formally, the problem is that the phenotypic (partial) regression of focal trait on partner phenotype (i.e. the estimator for Ψ) is not a valid estimate of the genetic regression (which it is when the relationship is truly causal). In a statistical sense, this is exactly the same issue that drives a need for cautious interpretation of non-zero selection differentials highlighted in section 4.2. The same problem can also apply in reverse; a small, non-significant estimate of Ψ obtained by regressing focal trait (z) on partner trait (x) may lead one to conclude there can be no IGEs. However, it is possible that IGEs are present but are not driven by the assumed trait x – a possibility termed 'cryptic IGEs' (Bailey and Hoskins 2014). In fact, it is even possible that there is a real dependence of z on x as hypothesized, but that this is being masked by variation in some other parameter. Where data on multiple partner traits (and/or nuisance variables) are available, multiple regression analysis presents a route to improve inference of causality (Bailey and Hoskins 2014), although generic concerns about conflating causation with correlation should always give us pause for thought.

4.3.2 Using Variance Partitioning Models

Developed initially for applications in livestock production, this strategy for estimating IGEs probably seems less intuitive for behavioural researchers new to quantitative genetics. However, it deserves attention from empiricists as it arguably offers wider applicability. It is also the more common approach that has been taken in practice, at least for studies of wild and captive vertebrate populations. This is in part because such populations are often characterized by complex pedigrees which are best analysed using a particular form of linear mixed effect model known as the 'animal model' (Wilson et al. 2009). Usefully, the standard animal model routinely employed to estimate DGEs is readily extended to include IGEs, and can be implemented in standard mixed model packages.

How does this work in practice? Recall that, under the variance partitioning approach, IGEs still arise from one (or more) interactor trait expressed by a social partner (see Eq. 4.10 for the dyadic case). However, this interactor trait is now unknown and unmeasured. Nonetheless, if pedigree information is available for both focal and partner individuals, it remains possible to estimate the variance in IGE as among-social partner genetic variance for some focal trait. This follows from the idea that a trait subject to IGEs can really be thought of as having genetic control that is 'shared' between multiple individuals. So in the dyadic case, a focal individual's (i) phenotype potentially depends on two genotypes – its own and that of its partner (j). In fact, we could simply reassign the phenotype of the focal individual to its partner and apply any appropriate statistical method suitable to estimate V_A. With focal phenotypes assigned to partners, the answer would now be an estimate of the indirect genetic variance (i.e. $V_{Az'}$ in Eq. 4.12). However, in practice it is better to model the trait as being dependent on both genotypes simultaneously since this allows joint estimation of direct and indirect genetic variances along with the DGE-IGE covariance that is so important for evolutionary dynamics. Within the animal model framework used to date, this involves fitting the direct (of the focal) and indirect (of the partner) breeding values as random effects although we note that other ways of indexing focal and partner genotypes for inclusion as random effects may be appropriate for particular data structures (e.g. by sire, line, or clone).

Relative to trait-based approaches, variance partitioning has several additional advantages. First, it is much easier to go beyond the dyadic case and analyse interactions in groups of larger size (following Eq. 4.11) (Bijma et al. 2007a). Typically, in doing so it is assumed that all individuals within a group interact equally while no interactions occur between members of different groups. However, this assumption can be relaxed, and it is possible to weight IGEs from specific partners according to, for instance, geographical proximity in tree competition studies (Cappa 2008) or connectedness in a social network for wild animals (suggested in Wilson (2014) but not yet attempted). Statistical inference is also more straightforward since comparison of, for instance, an animal model including a DGE only to one with DGE and IGE provides an explicit test for IGEs. In contrast, under a trait-based analysis both Ψ and h^2 of the partner trait must be estimated in separate analysis with the uncertainty in these two parameters subsequently combined in some way.

Although these features mean variance partitioning is preferred by most statistical modellers working with empirical data, this does not mean it is a trivial exercise; the availability of statistical methods is only useful if there are suitable data to analyse and here, as always, the devil is in the detail. How much data is actually needed for a powerful study? What breeding design would be best to use in an experimental study? How does power to detect IGE change with group size? Answering these questions is difficult, and beyond the scope of this chapter but we highlight that simple answers are hard to derive analytically (Bijma 2010) and data structures optimal for estimating IGEs may often be suboptimal for DGEs (and vice versa). A potentially useful strategy would be to use simulation-based methods of power and sensitivity analysis that are already available for the case of maternal genetic effects (Morrissey and Wilson 2010) and could be readily extended to more general IGE cases.

4.3.3 Is Knowledge of the Interactor Trait Critical to Your Study?

As already noted, the variance partitioning approach allows estimation of IGEs on a focal trait without having explicit knowledge of the partner behaviours. Conceptually, it is perhaps useful to think of IGEs as reflecting among-partner genetic variance for some latent trait that we don't directly observed, but can be conceptualized as 'impacting focal phenotype'. This is sometimes referred to as 'performance' but in reality it may be an amalgam of multiple unknown partner traits. This then is an important distinction between trait-based and variance partitioning methods; the former test for IGEs arising from a single partner trait (x), the latter integrate over the multivariate partner phenotype. Variance partitioning thus side-steps the 'cryptic IGE' issue discussed earlier. However, being agnostic about the source of the IGE necessarily means reduced insight into the behavioural processes mediating interactions. Whether this is problematic or not is rather dependent on the aims of a study. For instance, if we are interested in how IGEs from competitive interactions shape the evolution of some resource-dependent trait (e.g. growth) under strong selection, we may not be concerned about the behavioural details of the contests deciding resource acquisition. Conversely, if we want to know whether selection on growth could cause correlated evolution of aggression during contests over limited resources, it will be crucial.

Knowing – and measuring – both traits thus becomes important if we wish to understand their coevolution. This is especially true if both traits are likely to be under selection, a point that has previously been made in relation to maternal effects (see Hadfield 2012) where the two best-known models, those of Willham (1963) and Kirkpatrick and Lande (1989, 1992), provide exemplars of variance partitioning and trait-based approaches respectively. Willham's (1963) model is widely used in livestock scenarios where it may be reasonable to assume that natural selection on the maternal trait(s) causing IGEs on offspring phenotype is weak, at least relative to the strength of artificial selection on the focal trait (e.g. offspring growth rate). Conversely, in wild populations it is widely expected that expressing maternal care (the interactor trait) will be costly for a mother (i.e. selected against through maternal fitness), but cause effects on offspring phenotype (the focal trait) that are beneficial (i.e. selected for through offspring fitness). Here the use of the trait-based Kirkpatrick and Lande model would let us include the selection on maternal care and so capture the resulting coevolutionary dynamic.

One last point to note with respect to the trait-based versus variance partitioning distinction is that for the empiricist it is both possible and useful to cross-validate the conclusions of each using 'hybrid' modelling strategies (McAdam et al. 2014). For instance, if we have correctly identified a heritable partner trait x with a causal non-zero Ψ for focal trait z, then we predict: (i) that IGE will be detectable on y using a variance partitioning analysis and (ii) that there will be a direct genetic correlation between x and indirect 'performance' for z. The second prediction is readily tested in, for instance, a bivariate animal model of x and z with IGEs specified on the latter only. This only requires that there are pedigree links between focal and partner individuals, although power will be greater if both x and z are measured on all individuals (regardless of their focal versus partner designations).

4.4 Empirical Evidence for IGEs on Behavioural Traits

At this point, the reader may well be wondering just what evidence there actually is that IGEs are an important contributor to behavioural variation in natural systems. Unfortunately, despite the development of tractable modelling strategies as outlined above, theoretical progress towards understanding the evolutionary importance of Ψ and IGEs far exceeds empirical progress. In simple terms, we know IGEs could be really important for behavioural evolution, but we probably don't have enough empirical estimates yet to generalize about the extent to which they actually are. However, some insights can be gained from considering empirical studies to date.

4.4.1 Social Plasticity of Behaviour is Widespread

In Table 4.1 we present the available estimates of Ψ for behavioural traits as estimated within the trait-based IGE framework. The most striking feature of this table is the general lack of studies formally estimating Ψ. Current estimates of Ψ are restricted to only two taxonomic groups (arthropods and fish), with the former representing only a single species (the guppy, *Poecilia reticulata*). Moreover, all current estimates of Ψ are based on laboratory studies and most estimates have been obtained using either different inbred lines or populations. An obvious exception is work on *Drosophila serrata* using experimental populations to examine how Ψ evolves (Chenoweth et al. 2010). Table 4.2 shows that Ψ has been estimated for a diversity of behaviours known to be strongly dependent on social interactions for their expression. The most thoroughly examined are behaviours involved in antipredator response, including a suite of inspection, proximity and schooling behaviours in guppies (Bleakey and Brodie 2007; Edenbrow et al. 2017) and the startle response of *Drosophila melanogaster* (Bailey and Hoskins 2014). Ψ has also been estimated for aggressive behaviour, including precannibalistic aggression in the Socorro isopod (Bleakey et al. 2013) and the number of aggressive lunges in male *D. melanogaster* (Saltz 2013), as well as for female mounting latency in the field cricket *Teleogryllus oceanicus* (Bailey and Zuk 2012).

A second feature of Table 4.1 is the large variability in estimates of Ψ. For studies providing a single estimate of Ψ, there are equally as many positive (Saltz 2013; Chenoweth et al. 2010) and negative (Bleakey et al. 2013; Bailey and Hoskins 2014) values. Moreover, in studies where multiple estimates of Ψ are provided (due to the use of multiple inbred lines or populations), values often ranged from negative to positive for the same behaviour being examined (Bleakey and Brodie 2007; Bailey and Zuk 2012; Edenbrow et al. 2017). Even when consistent positive estimates of Ψ were documented for a given behaviour, such as for mean distance from predator and the proportion of time spent

Table 4.1 Empirical studies formally quantifying the coefficient of interaction (also known as Ψ) for behavioural traits.

Taxonomic group	Species	Common name	Design	Behaviour	Ψ	Reference
Arthropod	*Drosophila melanogaster*	Fruit fly	Inbred lines	Aggressive lunges	0.098	Saltz (2013)
			Inbred lines	Startle response	−0.49	Bailey and Hoskins (2014)
	Drosophila serrata	Fruit fly	Experimental evolution[a]	Adjusting cuticular hydrocarbons in response to female	Positive	Chenoweth et al. (2010)
	Thermosphaeroma thermophilum	Socorro isopod	Two populations	Latency to attack	−0.05	Bleakey et al. (2013)
	Teleogryllus oceanicus	Field cricket	Multiple populations[b]	Female mounting latency	−0.63–0.40[b]	Bailey and Zuk (2012)
Fish	*Peocilia reticulata*	Guppy	Inbred lines	Antipredator behaviours[c]	−1.14–0.93	Bleakey and Brodie (2007)
			Multiple populations[d]	Distance[e]	0.85–1.58	Edenbrow et al. (2017)
				Time	0.33–2.00	Edenbrow et al. (2017)
				Co-ordination	−0.26–0.34	Edenbrow et al. (2017)

a) A total of 16 generations of experimental evolution in replicate populations with varying intensities of natural and sexual selection. Ψ was shown to significantly evolve in response to sexual selection for two methylalkanes ($2MeC_{28}$ and $2MeC_{30}$).

b) A total of four different populations (ancestral and contemporary Kauai, Mission Beach, and Hilo) were shown to have significant values of Ψ for female choosiness to an artificial acoustic signal.

c) A range of antipredator behaviours (time in proximity to predator; time orientated to predator, time agitated by presence of a predator, time spent schooling, number of inspections of predator) showed significant values of Ψ. Ψ was always positive and strongest when examining the same behaviours in focal individual and partners.

d) Guppies from four high-predation and four low-predation populations in Trinidad were used.

e) A range of antipredator behaviours were examined: distance = mean distance from predator; time = proportion of time spent closest to predator; and co-ordination = mean distance between focal and partner fish.

Table 4.2 Empirical studies formally documenting the existence of significant IGEs for behavioural traits.

Taxonomic group	Species	Common name	Lab or field	TB or VB	Design	Behaviour	V_{Az}	$V_{Az'}$	$Cov_{Az,z'}$	T^2	Reference
Platyhelminth	Macrostomum lignano	Flatworm	Lab	Both	Inbred lines	Time to first copulation	—	—	—	—	Marie-Orleach et al. (2017)
Arthropod	Drosophila melanogaster	Fruit fly	Lab	TB	Inbred lines[a]	Startle response	—	—	—	—	Bailey and Hoskins (2014)
			Lab	TB	Inbred lines	Aggressive lunges	—	—	—	—	Saltz (2013)
			Lab	VP	Half-sib	Copula duration	15.88 (1.15)	1.33 (0.57)	−0.13 (0.28)	0.18 (0.07)↑	Edward et al. (2014)
	D. serrata	Fruit fly	Lab	TB	Half-sib	Adjusting cuticular hydrocarbons in response to female	—	—	—	—	Petfield et al. (2005)
	Nicrophorus vespilloides	Burying beetle	Lab	VP	Cross fostering	Larval wandering behaviour	323.81	706.58	−0.22 (0.26)	0.01 (0.04)↓	Head et al. (2012)
	Enchenopa binotata	Treehopper	Lab	VP	Full-sib split family	Mate preference	—	—	—	—	Rebar and Rodriguez (2013)
	Gryllus bimaculatus	Field cricket	Lab	VP	Pedigree population	Aggression	0.05 (0.02)	0.03 (0.02)	−0.83 (0.37)	0.02 (0.03)↓	Santostefano et al. (2017)
Fish	Peocilia reticulata	Guppy	Lab	TB	Inbred lines	Anti-predator behaviours	—	—	—	—	Bleakey and Brodie 2007
	Gadus morhua	Atlantic cod	Lab	VP	Half-sib	Second dorsal fin erosion[b]	16.44 (3.82)	0.19 (0.10)	0.30 (0.25)	1.37 (0.60)↑	Nielson et al. (2014)
			Lab	VP	Half-sib	Third dorsal fin erosion[b]	0.09 (1.08)	0.03 (0.03)	0.78 (0.44)	0.48 (0.39)↑	Nielson et al. (2014)
	Oryzias latipes	Medaka	Lab	TB	Artificial selection	Aggression[c]	—	—	—	—	Ruzzante and Doyle (1991)

(Continued)

Table 4.2 (Continued)

Taxonomic group	Species	Common name	Lab or field	TB or VB	Design	Behaviour	V_{Az}	$V_{Az'}$	$Cov_{Az,z'}$	T^2	Reference
Birds	*Gallus gallus*	Laying hen	Lab	VP	Pedigree population	Mortality due to pecking	960	132	—	0.20↑	Bijma et al. (2007b)
	Coturnix japonica	Japanese quail	Lab	VP	Artificial selection	Feeding efficiency[d]	—	—	—	—	Muir (2005)
	Larus canus	Common gull	Field	VP	Pedigree population	Timing of egg laying	4.52 (1.2)	1.49 (0.53)	−0.53 (0.22)	—	Brommer and Rattiste (2008)
Mammal	*Neovision vison*	Mink	Lab	VP	Artificial selection[e]	Total bite marks[b]	2.95 (0.90)	1.60 (0.32)	0.90 (0.15)	0.61 (0.08)↑	Alemu et al. (2014)
			Lab	VP	Artificial selection[e]	Total bite marks[b]	5.20 (1.12)	1.13 (0.31)	0.89 (0.10)	0.54 (0.11)↓	Alemu et al. (2015)
	Sus scrofa	Pig	Lab	VP	Artificial selection[f]	Aggression	—	—	—	—	Camerlink et al. (2014), Camerlink et al. (2015)
			Lab	VP	Pedigree population	Feed intake	16 950 (3247)	596 (220)	0.38 (0.22)	0.70 (0.17)↑	Bergsma et al. (2008)
	Bos taurus	Cattle	Field	VP	Pedigree population	Social dominance	0.28 (0.22, 0.33)	0.22 (0.18, 0.27)	−0.98 (−1.02, −0.94)	0.01 (−0.01, 0.03)↓	Sartori and Mantovani (2013)

Peromyscus maniculatus	Deer mouse	Lab	VP	Pedigree population	Rearing rate	0.02 (0.01)	0.01 (0.01)	0.79 (0.17)	0.61 (0.15)↑	Wilson et al. (2009)
		Lab	VP	Pedigree population	Reciprocal latency to fight	0.03 (0.02)	0.04 (0.01)	0.86 (0.16)	0.56 (0.15)↑	Wilson et al. (2009)
Cervus elaphus	Red deer	Field	VP	Pedigree population	Social dominance	0.36 (0.12)	0.36 (0.11)	−0.91 (0.07)	—	Wilson et al. (2011)

TB refers to a trait-based approach and VP to a variance partitioning approach. V_{Az} is the direct genetic component, $V_{Az'}$ is the social genetic component (the IGE), $Cov_{Az,z'}$ is the genetic correlation between the direct and indirect genetic effects and T^2 is the total heritable variation. ↑ indicates that the variance explained by T^2 (which includes IGEs) is higher that h^2 estimates based on conventional models (without IGEs), whereas ↓ indicates the opposite pattern. Standard errors for these estimates are provided in parenthesis (underneath estimates), with the exception of *Bos taurus* where these estimates are 95% credible intervals.

a) Thirteen single nucleotide polymorphisms (SNPs) from a genome-wide association (GWA) study were shown to underlie this IGE. The most significant was the *slowpoke* protein-coding gene.

b) These phenotypic measures are interpreted as a proxy for aggression and social dominance.

c) Fish selected for growth in a high interaction (HI) environment were less aggressive than those selected for growth in a low interaction (LI) environment, both when food was present or absent during testing. The agnostic behaviours examined include nips, chases and sudden changes and an index of agnostic interactions was used to compare selection lines.

d) Birds selection for six-week weight based only on direct effects (D-BLUP) or on both direct and indirect effects (C-BLUP). C-BLUP birds evolved to be significantly more efficient at feeding than D-BLUP selected birds. Feeding efficiency was measured as the amount of food wasted during feeding. C-BLUP selection regime was also more efficient in generating an evolutionary response in body weight than D-BLUP selection regime.

e) Artificial selection was followed by pedigree analysis using the animal model.

f) Pigs experienced one generation of divergent artificial selection on IGE for growth (IGEg). Pigs selected for a positive effect of IGEs on growth performed significantly less non-reciprocal biting and less aggression at reunion with familiar group members after 24 h of separation.

closest to a predator in guppies (Edenbrow et al. 2017), the magnitude of these estimates varied greatly (over six times for the latter behaviour). With the exception of aggressive lunges in *D. melanogaster* (Saltz 2013) and latency to attack in Socorro Isopods (Bleakey et al. 2013), the absolute magnitude of Ψ is moderate to high, suggesting that social plasticity is generally important for behaviour. As a slight caveat, we note that while McGlothlin and Brodie (2009) advocated standardizing traits to zero mean and unit variance to allow meaningful comparisons of Ψ, not all studies report whether this was done. Nonetheless, the importance of social plasticity as demonstrated through formal estimation of Ψ is also in accordance with the wider behavioural literature in which the dependence of focal behaviours on partner traits is extensively documented. Obvious examples of this include responsiveness of female mating behaviour to variation in male ornaments (Hunt et al. 2005), and focal contest behaviours being moderated by opponent resource holding potential (e.g. size, weaponry) in animal contests (Hardy and Briffa 2013).

4.4.2 Does Social Plasticity Generate IGEs on Focal Behaviour?

As noted above, a non-zero estimate of Ψ does not permit us to conclude that there are IGEs, only that there could be. However, comparisons of Ψ among populations or lineages with common ancestry do allow somewhat stronger inferences. This is because IGEs are likely to contribute to the evolution of social plasticity. For example, although direct estimates of Ψ were not provided, Petfield et al. (2005) showed that male *D. serrata* rapidly adjust their cuticular hydrocarbon (CHC) profile to match the CHC profile of their mating partner. This male response was shown to be stronger after 16 generations of experimental evolution in populations with sexual selection present versus populations where sexual selection was removed, indicating that Ψ is positive and able to evolve in direct response to sexual selection, at least under laboratory conditions (Chenoweth et al. 2010). A similar ability of Ψ to evolve has also been suggested by Bailey and Zuk (2012) for female mate choice in contemporary and ancestral Kauai populations of the field cricket *T. oceanicus*. Females from the ancestral Kauai decrease their choosiness in the presence of a male calling song (negative Ψ), whereas the reverse is true for females from the contemporary Kauai population (positive Ψ). It was argued that this difference in Ψ results from high predation by the acoustically orienting parasitoid (*Ormia ochracea*) that has rendered up to 90% of males unable to call due to a *flatwing* mutation in the contemporary Kauai population. While this finding suggests that Ψ can also evolve under selection in natural populations, it should be interpreted with caution as it is based on an unreplicated evolutionary event and this difference in Ψ could easily arise from genetic drift (Bailey and Zuk 2012).

Evidence that Ψ has evolved does not prove the presence of IGES in contemporary populations, but does imply that there is (or has been) genetic variance for social plasticity (i.e. a direct genotype-by-(social) environment interaction). In fact, if the social environment is a heritable partner trait, then this logically suggests not only the presence of additive IGEs on the focal behaviour but also the possibility of DGExIGE interactions (an interesting but rarely considered possibility). Regardless, the point is that IGEs occur only if the partner trait is heritable. Formal quantitative genetic analyses are required to confirm this criterion and Table 4.2 presents an overview of empirical studies that have documentED significant IGEs for behaviours. As in Table 4.1, we arrange these studies according to taxon, whether the work was conducted in the field or laboratory and the type of experimental design used. We also indicate whether the study takes a trait-based (TB) or variance partitioning (VP) approach, as well as estimates of the key genetic parameters (V_{Az}, $V_{Az'}$, $Cov_{Az, z'}$ and T^2) for IGEs when available. We have already defined V_{Az} and $V_{Az'}$ as the direct and IGE variances, respectively, and $Cov_{Az, z'}$ as the DGE-IGE covariance in Eq. 4.12. T^2 is simply an estimate of the total heritability for a focal trait, estimated by dividing the variance in TBVs (V_{TBVz}) by the phenotypic variance (V_P). Thus, the comparison of h^2 to T^2 reveals the proportion of the contribution of IGEs to the heritable variance that determines the potential of the population to respond to selection.

Perhaps surprisingly given the greater data requirements, IGEs for behaviour have actually been estimated across a broader taxonomic range than is represented by estimates of Ψ. The only major taxa where IGEs for behaviour have not been documented are amphibians and reptiles. However, given the importance of the social environment to behaviour in these taxa (e.g. Hettyey and Pearman 2003; Riley et al. 2017), it is unlikely that IGEs do not exist but rather that they simply have not been examined. Many of the studies in Table 4.3 focus on species used in agriculture and aquaculture, highlighting the important role that IGEs for behaviour can have not just on animal welfare but also on the effectiveness of selection on commercially important target traits. Most have been conducted on captive populations (with animals housed in the laboratory, pens or tanks) using a variance partitioning rather than trait-based approach. A large number of different specific experimental designs have been used, ranging from use of inbred lines, through classic quantitative genetic approaches (e.g. full-sib half-sib breeding designs) through to analyses of complex, multigenerational pedigree structures as found in wild vertebrates. Finally, although IGEs have been examined for a range of different behaviours, including those associated with reproduction, feeding and antipredator responses, behaviours associated with aggression and social dominance feature most prominently.

While not all studies shown in Table 4.2 provide estimates of all the key genetic parameters, several clear patterns do emerge. First, in the VP studies that predominate, direct genetic variance (V_{Az}) for behaviour typically exceeds indirect (social) genetic variance ($V_{Az'}$): $V_{Az} > V_{Az'}$ in 11 studies, $V_{Az} < V_{Az'}$ in two studies and $V_{Az} = V_{Az'}$ in a single study. Moreover, in those studies where $V_{Az} > V_{Az'}$, V_{Az} was on average 13.78 times larger than $V_{Az'}$, whereas in studies where $V_{Az} < V_{Az'}$, $V_{Az'}$ was only 1.76 greater than V_{Az}. There does not appear to be any taxonomic bias to the observed asymmetry between V_{Az} and $V_{Az'}$. Second, of the 14 estimates of $Cov_{Az, z'}$, seven were negative and seven were positive in sign. There appear to be differences in the sign of $Cov_{Az, z'}$ across taxon, with all available estimates being negative in arthropods (four studies) and birds (one study), positive in fish (two studies) and a mixture in mammals (five positive, two negative). The absolute magnitude of $Cov_{Az, z'}$ appears to be stronger in mammals than the other taxon where it has been estimated but this is probably due to the fact that two of the mammalian studies have focused on IGEs for social dominance in dyadic contests (a specific scenario in which we expect a perfect negative correlation between DGEs and IGEs) (Wilson et al. 2011; Sartori and Mantovani 2013). More generally, however, the balance of positive and negative covariances is consistent with the expectation that IGEs can both accelerate selection responses (always the case if $Cov_{Az, z'} > 0$) and dampen them (likely, but not inevitable, when $Cov_{Az, z'} < 0$).

Table 4.3 Theoretical models examining the evolutionary implications of indirect genetic effects (IGEs) for the evolution of phenotypic traits.

Evolutionary implications	Model type	Reference
IGEs alter the rate of evolution of interacting phenotypes	Analytical	Moore et al. (1997), Agrawal et al. (2001), Bijma et al. (2007a), McGlothlin et al. (2010)
IGEs alter the direction of evolution of interacting phenotypes	Analytical	Moore et al. (1997), Agrawal et al. (2001), Bijma et al. (2007a), McGlothlin et al. (2010), Bijma and Wade (2008), Trubenová and Hagar (2014)
IGEs allow traits lacking direct genetic variance to evolve	Analytical	Moore et al. (1997), Agrawal et al. (2001), Bijma et al. (2007a)
IGEs have a much larger effect on evolutionary responses in a structured population	Analytical	Agrawal et al. (2001)
IGEs can generate covariance between the breeding value of one individual and the phenotype of its social partner. This generates and drives social selection that can influence the evolution of social behaviours and altruism	Analytical	Agrawal et al. (2001), Bijma et al. (2007a), McGlothlin et al. (2010), Bijma and Wade (2008), Trubenová and Hagar (2014), Wolf et al. (1999)
When IGEs are present, only multilevel selection influences the response to selection. When IGEs are absent, multilevel selection and relatedness together influence the response to selection	Analytical	Bijma and Wade (2008)
IGEs allow runaway sexual selection to occur without direct genetic covariance between male traits and female preference	Analytical	Bailey and Moore (2012)
When Ψ is able to evolve, IGEs will have different effects on the rate and direction of evolution of interacting phenotypes compared to traditional analytical models	Numerical	Kazancioğlu et al. (2012)
The inclusion of IGEs into models captures more of the genetic variability in infectivity than models without IGEs	Numerical	Lipschutz-Powell et al. (2012)
BLUP selection regimes including IGEs increases the rate of inbreeding compared to conventional selection regimes lacking IGEs	Numerical	Khaw et al. (2014)
IGEs can modify direct genetic effects on trait values	Numerical	Trubenová and Hagar (2012)

Table 4.3 (Continued)

Evolutionary implications	Model type	Reference
IGEs can directly create selection pressure at the group level and lead to the evolution of social traits that would not evolve in the absence of social interactions	Numerical	Trubenová and Hagar (2014)
The evolutionary consequences of IGEs are limited when social interactions between individuals contain a feedback loop	Numerical	Trubenová et al. (2015)
Artificial selection programme including both direct and IGEs was more efficient in generating an evolutionary response than one without IGEs	Numerical	Liu and Tang (2016)
The size of the social effect and number of group members had the largest effect on how IGEs influenced the response to selection	Numerical	Trubenová and Hagar (2012), Liu and Tang (2016)

We refer to analytical models in the general sense in having a closed form solution. In the case of IGEs, these largely reflect quantitative genetic models. We refer to numerical models in the general sense of requiring a numerical time-stepping procedure to obtain the model's behaviour. In the case of IGEs, time is generally assessed across generations using a simulation approach.

4.5 What are the Evolutionary Consequences of IGEs?

We have argued that IGEs are likely to be present for behavioural traits expressed in social interactions and, though limited in number, the empirical studies highlighted above support this view. IGEs represent an additional source of genetic variance for behaviours that change our expectations, relative to those derived from DGE-only models (e.g. the breeder's equation), of phenotypic responses to selection. But what exactly are the evolutionary consequences? Unfortunately, this is hard to sum up succinctly because theoretical predictions differ greatly across biological scenarios, as well as with the details of specific models. Nonetheless, different theoretical treatments do agree on the major possible outcomes and we highlight these in Table 4.3. With respect to a focal trait of interest, IGEs will: (i) normally alter the rate of evolution, although both increases and decreases are possible; (ii) sometimes alter the direction of evolution, even in the univariate case whereby direct selection to increase the phenotypic mean of a trait could actually result in a decrease in this trait; and (iii) potentially facilitate the evolution of traits that actually lack direct genetic variance (i.e. appear non-heritable in the classic sense). However, the consequences of IGEs necessarily extend beyond the focal trait since they (iv) can lead to a coevolutionary dynamic between the (focal) trait under selection and the social environment provided by the phenotype(s) of interacting conspecific(s). Beyond this, the predicted outcomes of theoretical models become far more complicated.

In this section, we consider only a few of these models to help highlight some of the possible evolutionary consequences of IGEs. A full discussion of all available theoretical models is far beyond the scope of this chapter but we again direct the reader to Table 4.3 for a more complete overview of existing models and their predicted evolutionary outcomes.

4.5.1 What about the Role of Social Selection?

Social interactions between conspecifics can have a dramatic effect on an individual's fitness and whenever the variation in fitness covaries with the traits used in the social interactions, 'social selection' will exist and target the interacting phenotypes (see Figure 4.1) (Wolf et al. 1999). Social selection therefore differs from other forms of 'non-social' selection (e.g. natural or sexual selection) because an individual's fitness is not determined entirely by its own phenotype but also, in part, by the phenotype of its social partner (see Figure 4.1). The multiple regression-based approach outlined in section 4.2 can be readily extended to estimate and partition the social (β_S) and non-social (β_N) selection gradients for a set of correlated phenotypic traits (see Figure 4.1) (Wolf et al. 1999).

Early analytical models largely treated IGEs and social selection in isolation (Moore et al. 1997; Wolf et al. 1999). However, IGE models including social selection have shown that while both processes result from interactions among conspecifics, they generate evolutionary change in different ways (McGlothlin et al. 2010; Westneat 2012; Trubenová and Hagar 2014). Social selection will generate evolutionary change whenever the breeding value of one individual covaries with the phenotype of its social partner (McGlothlin et al. 2010). This can be best illustrated using the equation for the predicted evolutionary change in a focal trait ($\Delta \bar{z}$) in response to social and non-social selection:

$$\Delta \bar{z} = \mathbf{C}_{Az} \beta_N + \mathbf{C}_{Az'} \beta_S \tag{4.13}$$

This equation shows that the evolutionary change due to non-social selection is proportional to the matrix of covariances between the focal individuals' TBV and their own phenotypic values ($\mathbf{C}_{Az}$), whereas evolutionary change attributable to social selection is proportional to the matrix of covariances between the focal individuals' breeding values and the phenotypic values of their social partners ($\mathbf{C}_{Az'}$). Social selection will therefore only drive evolutionary change in cases where $\mathbf{C}_{Az'}$ differs from zero, as will occur whenever individuals are associated non-randomly based on their genotypes (e.g. due to relatedness) and/or if phenotypic expression is influenced by the interaction between individuals (e.g. due to IGEs) (McGlothlin et al. 2010).

Social selection can also generate similar feedback loops to IGEs that accelerate the rate of evolution but on different timescales (McGlothlin et al. 2010). While IGEs create feedback by influencing phenotypic expression within a generation (Moore et al. 1997), social selection creates feedback across generations because interacting phenotypes act simultaneously as both the targets and agents of selection (Wolf et al. 1999). That is, as the traits causing social selection evolve, the next generation experiences a very different social environment (that also alters social selection), creating a feedback in the evolution of the interacting phenotype (Wolf et al. 1999; McGlothlin et al. 2010). It has been argued that these feedback loops are central to the evolution of a range of social behaviours, including altruism, co-operation and aggression (McGlothlin et al. 2010; Trubenová and Hagar 2014).

4.5.2 What Happens When Ψ is Also Able to Evolve?

The traditional quantitative genetic framework for studying IGEs outlined in section 4.3 assumes that Ψ is fixed over both ecological and evolutionary timescales (Moore et al. 1997). However, the response of a focal individual's phenotype to its social partner is likely to be a trait with underlying genetic variation and subject to strong selection and therefore has the potential to evolve (Moore et al. 1997; McGlothlin and Brodie 2009). Indeed, as we outline in section 4.5 above, there is direct evidence from experimental populations of *D. serrata* (Chenoweth et al. 2010) and indirect evidence from natural populations of *T. oceanicus* (Bailey and Zuk 2012) that Ψ is able to evolve. Despite this evidence, only a single model has investigated whether permitting Ψ to evolve changes the evolutionary dynamics of traits involved in social interactions (Kazancioğlu et al. 2012).

To investigate this issue, Kazancioğlu et al. (2012) used simulation models to examine three different biological scenarios (following Moore et al. 1997) where interactions were between: (i) two different traits with non-reciprocal effects, (ii) two different traits with reciprocal effects, and (iii) a single trait with reciprocal effects. In each scenario, two different models were compared: a traditional model where Ψ was assumed to be static and a second model where Ψ is under directional selection to either increase or decrease and therefore has the ability to evolve. Under all scenarios, Kazancioğlu et al. (2012) found that the evolution of Ψ qualitatively altered predictions about the rate and direction of evolution of the interacting phenotypes. More specifically, when Ψ was fixed and had a positive value, trait values increased in each biological scenario. The same increase in trait values was also observed for each biological scenario when Ψ was selected to increase, although the rate of increase in trait values was greater than when Ψ was fixed. In contrast to these patterns, however, selection to decrease Ψ reversed the evolutionary dynamics and resulted in a decrease in trait values in all biological scenarios.

Collectively, this work clearly shows that a more complete understanding of how Ψ evolves is crucial to accurately predict the evolution of traits involved in social interactions. An essential first step in this process is further empirical research measuring Ψ and examining the range of biological conditions that facilitate its evolution. It is likely that experimental evolution studies, such as that conducted by Chenoweth et al. (2010), will prove particularly valuable in this regard although further theoretical work is also needed. The model developed by Kazancioğlu et al. (2012) assumed the evolution of Ψ through fixed selection coefficients. However, selection acting on Ψ is likely to be dynamic and change over ecological and evolutionary time. Indeed, if the conditions that generate selection on Ψ are determined by traits that comprise the social environment (e.g. aggression or dominance), then this seems almost inevitable. Incorporating this additional complexity into IGE models is likely to greatly enhance our ability to predict the long-term evolutionary dynamics of interacting phenotypes.

4.5.3 Can IGEs Influence Other Important Evolutionary Processes?

Many important processes in evolutionary biology (e.g. population divergence, speciation, sexual conflict) are mediated by social interactions, providing the potential for

IGEs to have widespread implications. This is particularly true for sexual selection where interactions within and between the sexes generate the variance in fitness needed for this process to operate and drive the evolution of male sexual traits and female preference for them.

One of the most influential models in sexual selection theory to account for the evolution of exaggerated male ornaments is Fisher's runaway process (Fisher 1958). Central to this model is a positive genetic correlation between the male sexual trait and female preference for the trait. This genetic correlation fuels the sexual coevolution of male trait and female preference, leading to their continued exaggeration until countered by natural selection. However, by incorporating IGEs into quantitative genetic models of sexual selection, Bailey and Moore (2012) showed that a direct genetic correlation is not essential for the runaway process if IGEs exist. This can be illustrated by examining the equation for the mean change in female preference ($\Delta \bar{p}$):

$$\Delta \bar{p} = \frac{1}{2}[B_{PT}\beta_{M_{NS}} + B_{PT}\beta_{M_{SS}}] + \Psi\frac{1}{2}[G_T\beta_{M_{NS}} + G_T\beta_{M_{SS}}] \qquad (4.14)$$

where B_{PT} is the genetic covariance between the male trait (T) and female preference (P), G_T is the genetic variance in the male trait, $\beta_{M_{NS}}$ and $\beta_{M_{SS}}$ represents natural and sexual selection acting on the male trait, respectively and $1/2$ reflects the sex-limited expression of both T and P. The first part of this equation demonstrates the standard quantitative genetic view of the runaway process whereby the change in mean female preference is a function of indirect selection acting on females through genetic covariance with males. However, the second half shows that when IGEs exist, female preference can also change due to selection acting on the male trait that is filtered through Ψ, enabling female preference to evolve even in the absence of any genetic covariance between these traits (Bailey and Moore 2012). That is, even if there is no genetic covariance between female preference and the male trait and the only form of selection targeting the male trait is natural selection (i.e. $B_{PT} = 0$ and $\beta_{M_{SS}} = 0$ in Eq. 4.14), it is still possible for $\Delta \bar{p}$ to change as a function of $\Psi\frac{1}{2}[G_T\beta_{M_{NS}}]$. Consequently, the social environment is predicted to play a key role in runaway sexual selection. Even a small advantage for the male trait in natural selection will result in a positive $\Delta \bar{p}$ provided that $\Psi > 0$. This means that the social environment can provide a mechanism for initiating the runaway process when there is only natural selection acting on the male trait and before any genetic covariance between T and P has been established. Once initiated, the social environment is also expected to enhance the runaway process when Ψ is positive but retard it when Ψ is negative (Bailey and Moore 2012).

Perhaps one of the most important features of this model is that it generates a number of testable predictions on how IGEs should influence the operation of sexual selection, including: (i) the genetic covariance between male traits and female preference may be rare or absent in many systems, (ii) social flexibility in female preference will be common in populations experiencing strong sexual selection, (iii) variation in the social environment should be associated with rapid divergence in male sexual traits, and (iv) male sexual traits should be more elaborate than previously predicted based on runaway sexual selection (Bailey and Moore 2012). There is some evidence to support prediction (i) with a recent meta-analysis showing that less than 20% of species examined exhibited a significant positive genetic correlation between male traits and female preference (Greenfield et al. 2014), although it should be noted that this pattern is also expected given the low power of the experimental designs used in these studies (Sharma et al. 2016). Unfortunately, the remaining predictions have received considerably less empirical attention.

4.5.4 What are the Longer Term Consequences of IGEs?

A common outcome of most theoretical models is that IGEs will influence the evolutionary dynamics of phenotypic traits involved in the social interactions (see Table 4.3). Despite this, however, very few empirical studies have examined the longer term evolutionary consequences of IGEs. An obvious and well-cited exception to this is work on maternal genetic effects in mice (Falconer 1963). Despite applying a regime of divergence artificial selection on litter size in mothers, the observed response in this trait was actually in the opposite direction to selection. This evolutionary reversal is caused by the fact that large mothers have more offspring but offspring from these litters are much smaller in size (due to sibling competition for resources). As a result, these small daughters have fewer offspring per litter, but these offspring grow to be large in size. Thus, a negative covariance between the environment provided by mothers and how this affects the reproduction of daughters drives this maladaptive reversal in the response to selection on litter size (Falconer 1963).

To our knowledge, however, only a single study has examined the longer term consequences of IGEs that involve unrelated individuals (Muir 2005). In this study, Muir (2005) used a regime of artificial selection on six-week body weight in Japanese quail (*Coturnix japonica*) where selection was based on an optimal index including both DGEs and IGEs (C-BLUP) or only DGEs (D-BLUP). After 23 generations of selection, C-BLUP lines produced a significant positive response to selection, whereas D-BLUP lines did not. Moreover, mortality rates decreased over this period in C-BLUP lines but increased in D-BLUP lines. This demonstrates that failure to include IGEs in breeding programmes is not only detrimental to the response to selection but also has an adverse effect on the well-being of animals. We clearly need more empirical studies like this that move beyond simply quantifying key IGE parameters to examining how and why they influence the evolutionary process, especially involving behaviours.

4.6 Conclusions and Future Directions

The relative contribution of genes and the environment to phenotypic variation (the so-called 'nature–nurture' debate) has been a focal point in evolutionary quantitative genetics for decades. We show in this chapter that attributing observed levels of phenotypic variation to either genes or the environment is an approach that, while convenient, lacks nuance and can (at least in theory) give a misleading view of evolutionary potential under selection. As many behaviours are only expressed during social interactions (e.g. aggression, dominance, mate choice) and are important determinants of fitness, it is particularly likely that IGEs play an important role in behavioural evolution.

When a given individual interacts with conspecifics in the population, those conspecifics provide a social environment that often contributes to the phenotype of the focal individual. If the outcome of social interactions depends on a partner trait and causes fitness variation amongst (focal) individuals then social selection occurs (see Figure 4.1) (Wolf et al. 1999). If partner traits determining focal phenotype vary (in part) because of genetic differences, IGEs also exist (see Figure 4.1) (Wolf et al. 1998). The concepts of social selection and IGEs are clearly not identical because one can occur without the other. However, when both are present there is a causal pathway from the genotype of one individual (the partner) to the fitness of the other (the focal) which has consequences for the evolutionary dynamics of both intermediate traits (i.e. that of the partner and that of the focal). This means that the partner trait can evolve, changing the net social environment in a way that may either exacerbate or dampen the change in mean focal trait normally expected under traditional (DGE) quantitative genetic models (Moore et al. 1997).

Many of these general insights were derived using analytical solutions to models intended to capture biologically simple scenarios, most notably dyadic interactions (e.g. Moore et al. 1997; Agrawal et al. 2001; McGlothlin et al. 2010). Although analytical models are ideal in this context they quickly become intractable for exploring more complex, and biologically realistic, social interactions. Here numerical modelling can probably help. For instance, under the trait-based view, allowing variation in Ψ has revealed some surprising consequences of IGEs (Trubenová and Hagar 2012, 2014), while negative feedback loops between focal and partner can lead to phenotypic means oscillating across generations (Trubenová et al. 2015). We may also benefit from more models – both analytical and numerical – that explore more specific types of social interaction. Work to date on sexual selection (Bailey and Moore 2012), sexual conflict (Westneat 2012) and infectivity and disease prevalence (Lipschutz-Powell et al. 2012) is valuable, but IGEs are likely to also be important in a much broader range of social interactions, such as sperm competition, foraging behaviour, cannibalism, and all types of resource competition.

However, we note that the development of models continues to outpace empirical research. Theoretical papers rarely use empirical estimates of Ψ and/or key genetic parameters (e.g. V_{Az}, $V_{Az'}$, $Cov_{Az, z'}$ and T^2) to seed their simulations and calculations (e.g. Trubenová et al. 2015). However, as Tables 4.1 and 4.2 clearly show, this is not surprising given the paucity of estimates available, at least for behavioural traits. If we hope to improve our understanding of the real-world evolutionary consequences of IGEs, we need more empirical work across a wider range of taxa, a more diverse set of behavioural traits and a more comprehensive set of social contexts. Meeting the high data requirements to allow precise estimation of quantitative genetic parameters is never trivial, and high-throughput behavioural phenotyping is especially challenging. It is perhaps unsurprising, then, that IGE parameters, and estimates of h^2 for behaviour more generally, are underrepresented in the literature relative to morphological and life history traits. Key parameters are also easier to estimate using inbred lines (or clones) than outbred (diploid) individuals, as the same genotype can be replicated in different individuals meaning that any changes in behaviour due to experience or carry-over effects can be avoided. Accordingly, many of the studies in Table 4.2 have taken this approach, and there is a lack of studies formally quantifying IGEs in wild (pedigree) populations. However, as studies on a number of bird (e.g. Bijma et al. 2007a; Brommer and Rattiste 2008) and mammal (e.g. Sartori and Mantovani 2013; Wilson et al. 2009, 2011) species demonstrate, this task is not impossible. Although we concede it is an ambitious goal, future studies would therefore benefit by placing greater emphasis on wild populations where IGEs can be assessed during 'natural' social interactions.

References

Agrawal, A.F., Brodie, E.D. III, and Wade, M.J. (2001). On indirect genetic effects in structured populations. *American Naturalist* 158 (3): 308–323.

Alemu, S.W., Bijma, P., Møller, S.H. et al. (2014). Indirect genetic effects contribute substantially to heritable variation in aggression-related traits in group-housed mink (*Neovison vison*). *Genetics Selection Evolution* 44: 30.

Alemu, S.W., Berg, P., Janss, L., and Bijma, P. (2015). Estimation of indirect genetic effects in group-housed mink (*Neovison vison*) should account for systematic interactions either due to kin or sex. *Journal of Animal Breeding and Genetics* 133: 43–50.

Andersson, S. (2014). Indirect genetic effects from competition in the clonal herb *Sedum album* (Crassulaceae). *PLOS One* 9 (8): e106104.

Bailey, N.W. and Hoskins, J.L. (2014). Detecting cryptic indirect genetic effects. *Evolution* 68 (7): 1871–1882.

Bailey, N.W. and Moore, A.J. (2012). Runaway sexual selection without genetic correlations: social environments and flexible mate choice initiate and enhance the Fisher process. *Evolution* 66 (9): 2674–2684.

Bailey, N.W. and Zuk, M. (2012). Socially flexible female choice differs among populations of the Pacific field cricket geographic variation in the interaction coefficient psi (Ψ). *Proceedings of the Royal Society of London B* 279: 3589–3596.

Bergsma, R., Kanis, E., Knol, E.F., and Bijma, P. (2008). The contribution of social effects to heritable variation in finishing traits of domestic pigs. *Genetics* 178: 1559–1570.

Bertram, S.M., Harrison, S.J., Thomson, I.R., and Fitzsimmons, L.P. (2013). Adaptive plasticity in wild field cricket's acoustic signalling. *PLOS one* 8: e69247.

Bijma, P. (2010). Estimating indirect genetic effects: precision of estimates and optimum designs. *Genetics* 186: 1–16.

Bijma, P. (2014). The quantitative genetics of indirect genetic effects: a selective review of modelling issues. *Heredity* 112: 61–69.

Bijma, P. and Wade, M.J. (2008). The joint effects of kin, multilevel selection and indirect genetic effects on response to selection. *Journal of Evolutionary Biology* 21: 1175–1188.

Bijma, P., Muir, W.M., and van Arendonk, J.A.M. (2007a). Multilevel selection 1: quantitative genetics of inheritance and response to selection. *Genetics* 175: 277–288.

Bijma, P., Muir, W.M., Ellen, E.D. et al. (2007b). Multilevel selection 2: estimating the genetic parameters determining inheritance and response to selection. *Genetics* 175: 289–299.

Bleakey, B.H. and Brodie, I.I.I.E.D. (2007). Indirect genetic effects influence antipredator behaviour in guppies: estimates of the coefficient of interaction *Psi* and the inheritance of reciprocity. *Evolution* 63 (7): 1796–1806.

Bleakey, B.H., Welter, S.M., McCauley-Cole, K.M. et al. (2013). Cannibalism as an interacting phenotype: precannibalistic aggression is influenced by social partners in the endangered Socorro Isopod (*Thermosphaeroma thermophilum*). *Journal of Evolutionary Biology* 26 (4): 832–842.

Brommer, J.E. and Rattiste, K. (2008). "Hidden" reproductive conflict between mates in a wild bird population. *Evolution* 62 (9): 2326–2333.

Callander, S., Kahn, A.T., Hunt, J. et al. (2013). The effect of competitors on calling effort and lifespan in male field crickets. *Behavioural Ecology* 24: 1251–1259.

Camerlink, I., Turner, S.P., Bijma, P., and Bolhuis, J.E. (2014). Indirect genetic effects and housing conditions in relation to aggression behaviour in pigs. *PLOS One* 8 (6): e65136.

Camerlink, I., Ursinus, W.W., Bijma, P. et al. (2015). Indirect genetic effects for growth rate in domestic pigs alter aggressive and manipulative biting behaviour. *Behaviour and Genetics* 45: 117–126.

Cappa, E.P. (2008). Direct and competition additive effects in tree breeding: Bayesian estimation from an individual tree mixed model. *Silvae. Genetica* 57: 45–56.

Chenoweth, S.F., Rundle, H.D., and Blows, M.W. (2010). Experimental evidence for the evolution of indirect genetic effects: changes in the interaction effect coefficient, *psi* (Ψ), due to sexual selection. *Evolution* 64 (6): 1849–1856.

Edenbrow, M., Bleakey, B.H., Dardem, S.K. et al. (2017). The evolution of cooperation: interacting phenotypes among social partners. *American Naturalist* 189 (6): 630–643.

Edward, D.A., Poissant, J., Wilson, A.J., and Chapman, T. (2014). Sexual conflict and interacting phenotypes; a quantitative genetic analysis of fecundity and copula duration in *Drosophila melanogaster*. *Evolution* 68 (6): 1651–1660.

Ellen, E.D., Peeters, K., Verhoeven, M. et al. (2015). Direct and indirect genetic effects in life-history traits of flour beetles (*Tribolium casteneum*). *Evolution* 70 (1): 207–217.

Falconer, D.S. (1963). Patterns of response in selection experiments with mice. *Cold Spring Harbor Symposia on Quantitative Biology* 20: 178–196.

Falconer, D.S. and Mackay, T.F.C. (1996). *Introduction to Quantitative Genetics*. London: Longman.

Fisher, R.A. (1958). *The Genetical Theory of Natural Selection*. New York: Dover Press.

Greenfield, M.D., Alem, S., Limousin, D., and Bailey, N.W. (2014). The dilemma of Fisherian sexual selection: mate choice for indirect benefits despite rarity and overall weakness of trait-preference genetic correlation. *Evolution* 68: 3524–3536.

Hadfield, J. (2012). The quantitative genetic theory of parental effects. In: *Evolution of Parental Care* (ed. N.J. Royle, P.T. Smiseth and M. Kölliker). Oxford: Oxford University Press.

Hardy, I.C.W. and Briffa, M. (2013). *Animal Contests*. Cambridge: Cambridge University Press.

Head, M.L., Berry, L.K., Royle, N.J., and Moore, A.J. (2012). Paternal care: direct and indirect genetic effects of fathers on offspring performance. *Evolution* 66 (11): 3570–3581.

Hedrick, A.V., Perez, D., Lichti, N., and Yew, J. (2002). Temperature preference of male field crickets (*Gryllus integer*) alter their mating calls. *Journal of Comparative Physiology A* 188: 799–805.

Hettyey, A. and Pearman, P.B. (2003). Social environment and reproductive interference affect reproductive success in the frog *Rana latastei*. *Behavioural Ecology* 14: 294–300.

Hunt, J. and Simmons, L.W. (2002). The genetics of maternal care: direct and indirect genetic effects on phenotype in the dung beetle *Onthophagus taurus*. *Proceedings of the National Academy of Sciences USA* 99: 6828–6832.

Hunt, J., Brooks, R., and Jennions, M.D. (2005). Female mate choice as a condition-dependent life-history trait. *American Naturalist* 166: 79–92.

Kasumovic, M.M., Hall, M.D., and Brooks, R.C. (2012). The juvenile social environment introduces variation in the choice and expression of sexually selected traits. *Ecology and Evolution* 2: 1036–1047.

Kazancioğlu, E., Klug, H., and Alonzo, S.H. (2012). The evolution of social interactions changes predictions about interacting phenotypes. *Evolution* 66 (7): 2056–2064.

Khaw, H.L., Ponzoni, R.W., and Bijma, P. (2014). Indirect genetic effects and inbreeding: consequences of BLUP selection for socially affected traits on rate of inbreeding. *Genetics Selection Evolution* 46: 39.

Kirkpatrick, M. and Lande, R. (1989). The evolution of maternal characters. *Evolution* 43: 485–503.

Kirkpatrick, M. and Lande, R. (1992). The evolution of maternal characters: errata. *Evolution* 46: 284.

Lande, R. (1979). Quantitative genetic analysis of multivariate evolution, applied to brain:body size allometry. *Evolution* 33 (1): 402–416.

Lande, R. and Arnold, A.J. (1983). The measurement of selection on correlated characters. *Evolution* 37 (6): 1210–1226.

Lande, R. and Kirkpatrick, M. (1990). Selection response in traits with maternal inheritance. *Genetics Research* 55 (3): 189–197.

Lipschutz-Powell, D., Woolliams, J.A., Bijma, P., and Doeschl-Wilson, A.B. (2012). Indirect genetic effects and the spread of infectious disease: are we capturing the full heritable variation underlying disease prevalence? *PLOS One* 7 (6): e39551.

Liu, J. and Tang, G. (2016). Investogating the contribution of social genetic effect to longer selection response in a ten generations breeding programme simulated. *Italian Journal of Animal Science* 15 (4): 610–616.

Lynch, M. and Walsh, B. (1998). *Genetics and Analysis of Quantitative Traits*. New York: Sinauer.

Marie-Orleach, L., Voft-Burri, N., Mouginot, P. et al. (2017). Indirect genetic effects and sexual conflicts: partner genotype influences multiple morphological and behavioural reproductive traits in flatworms. *Evolution* 71 (5): 1232–1245.

McAdam, A.G., Garant, D., and Wilson, A.J. (2014). The effect of other's genes: maternal and other indirect genetic effects. In: *Quantitative Genetics in the Wild* (ed. A. Charmantier, D. Garant and L.E.B. Kruuk). Oxford: Oxford University Press.

McGlothlin, J.W. and Brodie, E.D. III, (2009). How to measure indirect genetic effects: the congruence of trait-based and variance-partitioning approaches. *Evolution* 63 (7): 1785–1795.

McGlothlin, J.W., Moore, A.J., Wolf, J.B., and Brodie, E.D. III, (2010). Interacting phenotypes and the evolutionary process. III. Social evolution. *Evolution* 64 (9): 2558–2574.

Moore, A.J., Brodie, E.D. III, and Wolf, J.B. (1997). Interacting phenotypes and the evolutionary process: I. Direct and indirect genetic effects of social interactions. *Evolution* 51 (5): 1352–1362.

Morrissey, M.B. and Wilson, A.J. (2010). PEDANTICS: an R package for pedigree-based genetic simulation and pedigree manipulation, characterization and viewing. *Molecular Ecology Resources* 10 (4): 711–719.

Morrissey, M.B., Kruuk, L.E.B., and Wilson, A.J. (2010). The danger of applying the breeder's equation in observational studies of natural populations. *Journal of Evolutionary Biology* 23: 2277–2288.

Muir, W.M. (2005). Incorporation of competitive effects in forest tree or animal breeding programs. *Genetics* 170: 1247–1259.

Nielson, H.M., Monsen, B.B., Ødegard, J. et al. (2014). Direct and social genetic parameters for growth and fin damage in Atlantic cod (*Gadus morhua*). *Genetic Selection Evolution* 46: 5.

Petfield, D., Chenoweth, S.F., Rundle, H.D., and Blows, M.W. (2005). Genetic variance in female condition predicts indirect genetic variance in male sexual display traits. *Proceedings of the National Academy of Sciences USA* 102 (17): 6045–6050.

Rapkin, J., Archer, C.R., Grant, C. et al. (2017). Little evidence for intralocus sexual conflict over the optimal intake of nutrients for lifespan and reproduction in the black field cricket *Teleogryllus commodus*. *Evolution* 71: 2159–2177.

Rebar, D. and Rodríguez, R.L. (2013). Genetic variation in social influence on mate preferences. *Proceedings of the Royal Society of London B* 280: 20130803.

Reimert, I., Rodenburg, T.S., Ursinus, W.W. et al. (2014). Selection based in indirect genetic effects for growth, environmental enrichment and coping style affect the immune status of pigs. *PLOS One* 9 (10): e108700.

Riley, J.L., Noble, D.W.A., Byrne, R.W., and Whiting, M.J. (2017). Early social environment influences the behaviour of a family-living lizard. *Royal Society Open Science* 4: 161082.

Ruzzante, D.E. and Doyle, R.W. (1991). Rapid behavioural changes in medaka (*Oryzias latipes*) caused by selection for competitive and noncompetitive growth. *Evolution* 45 (8): 1936–1946.

Saltz, J. (2013). Genetic composition of social groups influences male aggressive behaviour and fitness in natural genotypes of *Drosophila melanogaster*. *Proceedings of the Royal Society of London B* 280: 20131926.

Santostefano, F., Wilson, A.J., Niemelä, P.T., and Dingenmanse, N.J. (2017). Indirect genetic effects: a key component of the genetic architecture of behaviour. *Scientific Reports.* 7: 10235.

Sartori, C. and Mantovani, R. (2013). Indirect genetic effects and the genetic bases of social dominance: evidence from cattle. *Heredity* 110: 3–9.

Sharma, M.D., Wilson, A.J., and Hosken, D.J. (2016). Fisher's sons' effect in sexual selection: absent, intermittent or just low experimental power? *Journal of Evolutionary Biology* 29: 2464–2470.

Trubenová, B. and Hagar, R. (2012). Phenotypic and evolutionary consequences of social behaviour: interactions among individuals affect direct genetic effects. *PLOS One* 7 (11): e46273.

Trubenová, B. and Hagar, R. (2014). Social selection and indirect genetic effects in structured populations. *Evolutionary Biology* 41: 123–133.

Trubenová, B., Novak, S., and Hagar, R. (2015). Indirect genetic effects and the dynamics of social interactions. *PLOS One* 10 (5): e0126907.

Westneat, D.F. (2012). Evolution in response to social selection: the importance of interactive effects of traits on fitness. *Evolution* 66 (3): 890–895.

Willham, R.L. (1963). The covariance between relatives for characters composed of components contributed by related individuals. *Biometrics* 15: 18–27.

Wilson, A.J. (2014). Competition as a source of constraint on life history evolution in natural populations. *Heredity* 112: 70–78.

Wilson, A.J., Gelin, U., Perron, M.-C., and Réale, D. (2009). Indirect genetic effects and the evolution of aggression in a vertebrate system. *Proceedings of the Royal Society of London B* 276: 533–541.

Wilson, A.J., Morrissey, M.B., Adams, M.J. et al. (2011). Indirect genetic effects and evolutionary constraint: an analysis of social dominance in red deer, *Cervus elaphus*. *Journal of Evolutionary Biology* 24: 772–783.

Wolf, J.B. (2003). Genetic architecture and evolutionary constraint when the environment contains genes. *Proceedings of the National Academy of Sciences USA* 100 (8): 4655–4660.

Wolf, J.B. and Wade, M.J. (2009). What are maternal effects (and what are they not)? *Philosophical Transactions of the Royal Society of London B Biological Sciences* 364: 1107–1115.

Wolf, J.B. and Wade, M.J. (2016). Evolution genetics of maternal effects. *Evolution* 70 (4): 827–839.

Wolf, J.B., Brodie, E.D. III, Cheverud, J.M. et al. (1998). Evolutionary consequences of indirect genetic effects. *Trends in Ecology & Evolution* 13 (2): 64–69.

Wolf, J.B., Brodie, E.D. III, and Moore, A.J. (1999). Interacting phenotypes and the evolutionary process. II. Selection resulting from social interactions. *American Naturalist* 153 (3): 254–266.

5

Genes and Behaviour

Chelsea A. Weitekamp and Laurent Keller

Department of Ecology and Evolution, University of Lausanne, 1015 Lausanne, Switzerland

Behaviour is the response of an animal to stimuli in its internal or external environment, ranging from simple reflexive behaviours to those that are more complex and goal directed, such as foraging, finding a mate, or engaging in aggressive interactions. However, even reflexive behaviours can be modified by experience. For example, in the zebrafish, *Danio rerio*, the decision to escape or swim is influenced by social status, achieved through a shift in the excitability of neural circuits (Miller et al. 2017). Therefore, a behavioural act requires an individual not only to process sensory information and respond with motor output, but also to integrate its current internal motivational state and memory of past experiences (Bendesky and Bargmann 2011; O'Connell and Hofmann 2011). As such, the genes that affect behaviour can act to influence many different layers of the nervous system, ranging from sensory perception to the connectivity and modulation of neural circuits (Marder 2012; McGrath 2013). This feature of behaviour, the ability to be modified at many different levels, may contribute to the high evolvability of behavioural traits (Blomberg et al. 2003).

To gain an understanding of how animal behaviour evolves requires an integrative approach that examines how behavioural traits are inherited and also characterizes the genetic variants underlying behaviour and their specific effects on neural processing. In this chapter, we present a current understanding of the relationship between genes (of large effect) and behaviour. We first outline how most phenotypic traits, including behaviour, are controlled by many variants of small effect (see also Chapters 1 and 2). We then describe several well-studied examples of single genes that mediate behaviour, as well as 'supergenes' that can control behavioural divergence within species. Next, we discuss how certain classes of genes may be more likely to influence the evolution of behaviour. Finally, we consider whether the genetic architecture of behavioural traits is unique in relation to other phenotypic traits. We conclude the chapter by suggesting that an integrative approach to the study of genes and behaviour will lend the most insight into the forces underlying behavioural and genetic diversity.

Genes and Behaviour: Beyond Nature-Nurture, First Edition.
Edited by David J. Hosken, John Hunt and Nina Wedell.
© 2019 John Wiley & Sons Ltd. Published 2019 by John Wiley & Sons Ltd.

5.1   Genetic Architecture of Phenotypic Traits

Evolutionary change requires variation in phenotypic traits to have a genetic component. The proportion of variance of a phenotypic trait in a population that is due to genetic factors can be described by the degree of heritability (Visscher et al. 2008). Heritability has been best quantified in humans. A meta-analysis of twin studies identified no traits with heritability estimates of zero, suggesting that all traits are heritable (Polderman et al. 2015). The narrow-sense heritability for one of the most well-studied traits, human height, is around 80%, meaning that 80% of the variance is due to additive genetic factors (Silventoinen et al. 2003). Many behavioural human diseases are also highly heritable, with heritability for schizophrenia and autism spectrum disorder at 80% and 50%, respectively (Purcell et al. 2009a; Gaugler et al. 2014). There is great interest in identifying the causal genetic variants that underlie the heritable variation in phenotypic traits; finding these variants has important implications for identifying disease risk, developing drug treatments, improving efficiency in agriculture, and appreciating the effect of selection on maintaining genetic and phenotypic diversity (Robinson et al. 2014).

Quantitative trait locus (QTL) mapping in inbred laboratory model organisms was the first approach commonly used to examine the genetic architecture of traits. The results from early QTL studies, often biased by low statistical power, suggested that a few large-effect QTL could explain a high proportion of trait variation, fuelling hope that large-effect variants underlying disease could be identified in humans (Flint and Mackay 2009). However, subsequent fine-mapping studies revealed that multiple QTLs, often with opposite effects, were contained within the originally identified single QTLs (Flint and Mackay 2009). Furthermore, even a well-defined QTL contains some 300–500 genes (Mackay 2004; Mott and Flint 2008). The picture that eventually emerged suggests that allelic effects follow an exponential distribution, with few loci of large effect and many loci of small effect (Orr 1998; Flint and Mackay 2009; Rockman 2012).

A genetic architecture with many variants of small effect has also been found using the more recently developed tool of genome-wide association studies (GWAS) (Manolio et al. 2009). Using outbred populations, this approach identifies single nucleotide polymorphisms associated with phenotypic traits and overcomes many of the challenges associated with QTL analyses that typically only identify regions containing tens of hundreds of genes. GWAS with large sample sizes (250 000 individuals) have shown that the genetic architecture for human height is characterized by a very large number of causal variants (Wood et al. 2014). The effect sizes of the thousands of loci implicated in human height are typically small, with any given allele typically accounting for less than 1 mm difference in height (Wood et al. 2014). Behavioural human diseases appear to have a similar genetic architecture with many genes of small effect. For example, both schizophrenia and autism spectrum disorder are associated with a large number of common genetic variants (Purcell et al. 2009a; Gaugler et al. 2014).

Somewhat surprisingly, even very large-scale GWAS still often explain only a limited proportion of the heritability of a given trait. This has led to the suggestion that non-additive genetic variance contributes to the 'missing heritability' (Eichler et al. 2010; Zuk et al. 2012a). However, most evidence suggests that the missing heritability probably lies in unidentified variants of very small effect which studies lack the power to detect (Rockman 2012). The proportion of accounted-for heritability has been shown to increase with ever larger sample sizes and inclusion of more single nucleotide polymorphisms (SNPs). For example, the common variants identified across independent studies of human height now account for 60% of the heritability (Wood et al. 2014). Similarly, a meta-analysis based on over 14.5 million pairs of human twins found that for two-thirds of all traits, twin resemblance fit a simple model of additive genetic variance (Polderman et al. 2015). Deviation from the model for the remaining one-third of traits apparently stems from the effect of shared environmental factors, rather than a prevalence of non-additive genetic variance (Polderman et al. 2015). It is also possible that trait heritability is overestimated due to the presence of epistasis, so the concern over the problem of 'missing heritability' may have been overstated (Zuk et al. 2012b). Unfortunately, the statistical power to detect non-additive genetic effects via genome-wide scans is very low given the nearly infinite number of models that can be fit. On smaller scales, the importance of non-additive genetic effects is well documented (Greenspan 2001; Meffert et al. 2002). For example, gene knockout studies have shown that the genetic background influences the resulting phenotype (Holmes et al. 2003; Dowell et al. 2010).

While it appears that most phenotypic traits are characterized by a genetic architecture composed of many variants of small effect acting additively, as described above, there are also many cases where traits are under the control of variants of large effect. For example, two variants with large effects on human personality are due to inversions (Giglio et al. 2001; Stefansson et al. 2005; Huddleston and Eichler 2016). The 17q21.31 inversion in humans contains several neurological-related genes, including microtubule-associated protein tau (*MAPT*) and corticotropin-releasing hormone receptor 1 (*CRHR1*) (Stefansson et al. 2005). Two haplotypes exist within human populations, H1 (direct) and H2 (inverted), and the two do not recombine over ~1.5 Mb (Steinberg et al. 2012). A historical double recombination event may have occurred between the two haplotypes (Steinberg et al. 2012). Eight structural subtypes, five in the H1 lineage and three in the H2 lineage, have also been identified as a result of duplications and complex rearrangements within the inverted region (Steinberg et al. 2012). The primary H1 haplotype has been implicated in many neurodegenerative diseases (de Jong et al. 2012; Puig et al. 2015). One subtype of the H2 haplotype, H2D, increases the risk of microdeletion syndrome and has greatly increased in frequency in individuals of European descent, mostly likely because women carrying this subtype tend to have more children (Stefansson et al. 2005). It has been suggested that inversions and other structural variants may be an important and unappreciated cause underlying variation in human phenotypes (Huddleston and Eichler 2016).

Below, we discuss further examples of single genes and large non-recombining chromosomal regions with large effects on behavioural phenotypes.

5.2 Effects of Single Genes on Behaviour

Most genes with a large effect on behavioural phenotypes have been identified through mutagenesis experiments and association mapping in classic model laboratory organisms (Greenspan 2009). Another successful approach has been to conduct interspecific QTL mapping, which involves crosses between closely related species that differ in characteristic behavioural phenotypes (Schielzeth and Husby 2014). While powerful, an issue with this method is that effect size is often overestimated when traits are mapped in two related species, probably as a by-product of differences in the structure of the mapping populations (Flint et al. 2005). QTL mapping in *Peromyscus* mice revealed that a simple genetic architecture may underlie tunnel burrowing behaviour (Weber et al. 2013). Oldfield mice, *Peromyscus polionotus*, build complex burrows with long entrances and escape tunnels. In contrast, the sister species, the deer mouse, *Peromyscus manic- ulatus*, build simple burrows with no escape tunnel. Genetic crosses between the two species resulted in offspring that build tunnels similar to oldfield mice, suggesting that the alleles that affect burrowing behaviour segregate in a dominant manner (Weber et al. 2013). QTL mapping on a recombinant backcross generation revealed that three addi- tive genetic variants (QTLs) associated with the length of the entrance tunnel explain more than half of the genetic variation for the trait. A single variant was found to be associated with whether or not an individual builds an escape tunnel (Weber et al. 2013).

The same approach can be used on within-species crosses when there is strong vari- ation among populations. For example, two independent studies using QTL mapping have examined the genetic architecture associated with schooling behaviour in fishes. By crossing strongly schooling marine and weakly schooling benthic populations of the threespine stickleback, *Gasterosteus aculeatus*, a genetic variant associated with school- ing position was identified (Greenwood et al. 2013). Interestingly, this variant is also associated with lateral line anatomy, a peripheral neurosensory system important to positioning during social group formation in fishes (Greenwood et al. 2013). Specifically, the identified region contains genes that underlie variation in the number and pattern of neuromasts, as well as other candidate genes implicated in lateral line development or social behaviour. Similarly, in the Mexican tetra, *Astyanax mexicanus*, crosses between a sighted surface-dwelling form that schools and a blind cave-dwelling form that does not revealed that the genetic variants related to loss of schooling behaviour are, in part, associated with loss of vision, rather than changes in lateral line anatomy, as was found for benthic stickleback (Kowalko et al. 2013).

While QTL mapping studies have uncovered many genetic variants associated with behaviour, the causal gene(s) and functional changes in DNA sequence that contribute to behavioural variation have only rarely been identified. One exception comes from a study of parental care in *Peromyscus* mice (Bendesky et al. 2017). A large-effect QTL associated with nest building, a component of parental care, was identified by conducting an interspecies cross between two *Peromyscus* species that differ in parental care behaviour. By examining expression differences in the hypothalamus for the approximately 500 genes identified within the QTL, the neuromodulator arginine vasopressin (AVP) was identified as a main candidate. Increased expression of AVP was associated with less nest building. Pharmacology and chemogenetic experiments showed that vasopressin neurons within the hypothalamus are critical to parental nest-building behaviour, suggesting that differences in this gene contribute to the evolution of parental care in mice (Bendesky et al. 2017).

A comparative approach that examines variation in RNA or protein expression across species that differ in behavioural phenotypes is another way in which specific large-effect genes have been identified. For example, variation in the distribution of AVP 1a receptors in the male brain have been associated with differences in pair-bonding behaviour between vole species (Insel et al. 1994). This gene was originally examined as a candidate because central administration of vasopressin was demonstrated to have wide-ranging effects on reproductive and parental care behaviours across species (Winslow et al. 1993). Findings such as these suggested that there may be many genes underlying behaviour with evolutionarily conserved functions, prompting the idea of a 'genetic toolkit' for behaviour, in which the same gene is either conserved or repeatedly co-opted in the evolution of shared behavioural phenotypes (Fitzpatrick et al. 2005; Toth et al. 2007; Rittschof and Robinson 2016). Recent comparative genomics approaches, however, have largely failed to identify new single genetic variants associated with behaviour across taxa (Rittschof et al. 2014; Kapheim et al. 2015). Given the large number of loci found which contribute to phenotypic variation, as well as the general conservation of physiological processes, there remain relatively few examples of the same genes contributing to shared behaviours in different species (Flint and Mackay 2009). Nevertheless, there are several striking examples of single genes with large effects on behaviour both between and within species; examples are described below (see also Figure 5.1).

5.2.1   The Foraging Gene and Food-Search Behaviour

The *foraging* (*for*) gene, which underlies a naturally occurring polymorphism in the food-searching strategy of *Drosophila melanogaster* larvae and adults, was one of the first large-effect behavioural genes to be identified (de Belle et al. 1989). The gene *for* encodes for a cyclical guanosine monophosphate (cGMP)-dependent protein kinase G (PKG) and has two naturally occurring variants (Osborne et al. 1997). The rover pheno-type moves further distances in search of food compared to the sitter phenotype. Rover individuals have higher PKG activity and higher *for* gene expression relative to sitters

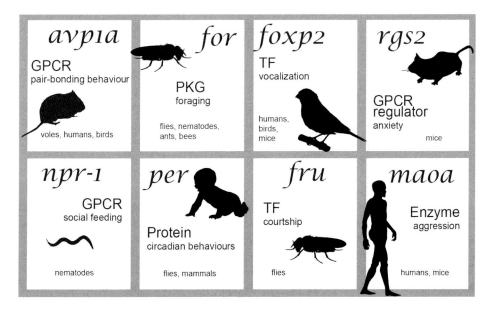

Figure 5.1 Examples of single genes with large effects on behavioural phenotypes. Information listed for each gene includes the product which it encodes, its effect on behaviour, and the species in which it has been well studied. GPCR, G-protein coupled receptor; PKG, cGMP-dependent protein kinase; TF, transcription factor.

(Osborne et al. 1997). The *for* gene also has pleiotropic effects on other food-related traits, including metabolism and insulin signalling (Kent et al. 2009).

The *for* ortholog also affects food-searching strategies in other invertebrates. In the honey bee, *Apis mellifera*, gene expression of *for* increases in association with the age-related transition from in-hive activities to foraging behaviour (Ben-Shahar et al. 2002). The opposite pattern appears to be true for ants, whereby foraging behaviour is associated with lower levels of *for* gene expression (Ingram et al. 2005; Lucas et al. 2015), although the relationship between age, foraging behaviour, and *for* activity may be complex (Oettler et al. 2015). The specific tissues, cells, and gene networks within which *for* exerts its behavioural effects remain to be elucidated (Allen et al. 2017).

5.2.2 Arginine Vasopressin Receptor and Pair-Bonding Behaviour

The nonapeptides oxytocin and AVP have wide-ranging effects on social behaviour across species (Goodson 2013). They function as hormones in the periphery and as neuromodulators in the central nervous system. In the brain, differences in receptor expression, ligand binding, and microsatellite length have been shown to predict differences in pair bonding, social flocking, parental care, and various other social behaviours (Goodson 2013). The vasopressin 1a receptor (V1aR, encoded by *avpr1a*) has been associated with pair bonding in several species, including humans (Walum et al. 2008), but has been particularly well studied in the socially monogamous prairie vole *Microtus ochrogaster* (Young et al. 2011). V1aR exhibits high intraspecific variation in this species. The effects of V1aR on behaviour depend on the brain region/neural circuit in which it is acting. For example, administration of a V1aR antagonist in the lateral septum or ventral pallidum, regions involved in the reward circuit, prevents partner preference formation in males (Lim and Young 2004). *Avpr1a* expression in a spatial memory circuit, but not in the lateral septum or ventral pallidum, is associated with male sexual fidelity (Ophir et al. 2008).

While prairie voles are socially monogamous, nearly one-fourth of offspring are sired by males that engage in extra-pair fertilizations. These males have larger home ranges and more frequently intrude in other territories compared to the majority of males that adopt a 'resident' strategy (Okhovat et al. 2015). Interestingly, levels of V1aR in spatial memory-related brain regions are associated with space use and site fidelity. Also in these brain regions, single nucleotide polymorphisms in *avpr1a* were found to predict individual differences in V1aR abundance. These genetic markers appear to be under balancing selection, reflecting the fitness trade-offs associated with either closely maintaining a pair bond or engaging in extra-pair mating (Okhovat et al. 2015).

5.2.3 Neuropeptide Y Homolog, Sensory Neurons, and Social Feeding Behaviour

The nematode *Caenorhabditis elegans* typically lives within decaying fruit where oxygen levels are low due to microbial respiration (Laurent et al. 2015). If an individual detects an increase in oxygen levels, suggesting that it is approaching the surface, it exhibits avoidance behaviour and reverses direction (McGrath et al. 2009). Failure to find an environment with lower oxygen leads to an aroused state with a suite of related behaviours, including forming social aggregations on food (Busch et al. 2012; Laurent et al. 2015). A laboratory-cultivated strain, N2, shows only a weak response to increased oxygen and maintains solitary feeding habits in the laboratory (de Bono and Bargmann 1998; Gray et al. 2004). During the early phases of domestication, this strain adapted to the laboratory environment and acquired a single nucleotide substitution in the homolog of the neuropeptide Y gene *npr-1* (McGrath et al. 2009). The high activity version of the allele present in the N2 strain decreases aversion to oxygen levels when consuming bacterial food, leading to modification of aggregation behaviour and differences in adult body size, fecundity, and physiology (Gray et al. 2004; Milward et al. 2011; Andersen et al. 2014). The effect of *npr-1* on these behaviours has been linked to a single pair of inter/motor neurons, called RMG neurons (Macosko et al. 2009). Multiple distributed sensory inputs, including pheromone and oxygen detection, are co-ordinated through gap junctions with the common target neurons (Jang et al. 2017). NPR-1 inhibits RMG activity in the solitary N2 strain, which serves to uncouple the circuit while maintaining the function of the input sensory neurons (Macosko et al. 2009; Bargmann and Marder 2013). Similar neural circuits linking variation in neuropeptide Y homologs with social feeding behaviour may occur in other species, such as *D. melanogaster* (Wu et al. 2003).

Foraging strategies in *C. elegans* are also altered by conspecific pheromones. Heritable variation in pheromone sensitivity is linked to a G-protein coupled pheromone receptor, *srx-43*, that acts on sensory neurons to suppress exploratory foraging (Greene et al. 2016). The genomic region associated with *srx-43* is under balancing selection as the two different haplotypes confer bidirectional effects on fitness dependent on food distribution and pheromone detection via *srx-43* (Greene et al. 2016).

5.3 Effects of Supergenes on Behaviour

Phenotypic associations between multiple traits among individuals within populations occur commonly across taxa and across traits, including behaviour (Saltz et al. 2017). For example, aggression is correlated with colour variation in many species (Ducrest et al. 2008). In the African cichlid fish, *Astatotilapia burtoni*, yellow male morphs are more aggressive than their blue counterparts (Dijkstra et al. 2017). When such trait correlations are adaptive, recombination can be disruptive and can impose a cost on fitness. This cost is avoided by a genetic architecture that preserves favourable combinations of alleles (Darlington and Mather 1949; Dobzhansky 1970; Thompson and Jiggins 2014). The maintenance of correlated traits can be achieved by genomic rearrangements that are inherited as a single locus. Such solutions have evolved repeatedly and underlie, for example, the evolution of sex chromosomes (Charlesworth 1996).

Supergenes are defined as multiple tightly linked loci that each affect discrete developmental or behavioural phenotypes (Schwander et al. 2014; Thompson and Jiggins 2014). Butterfly mimicry is a classic example of a supergene maintaining a balanced polymorphism, whereby multiple morphs in the same species mimic several different toxic species, functioning to reduce predation (Joron et al. 2011). Any recombination of traits that would reduce phenotype matching would have negative effects on fitness. Recently, supergenes have also been found to maintain several behavioural polymorphisms, described below (Figure 5.2).

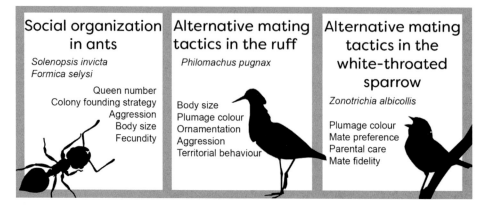

Figure 5.2 Examples of behavioural polymorphisms under the control of supergenes.

5.3.1 Social Organization in Ants

Social organization, defined as the pattern of relationships between individuals within a social group, including the way in which reproduction is partitioned, can vary even within species or populations (Ross and Keller 1995). This phenomenon has been well studied in the fire ant, *Solenopsis invicta*, where there is variation in the number of reproductive queens per colony. *S. invicta* is polymorphic in queen number, as well as in a suite of related traits (Keller 1993). Colony acceptance of multiple queens is directly linked to allelic variation for the gene *Gp-9* (B and b alleles), which encodes an odorant binding protein (Ross and Keller 1998). Colonies in which all workers are homozygous at this locus (BB) will only accept a single BB queen. In contrast, colonies in which at least 10% of workers are heterozygous (Bb) will accept multiple, but only heterozygous queens (Ross and Keller 2002; Gotzek and Ross 2008). It was later shown that *Gp-9* is in fact part of a large non-recombining supergene of approximately 13 Mb with an estimated 616 genes in tight linkage (Wang et al. 2013). Interestingly, the Alpine silver ant, *Formica selysi*, shows a similar polymorphism in the number of queens per colony. This too was found to be under the control of a non-recombining supergene, although the specific location and content of the supergene differ from that of *S. invicta* (Purcell et al. 2014).

5.3.2 Alternative Mating Tactics in Birds

The Eurasian sandpiper, *Philomachus pugnax*, is a lek-breeding wading bird with three alternative male morphs. Independent males, the most common morph, defend territories and court females. Satellite males are non-territorial, but co-display and steal matings when independents are distracted. The third morph is the rare female-mimicking sneaker male (called a faeder) (Jukema and Piersma 2006). The three male morphs have been suggested to be under the control of a single Mendelian locus with three alleles (Lank et al. 1995). This was revealed to be a supergene consisting of an estimated 125 genes contained within a 4.5 Mb inversion (Küpper et al. 2015; Lamichhaney et al. 2016). The satellite and faeder morph alleles are dominant to the ancestral independent morph sequence (Lamichhaney et al. 2016). Homozygosity for the inversion is lethal, and heterozygosity reduces survival. Satellite and faeder males have larger testes, suggesting that a higher reproductive success may offset the costs of carrying the inversion (Küpper et al. 2015).

In the white-throated sparrow, *Zonotrichia albicollis*, a supergene controls two alternative morphs in both males and females that differ in plumage colour and social behaviour (Tuttle et al. 2016). Tan morphs are monogamous, while white morphs are promiscuous and invest less in parental care (Tuttle 2003). The supergene is a large inversion over 100 Mb and contains an estimated 1137 genes. The inversion contains several genes that are well known for their role in the neural control of social behaviour and regulation of aggression, including serotonin and oestrogen receptors, as well as vasoactive intestinal peptide (Tuttle et al. 2016). White and tan morphs show strong disassortative mating – there are negative fitness effects for the rare cases of assortative mating. Interestingly, the white morph allele may be degrading, and for genes within the inversion, gene expression is lower compared to the tan morph, suggesting the white allele is similar to a neo-sex chromosome (Tuttle et al. 2016).

5.4 Evolvability of Behaviour-Associated Genes

Certain classes of genes may be more likely to contribute to the evolution of behaviour. Sensory genes, for example, are among the fastest evolving families of genes (McGrath 2013). A genetic change to a sensory receptor provides a simple path to modify a behaviour by changing the perception of a stimulus without negative effects on other aspects of the phenotype (Bendesky and Bargmann 2011). In ants, communication is mainly through pheromone signalling – a large and novel clade of odorant receptor genes that allow pheromone detection evolved in the ancestor of all ants, probably affecting rates of speciation (McKenzie et al. 2016). The visual system, too, is a prime target. For example, changes in visual sensitivity due to genetic modifications of opsins was a major driver of the rapid speciation of African cichlid fishes, in part due to the effects on mate preference behaviour (Kocher 2004; Terai et al. 2006).

Neuromodulators also appear to be highly evolutionarily labile – they can act at a distance from the target cell and are not always essential for neurotransmission (Bendesky and Bargmann 2011; Marder 2012). Interestingly, in a meta-analysis of vertebrates, the sites of ligand production in the brain were found to be less conserved than the spatial distribution of their receptors (O'Connell and Hofmann 2012). It may be that only small developmental changes are necessary to shift the sites of ligand production (Marín and Rubenstein 2003; O'Connell and Hofmann 2012). While the spatial distribution of receptors was found to be highly conserved across taxa, the density of receptors in specific brain regions is well known to influence behaviour. For example, differences in the density of oxytocin, vasopressin, and dopamine receptors may underlie many of the behavioural differences between monogamous prairie voles and promiscuous montane and meadow voles (Smeltzer et al. 2006).

5.5 Are Behavioural Traits Unique?

Phenotypic traits for which the genetic architecture has been well characterized are often morphological or, in the case of humans, disease related. The extent to which the genetic architecture underlying behavioural traits is unique is not well understood. Studies of outbred mice and rats suggest that the effect sizes for individual QTL may be lower for behavioural compared to physiological traits (Rat Genome Sequencing and Mapping Consortium et al. 2013; Parker et al. 2016). In addition, behavioural phenotypes often have lower heritability estimates compared to morphological and physiological traits (Roff and Mousseau 1987; Meffert et al. 2002; Parker et al. 2016). These low estimates could arise, in part, because measuring behavioural traits is challenging and their repeatability is often not examined (Croston et al. 2015; Greives et al. 2017). Experimental noise in behaviour assays can arise from the effects of age, nutrition, and stress, as well as from abiotic sources, such as temperature (Boake 1994; Meffert et al. 2002). In addition, behavioural traits are often complex, interrelated, and labile across time and development, so it is difficult to dissect behavioural phenotypes into quantifiable components.

Behaviours that involve a social component create an additional level of complexity to studying the genetic architecture of behaviour due to the presence of indirect genetic effects (IGEs), which describe how the phenotype of a focal individual can be influenced by the genes expressed by its interacting partners (Moore et al. 1997; Schneider et al. 2016) (reviewed in Chapter 4). These IGEs make the social environment itself heritable and thus also open to the effects of selection (Wolf et al. 1998). IGEs can have additive genetic effects, as well as non-additive effects in the form of epistasis (Wolf 2000). Importantly, because selection can act on traits in the absence of additive genetic variance, these effects can obscure heritability estimates and alter evolutionary trajectories (Meffert 1995; Wolf et al. 1998; Meffert et al. 2002; García-González and Simmons 2007). IGEs are prevalent in species with parental care, whereby the phenotype of the offspring depends on the genotype of the parent, often of the mother, beyond the contribution of direct genetic inheritance (i.e. maternal effects) (Mousseau et al. 2009; McAdam et al. 2014) (see Chapter 7). These effects also play a large role in social insect colonies, whereby the genotype of nestmates affects individual and colony-level phenotypes (Pankiw et al. 2002; Linksvayer and Wade 2005; Linksvayer 2006). For example, in the fire ant, *S. invicta*, colonies switch their social organization and a suite of related behaviours when the proportion of colony members with a certain genotype passes above a critical threshold (Ross and Keller 2002) (see below for more detail). IGEs have also been found to affect a wide range of other behavioural phenotypes across species, including courtship (Petfield et al. 2005), aggression (Wilson et al. 2009), mate choice (Bailey and Zuk 2012) (see Chapter 6), and antipredator behaviour (Bleakley and Brodie 2009). Given that social interactions among animals are nearly ubiquitous, the role of IGEs in affecting trait evolution should be carefully considered. These may disproportionately affect the genetic architecture of behavioural traits, in particular.

Interestingly, most of the genes found to have a large effect on behavioural phenotypes come from crosses between different species or populations. These larger effects may be a by-product of differences in the structure of the mapping populations (Flint et al. 2005). Most QTL studies of behaviour conducted within single populations have revealed a complex genetic architecture, similar to that found for other phenotypic traits (Bendesky and Bargmann 2011). Indeed, it appears that variation in behaviour, like other phenotypic traits, is usually modulated by many common variants of small effect (Valdar et al. 2006; Flint and Mackay 2009). For example, studies on the behaviour of inbred strains of mice and rats revealed a large number of genetic variants of small effect, as well as the presence of complex non-additive genetic effects (Flint 2003).

5.6 Conclusion

Animal behaviour is often complex and is affected by genes, experiences, and the environment. In addition, many behavioural phenotypes show adaptive correlations with other phenotypic traits. These beneficial associations can be maintained through pleiotropy or linkage, and supergenes have properties of both (Saltz et al. 2017). Given the frequency of balanced polymorphisms that involve behaviour, it may be that supergenes are a common part of the genetic architecture underlying polymorphic behavioural phenotypes. The extent to which the genetic architecture of behavioural traits is unique, due to either the occurrence of supergenes, the presence of indirect genetic effects, or the way in which the nervous system responds to selection, remains to be determined. Future behavioural studies that are able to draw a link between heritability, fitness, genes, and neural circuits will be crucial to gain a more complete understanding of the genetic basis of behaviour.

Acknowledgements

We thank Jonathan Flint and Michel Chapuisat for comments.

References

Allen, A.M., Anreiter, I., Neville, M.C., and Sokolowski, M.B. (2017). Feeding-related traits are affected by dosage of the foraging gene in *Drosophila melanogaster*. *Genetics* 205 (2): 761–773.

Alves, J.M., Lima, A.C., Pais, I.A. et al. (2015). Reassessing the evolutionary history of the 17q21 inversion polymorphism. *Genome Biology and Evolution* 7 (12): 3239–3248.

Andersen, E.C., Bloom, J.S., Gerke, J.P. et al. (2014). A variant in the neuropeptide receptor npr-1 is a major determinant of *Caenorhabditis elegans* growth and physiology. *PLoS Genetics* 10 (2): e1004156.

Bailey, N.W. and Zuk, M. (2012). Socially flexible female choice differs among populations of the pacific field cricket: geographical variation in the interaction coefficient psi (ψ). *Proceedings of the Royal Society of London B: Biological Sciences* 279: 3589–3596. http://rspb.royalsocietypublishing.org/content/early/2012/05/24/rspb.2012.0631.short.

Bargmann, C.I. and Marder, E. (2013). From the connectome to brain function. *Nature Methods* 10 (6): 483–490.

de Belle, J.S., Hilliker, A.J., and Sokolowski, M.B. (1989). Genetic localization of foraging (for): a major gene for larval behavior in *Drosophila melanogaster*. *Genetics* 123 (1): 157–163.

Bendesky, A. and Bargmann, C.I. (2011). Genetic contributions to behavioural diversity at the gene–environment interface. *Nature Reviews Genetics* 12 (12): 809.

Bendesky, A., Kwon, Y.-M., Lassance, J.-M. et al. (2017). The genetic basis of parental care evolution in monogamous mice. *Nature* 544: 434–439.

Ben-Shahar, Y., Robichon, A., Sokolowski, M.B., and Robinson, G.E. (2002). Influence of gene action across different time scales on behavior. *Science* 296 (5568): 741–744.

Bleakley, B.H. and Brodie, E.D. (2009). Indirect genetic effects influence antipredator behavior in guppies: estimates of the coefficient of interaction *psi* and the inheritance of reciprocity. *Evolution* 63 (7): 1796–1806.

Blomberg, S.P., Garland, T., and Ives, A.R. (2003). Testing for phylogenetic signal in comparative data: behavioral traits are more labile. *Evolution* 57 (4): 717.

Boake, C.R.B. (1994). Evaluation of applications of the theory and methods of quantitative genetics to behavioral evolution. In: *Quantitative Genetic Studies of Behavioral Evolution* (ed. C.R.B. Boake), 305–325. Chicage: University of Chicago Press.

Boettger, L.M., Handsaker, R.E., Zody, M.C., and McCarroll, S.A. (2012). Structural haplotypes and recent evolution of the human 17q21.31 region. *Nature Genetics* 44 (8): 881–885.

de Bono, M. and Bargmann, C.I. (1998). Natural variation in a neuropeptide y receptor homolog modifies social behavior and food response in *C. elegans*. *Cell* 94 (5): 679–689.

Busch, K.E., Laurent, P., Soltesz, Z. et al. (2012). Tonic signaling from o_2 sensors sets neural circuit activity and behavioral state. *Nature Neuroscience* 15 (4): 581–591.

Charlesworth, B. (1996). The evolution of chromosomal sex determination and dosage compensation. *Current Biology* 6 (2): 149–162.

Croston, R., Branch, C.L., Kozlovsky, D.Y. et al. (2015). Heritability and the evolution of cognitive traits. *Behavioral Ecology* 26 (6): 1447–1459.

Darlington, C.D. and Mather, K. (1949). *The Elements of Genetics*. London: George Allen & Unwin.

Dijkstra, P.D., Maguire, S.M., Harris, R.M. et al. (2017). The melanocortin system regulates body pigmentation and social behaviour in a colour polymorphic cichlid fish. *Proceedings of the Royal Society B: Biological Sciences* 284 (1851): ii.

Dobzhansky, T. (1970). *Genetics of the Evolutionary Process*. New York: Columbia University Press.

Dowell, R.D., Ryan, O., Jansen, A. et al. (2010). Genotype to phenotype: a complex problem. *Science* 328 (5977): 469.

Ducrest, A.-L., Keller, L., and Roulin, A. (2008). Pleiotropy in the melanocortin system, coloration and behavioural syndromes. *Trends in Ecology & Evolution* 23 (9): 502–510.

Eichler, E.E., Flint, J., Gibson, G. et al. (2010). Missing heritability and strategies for finding the underlying causes of complex disease. *Nature Reviews Genetics* 11 (6): 446–450.

Fitzpatrick, M.J., Ben-Shahar, Y., Smid, H.M. et al. (2005). Candidate genes for behavioural ecology. *Trends in Ecology & Evolution* 20 (2): 96–104.

Flint, J. (2003). Analysis of quantitative trait loci that influence animal behavior. *Journal of Neurobiology* 54 (1): 46–77.

Flint, J. and Mackay, T.F.C. (2009). Genetic architecture of quantitative traits in mice, flies, and humans. *Genome Research* 19 (5): 723–733.

Flint, J., Valdar, W., Shifman, S., and Mott, R. (2005). Strategies for mapping and cloning quantitative trait genes in rodents. *Nature Reviews Genetics* 6 (4): 271–286.

García-González, F. and Simmons, L.W. (2007). Paternal indirect genetic effects on offspring viability and the benefits of polyandry. *Current Biology* 17 (1): 32–36.

Gaugler, T., Klei, L., Sanders, S.J. et al. (2014). Most genetic risk for autism resides with common variation. *Nature Genetics* 46 (8): 881–885.

Giglio, S., Broman, K.W., Matsumoto, N. et al. (2001). Olfactory receptor–gene clusters, genomic-inversion polymorphisms, and common chromosome rearrangements. *American Journal of Human Genetics* 68 (4): 874–883.

Goodson, J.L. (2013). Deconstructing sociality, social evolution and relevant nonapeptide functions. *Psychoneuroendocrinology* 38 (4): 465–478.

Gotzek, D. and Ross, K.G. (2008). Experimental conversion of colony social organization in fire ants (*Solenopsis invicta*): worker genotype manipulation in the absence of queen effects. *Journal of Insect Behavior* 21 (5): 337–350.

Gray, J.M., Karow, D.S., Lu, H. et al. (2004). Oxygen sensation and social feeding mediated by a c. elegans guanylate cyclase homologue. *Nature* 430 (6997): 317–322.

Greene, J.S., Brown, M., Dobosiewicz, M. et al. (2016). Balancing selection shapes density-dependent foraging behaviour. *Nature* 539 (7628): 254–258.

Greenspan, R.J. (2001). The flexible genome. *Nature Reviews Genetics* 2 (5): 383–387.

Greenspan, R.J. (2009). Selection, gene interaction, and flexible gene networks. *Cold Spring Harbor Symposia on Quantitative Biology* 74: 131–138.

Greenwood, A.K., Wark, A.R., Yoshida, K., and Peichel, C.L. (2013). Genetic and neural modularity underlie the evolution of schooling behavior in threespine sticklebacks. *Current Biology* 23 (19): 1884–1888.

Greives, T.J., Dochtermann, N.A., and Stewart, E.C. (2017). Estimating heritable genetic contributions to innate immune and endocrine phenotypic correlations: a need to explore repeatability. *Hormones and Behavior* 88: 106–111.

Holmes, A., Li, Q., Murphy, D.L. et al. (2003). Abnormal anxiety-related behavior in serotonin transporter null mutant mice: the influence of genetic background. *Genes, Brain and Behavior* 2 (6): 365–380.

Huddleston, J. and Eichler, E.E. (2016). An incomplete understanding of human genetic variation. *Genetics* 202 (4): 1251–1254.

Ingram, K.K., Oefner, P., and Gordon, D.M. (2005). Task-specific expression of the foraging gene in harvester ants. *Molecular Ecology* 14 (3): 813–818.

Insel, T., Wang, Z., and Ferris, C. (1994). Patterns of brain vasopressin receptor distribution associated with social organization in microtine rodents. *Journal of Neuroscience* 14 (9): 5381–5392.

Jang, H., Levy, S., Flavell, S.W. et al. (2017). Dissection of neuronal gap junction circuits that regulate social behavior in caenorhabditis elegans. *Proceedings of the National Academy of Sciences of the United States of America* 114 (7): E1263–E1272.

de Jong, S., Chepelev, I., Janson, E. et al. (2012). Common inversion polymorphism at 17q21.31 affects expression of multiple genes in tissue-specific manner. *BMC Genomics* 13 (1): 458.

Joron, M., Frezal, L., Jones, R.T. et al. (2011). Chromosomal rearrangements maintain a polymorphic supergene controlling butterfly mimicry. *Nature* 477 (7363): 203–206.

Jukema, J. and Piersma, T. (2006). Permanent female mimics in a lekking shorebird. *Biology Letters* 2 (2): 161–164.

Kapheim, K.M., Pan, H., and Li, C. (2015). Genomic signatures of evolutionary transitions from solitary to group living. *Science* 348 (6239): 1139–1143.

Keller, L. (1993). *Queen Number and Sociality in Insects*. Oxford: Oxford University Press.

Kent, C.F., Daskalchuk, T., Cook, L. et al. (2009). The drosophila foraging gene mediates adult plasticity and gene–environment interactions in behaviour, metabolites, and gene expression in response to food deprivation. *PLoS Genetics* 5 (8): e1000609.

Kocher, T.D. (2004). Adaptive evolution and explosive speciation: the cichlid fish model. *Nature Reviews Genetics* 5 (4): 288–298.

Kowalko, J.E., Rohner, N., and Rompani, S.B. (2013). Loss of schooling behavior in cavefish through sight-dependent and sight-independent mechanisms. *Current Biology* 23 (19): 1874–1883.

Küpper, C., Stocks, M., Risse, J.E. et al. (2015). A supergene determines highly divergent male reproductive morphs in the ruff. *Nature Genetics* 48 (1): 79–83.

Lamichhaney, S., Fan, G., Widemo, F. et al. (2016). Structural genomic changes underlie alternative reproductive strategies in the ruff (*Philomachus pugnax*). *Nature Genetics* 48 (1): 84–88.

Lank, D.B., Smith, C.M., Hanotte, O. et al. (1995). Genetic polymorphism for alternative mating behaviour in lekking male ruff *Philomachus pugnax*. *Nature* 378 (6552): 59–62.

Laurent, P., Soltesz, Z., Nelson, G.M. et al. (2015). Decoding a neural circuit controlling global animal state in *C. elegans*. *eLife* 4: 865–879.

Lim, M. and Young, L. (2004). Vasopressin-dependent neural circuits underlying pair bond formation in the monogamous prairie vole. *Neuroscience* 125 (1): 35–45.

Linksvayer, T.A. (2006). Direct, maternal, and sibsocial genetic effects on individual and colony traits in an ant. *Evolution* 60 (12): 2552.

Linksvayer, T.A. and Wade, M.J. (2005). The evolutionary origin and elaboration of sociality in the aculeate hymenoptera: maternal effects, sib-social effects, and heterochrony. *Quarterly Review of Biology* 80 (3): 317–336.

Lucas, C., Nicolas, M., and Keller, L. (2015). Expression of *foraging* and *gp-9* are associated with social organization in the fire ant *Solenopsis invicta*. *Insect Molecular Biology* 24 (1): 93–104.

Mackay, T. (2004). The genetic architecture of quantitative traits: lessons from *Drosophila*. *Current Opinion in Genetics & Development* 14 (3): 253–257.

Macosko, E.Z., Pokala, N., Feinberg, E.H. et al. (2009). A hub-and-spoke circuit drives pheromone attraction and social behaviour in *C. elegans*. *Nature* 458 (7242): 1171–1175.

Manolio, T.A., Collins, F.S., Cox, N.J. et al. (2009). Finding the missing heritability of complex diseases. *Nature* 461 (7265): 747–753.

Marder, E. (2012). Neuromodulation of neuronal circuits: back to the future. *Neuron* 76 (1): 1–11.

Marín, O. and Rubenstein, J.L.R. (2003). Cell migration in the forebrain. *Annual Review of Neuroscience* 26 (1): 441–483.

McAdam, A.G., Garant, D., and Wilson, A.J. (2014). The effects of others' genes: maternal and other indirect genetic effects. In: *Quantitative Genetics in the Wild* (ed. A. Charmantier, D. Garant and L.E.B. Kruuk), 84–103. Oxford: Oxford University Press.

McGrath, P.T. (2013). Varieties of behavioral natural variation. *Current Opinion in Neurobiology* 23 (1): 24–28.

McGrath, P.T., Rockman, M.V., Zimmer, M. et al. (2009). Quantitative mapping of a digenic behavioral trait implicates globin variation in *C. elegans* sensory behaviors. *Neuron* 61 (5): 692–699.

McKenzie, S.K., Fetter-Pruneda, I., Ruta, V., and Kronauer, D.J.C. (2016). Transcriptomics and neuroanatomy of the clonal raider ant implicate an expanded clade of odorant receptors in chemical communication. *Proceedings of the National Academy of Sciences USA* 113 (49): 14091–14096.

Meffert, L.M. (1995). Bottleneck effects on genetic variance for courtship repertoire. *Genetics* 139 (1): 365–374.

Meffert, L.M., Hicks, S.K., and Regan, J.L. (2002). Nonadditive genetic effects in animal behavior. *Te American Naturalist* (S6): S198–S213.

Miller, T.H., Clements, K., Ahn, S. et al. (2017). Social status-dependent shift in neural circuit activation affects decision making. *Journal of Neuroscience* 37 (8): 2137–2148.

Milward, K., Busch, K.E., Murphy, R.J. et al. (2011). Neuronal and molecular substrates for optimal foraging in caenorhabditis elegans. *Proceedings of the National Academy of Sciences of the United States of America* 108 (51): 20672–20677.

Moore, A.J., Brodie, E.D., and Wolf, J.B. (1997). Interacting phenotypes and the evolutionary process: I. Direct and indirect genetic effects of social interactions. *Evolution* 51 (5): 1352.

Mott, R. and Flint, J. (2008). Prospects for complex trait analysis in the mouse. *Mammalian Genome* 19 (5): 306–308.

Mousseau, T.A., Uller, T., Wapstra, E., and Badyaev, A.V. (2009). Evolution of maternal effects: past and present. *Philosophical Transactions of the Royal Society B: Biological Sciences* 364 (1520): 1035–1038.

O'Connell, L.A. and Hofmann, H.A. (2011). Genes, hormones, and circuits: an integrative approach to study the evolution of social behavior. *Frontiers in Neuroendocrinology* 32 (3): 320–335.

O'Connell, L.A. and Hofmann, H.A. (2012). Evolution of a vertebrate social decision-making network. *Science* 336 (6085): 1154–1157.

Oettler, J., Nachtigal, A.-L., Schrader, L. et al. (2015). Expression of the foraging gene is associated with age polyethism, not task preference, in the ant cardiocondyla obscurior. *PLoS One* 10 (12): e0144699.

Okhovat, M., Berrio, A., Wallace, G. et al. (2015). Sexual fidelity trade-offs promote regulatory variation in the prairie vole brain. *Science* 350 (6266): 1371–1374.

Ophir, A.G., Wolff, J.O., and Phelps, S.M. (2008). Variation in neural V1aR predicts sexual fidelity and space use among male prairie voles in semi-natural settings. *Proceedings of the National Academy of Sciences USA* 105 (4): 1249–1254.

Orr, H.A. (1998). The population genetics of adaptation: the distribution of factors fixed during adaptive evolution. *Evolution* 52 (4): 935.

Osborne, K.A., Robichon, A., Burgess, E. et al. (1997). Natural behavior polymorphism due to a cgmp-dependent protein kinase of *Drosophila*. *Science* 277 (5327): 834–836.

Pankiw, T., Tarpy, D.R., and Page, R.E. (2002). Genotype and rearing environment affect honeybee perception and foraging behaviour. *Animal Behaviour* 64 (4): 663–672.

Parker, C.C., Gopalakrishnan, S., Carbonetto, P. et al. (2016). Genome-wide association study of behavioral, physiological and gene expression traits in outbred cfw mice. *Nature Genetics* 48 (8): 919–926.

Petfield, D., Chenoweth, S.F., Rundle, H.D., and Blows, M.W. (2005). Genetic variance in female condition predicts indirect genetic variance in male sexual display traits. *Proceedings of the National Academy of Sciences USA* 102 (17): 6045–6050.

Polderman, T.J.C., Benyamin, B., de Leeuw, C.A. et al. (2015). Meta-analysis of the heritability of human traits based on fifty years of twin studies. *Nature Genetics* 47 (7): 702–709.

Puig, M., Casillas, S., Villatoro, S. et al. (2015). Human inversions and their functional consequences. *Briefings in Functional Genomics* 14 (5): 369–379.

Purcell, S.M., Wray, N.R., Stone, J.L. et al. (2009). Common polygenic variation contributes to risk of schizophrenia and bipolar disorder. *Nature* 460 (7256): 748.

Purcell, J., Brelsford, A., Wurm, Y. et al. (2014). Convergent genetic architecture underlies social organization in ants. *Current Biology* 24 (22): 2728–2732.

Rittschof, C.C. and Robinson, G.E. (2016). Behavioral genetic toolkits: toward the evolutionary origins of complex phenotypes. *Current Topics in Developmental Biology* 119: 157–204.

Rittschof, C.C., Bukhari, S.A., Sloofman, L.G. et al. (2014). Neuromolecular responses to social challenge: common mechanisms across mouse, stickleback fish, and honey bee. *Proceedings of the National Academy of Sciences USA* 111 (50): 17929–17934.

Robinson, M.R., Wray, N.R., and Visscher, P.M. (2014). Explaining additional genetic variation in complex traits. *Trends in Genetics* 30 (4): 124–132.

Rockman, M.V. (2012). The QTN program and the alleles that matter for evolution: all that's gold does not glitter. *Evolution* 66 (1): 1–17.

Roff, D.A. and Mousseau, T.A. (1987). Quantitative genetics and fitness: lessons from *Drosophila*. *Heredity* 58 (Pt 1): 103–118.

Ross, K.G. and Keller, L. (1995). Ecology and evolution of social organization: insights from fire ants and other highly eusocial insects on jstor. *Annual Review of Ecology and Systematics* 26: 631–656.

Ross, K.G. and Keller, L. (1998). Genetic control of social organization in an ant. *Proceedings of the National Academy of Sciences USA* 95 (24): 14232–14237.

Ross, K. and Keller, L. (2002). Experimental conversion of colony social organization by manipulation of worker genotype composition in fire ants (*Solenopsis invicta*). *Behavioral Ecology and Sociobiology* 51 (3): 287–295.

Saltz, J.B., Hessel, F.C., and Kelly, M.W. (2017). Trait correlations in the genomics era. *Trends in Ecology & Evolution* 32 (4): 279–290.

Schielzeth, H. and Husby, A. (2014). Challenges and prospects in genome-wide quantitative trait loci mapping of standing genetic variation in natural populations. *Annals of the New York Academy of Sciences* 1320 (1): 35–57.

Schneider, J., Atallah, J., and Levine, J.D. (2016). Social structure and indirect genetic effects: genetics of social behaviour. *Biological Reviews* 92 (2): 1027–1038.

Schwander, T., Libbrecht, R., and Keller, L. (2014). Supergenes and complex phenotypes. *Current Biology* 24 (7): R288–R294.

Silventoinen, K., Sammalisto, S., and Perola, M. (2003). Heritability of adult body height: a comparative study of twin cohorts in eight countries. *Twin Research* 6 (5): 399–408.

Smeltzer, M.D., Curtis, J.T., Aragona, B.J., and Wang, Z. (2006). Dopamine, oxytocin, and vasopressin receptor binding in the medial prefrontal cortex of monogamous and promiscuous voles. *Neuroscience Letters* 394 (2): 146–151.

Stefansson, H., Helgason, A., Thorleifsson, G. et al. (2005). A common inversion under selection in Europeans. *Nature Genetics* 37 (2): 129–137.

Steinberg, K.M., Antonacci, F., Sudmant, P.H. et al. (2012). Structural diversity and african origin of the 17q21.31 inversion polymorphism. *Nature Genetics* 44 (8): 872–880.

Terai, Y., Seehausen, O., Sasaki, T. et al. (2006). Divergent selection on opsins drives incipient speciation in Lake Victoria cichlids. *PLoS Biology* 4 (12): e433.

Thompson, M.J. and Jiggins, C.D. (2014). Supergenes and their role in evolution. *Heredity* 113 (1): 1–8.

Toth, A.L., Robinson, G.E., and LeRoux, M. (2007). Evo-devo and the evolution of social behavior. *Trends in Genetics* 23 (7): 334–341.

Tuttle, E.M. (2003). Alternative reproductive strategies in the white-throated sparrow: behavioral and genetic evidence. *Behavioral Ecology* 14 (3): 425–432.

Tuttle, E.M., Bergland, A.O., Korody, M.L. et al. (2016). Divergence and functional degradation of a sex chromosome-like supergene. *Current Biology* 26 (3): 344–350.

Valdar, W., Solberg, L.C., and Gauguier, D. (2006). Genome-wide genetic association of complex traits in heterogeneous stock mice. *Nature Genetics* 38 (8): 879–887.

Visscher, P.M., Hill, W.G., and Wray, N.R. (2008). Heritability in the genomics era – concepts and misconceptions. *Nature Reviews Genetics* 9 (4): 255–266.

Walum, H., Westberg, L., Henningsson, S. et al. (2008). Genetic variation in the vasopressin receptor 1a gene (AVPR1A) associates with pair-bonding behavior in humans. *Proceedings of the National Academy of Sciences USA* 105 (37): 14153–14156.

Wang, J., Wurm, Y., Nipitwattanaphon, M. et al. (2013). A y-like social chromosome causes alternative colony organization in fire ants. *Nature* 493 (7434): 664–668.

Weber, J.N., Peterson, B.K., and Hoekstra, H.E. (2013). Discrete genetic modules are responsible for complex burrow evolution in peromyscus mice. *Nature* 493 (7432): 402–405.

Wilson, A.J., Gelin, U., Perron, M.-C., and Réale, D. (2009). Indirect genetic effects and the evolution of aggression in a vertebrate system. *Proceedings of the Royal Society B: Biological Sciences* 276 (1656): 533–541.

Winslow, J.T., Hastings, N., Carter, C.S. et al. (1993). A role for central vasopressin in pair bonding in monogamous prairie voles. *Nature* 365 (6446): 545–548.

Wolf, J.B. (2000). Indirect genetic effects and gene interactions. In: *Epistasis and the Evolutionary Process* (ed. J.B. Wolf, E.D. Brodie III, and M.J. Wade), 158–176. Oxford: Oxford University Press.

Wolf, J.B., Brodie, E.D. III, Cheverud, J.M. et al. (1998). Evolutionary consequences of indirect genetic effects. *Trends in Ecology & Evolution* 13 (2): 64–69.

Wood, A.R., Esko, T., Yang, J. et al. (2014). Defining the role of common variation in the genomic and biological architecture of adult human height. *Nature Genetics* 46 (11): 1173–1186.

Wu, Q., Wen, T., Lee, G. et al. (2003). Developmental control of foraging and social behavior by the drosophila neuropeptide y-like system. *Neuron* 39 (1): 147–161.

Young, K.A., Gobrogge, K.L., Liu, Y., and Wang, Z. (2011). The neurobiology of pair bonding: insights from a socially monogamous rodent. *Frontiers in Neuroendocrinology* 32 (1): 53–69.

Zody, M.C., Jiang, Z., Fung, H.-C. et al. (2008). Evolutionary toggling of the mapt 17q21.31 inversion region. *Nature Genetics* 40 (9): 1076–1083.

Zuk, O., Hechter, E., Sunyaev, S.R., and Lander, E.S. (2012). The mystery of missing heritability: genetic interactions create phantom heritability. *Proceedings of the National Academy of Sciences USA* 109 (4): 1193–1198.

Rat Genome Sequencing and Mapping Consortiumet al. (2013). Combined sequence-based and genetic mapping analysis of complex traits in outbred rats. *Nature Genetics* 45 (7): 767–775.

6

Genes and Environments in *Drosophila* Sex

David J. Hosken[1], Amanda Bretman[2], Stephen F. Goodwin[3] and C. Ruth Archer[1]

[1] *Centre for Ecology & Conservation, University of Exeter, Penryn Campus, Penryn, TR10 9EZ, UK*
[2] *School of Biology, Faculty of Biological Sciences, University of Leeds, Leeds, LS2 9JT, UK*
[3] *Centre for Neural Circuits & Behaviour, University of Oxford, Oxford OX1 3SR, UK*

Drosophila melanogaster has become a model for the study of many phenomena, from insecticide resistance, to sperm evolution and use, spermatogenesis, development and HOX genes, to circadian clocks (Guss et al. 2001; Miller and Pitnick 2002; Jiang and White-Cooper 2003; Snook and Hosken 2004; Rostant et al. 2015). It has also become a workhorse for understanding behaviour, as have a number of its close relatives (Speith 1974; Hine et al. 2002; Ritchie et al. 2005; Snook et al. 2005; Lasbleiz et al. 2006; Kent et al. 2008; Taylor et al. 2008a,b; Okada et al. 2011; Saltz 2017). An understanding of *Drosophila* behaviour has been aided by the ease with which they can be reared in large numbers, and genetically manipulated, and has expanded in scale with the major sequencing efforts that have been undertaken across the clade (Drosophila 12 Genomes Consortium 2007). We are now dissecting *Drosophila* behaviour in ever increasing detail and this is accelerating thanks to the sharing of resources and data (e.g. FlyBase – an online database of *Drosophila* genes and genomes).

Drosophila researchers have investigated a broad range of behaviours from aggression to alcohol preference, territoriality. and foraging. Here we focus on a small subset of this huge body of work and primarily discuss genetic and environmental influences on *Drosophila* sexual behaviours, with more focus on social environmental effects and less on abiotic environments. We begin by briefly introducing some challenges in measuring behaviour and its influences before describing *Drosophila* courtship. We then move on to genetic effects, and then environmental and interactive effects influencing *Drosophila* sexual behaviour, before offering some concluding remarks.

6.1 Some Challenges

At present, most research linking genotype to phenotypes has overlooked how both are affected by and interact with the environment, an understandable but important omission. Incorporating all these links is challenging if only because behaviour is so variable that its study requires large sample sizes (Anholt and Mackay 2004). Further, in choosing phenotypes to assay, care is needed to ensure that they are either biologically

Genes and Behaviour: Beyond Nature-Nurture, First Edition.
Edited by David J. Hosken, John Hunt and Nina Wedell.
© 2019 John Wiley & Sons Ltd. Published 2019 by John Wiley & Sons Ltd.

meaningful or that clear genotype-phenotype links can be made subsequently. So, for example, while assaying total male attractiveness is relatively simple (Taylor et al. 2007) and biologically relevant (Prokop and Drobniak 2016), it is not always clear precisely which elements of the phenotype contribute most to attractiveness and therefore which genes should be the focus of any subsequent investigation. Additionally, careful experimental control is needed because of the considerable effects of environment, circadian rhythms, age, sex, genetic background, etc. on both behaviour and gene expression (Anholt and Mackay 2004).

6.2   Introducing *Drosophila*

Research on *D. melanogaster* has revealed that behaviours are variable, highly sexually dimorphic (ca. 50% of the *D. melanogaster* transcriptome shows sex biases), polygenic, and highly epistatic (Anholt and Mackay 2015). Furthermore, the extensive pleiotropy found for most behaviours shows that specific genes are not solely dedicated to controlling complex behaviour (Markow and O'Grady 2005; Anholt and Mackay 2004). For example, the genes that influence song in *D. melanogaster* are also involved in flight, regulation of translation, ion channel functioning, and sex determination. Indeed, *cacophony*, which was initially thought to only affect song (Yamamoto et al. 1997), is now known to be involved in autophagy, calcium ion transport, exocytosis, homeostasis of neuron cells and so on (FlyBase). Behaviours are also highly sensitive to environmental effects and transcriptome profiling of *D. melanogaster* reveals that at least 15% of the transcriptome varies across environments (Zhou et al. 2012). However, because of the polygenic nature of behaviour, underpinned as it is by large networks of interacting pleiotropic genes, quantitative genetic approaches to study behavioural phenotypes are also powerful and need to be employed because of the probabilistic links between genes and behaviour. This quantitative (statistical) genetic approach has also been widely employed to investigate the contribution of genes, environment, and their interaction on behaviour (Petfield et al. 2005; Narraway et al. 2010). Studies that look consistently find genotype-by-environment interactions for behaviours and this has important implications, most notably providing a means to maintain genetic variation in behaviour, but also potentially eroding the honesty of information conveyed in behavioural interactions (Hunt and Hosken 2014). More recently, there has been an increasing focus on interactions between the social environment and behaviour (Saltz and Foley 2011) (see below) and a more thorough linking of both with genetic data promises to provide novel insights into behavioural phenotypes.

6.3   The Behaviours

To gain a full understanding of the genetic and environmental drivers of behaviour, behavioural phenotypes need to be well characterized. Houle (2010) has argued that this is the most difficult element in establishing the genotype–phenotype link because phenomes are more complicated than genomes, and this is especially true of behavioural phenotypes, which are extremely complex, even if they contain certain 'fixed' elements. *Drosophila* mating behaviour contains stereotyped species-specific elements of courtship and mating, that include male behaviours like wing-flicking,

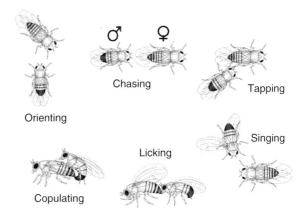

Figure 6.1 An example of the types of behaviours frequently seen in *Drosophila* courtship, starting from males orienting towards females and chasing them, until when courting is finally successful and copulation occurs.

tapping, wing vibration and genital 'licking', and female responses like ovipositor extrusion, kicking and abdomen elevation (Speith 1974; Markow and O'Grady 2005) (Figure 6.1). There are clear species differences with, for example, male *Drosophila yakuba* employing a double-wing sweep, while *Drosophila simulans* males only move one wing at a time. Furthermore, some taxa only mate when it is light, in some females also sing in duets with males and so on (for a full description of variation across species variation, see Markow and O'Grady 2005). These elements can also vary across genotypes within species (Rostant et al. 2017). As with much *Drosophila* work, *D. melanogaster* mating behaviour has been most extensively studied.

Courtship in *D. melanogaster* begins with males recognizing the presence of a female and orienting towards her. Tapping of the female abdomen then occurs and if the female moves, the male will follow. As the male orients around the female, looking towards her, wing extensions and vibrations produce a courtship song (and visual signals) (Shevtsova et al. 2011), males then attempt to lick the female genitalia and to copulate. If unsuccessful, there may be a behavioural pause after which the male begins orienting and singing once more and repeating the repertoire until copulation occurs or the female signals rejection by extending her ovipositor or flying or walking away (Speith 1974; Hall 1994; Yamamoto et al. 1997).

There is phenotypic variation in each of these elements of courtship and all contribute to the likelihood that copulation will occur. Thus, male attractiveness can be thought of as a composite trait that is the sum of these courtship elements and male odour (Ingleby et al. 2014). While this meta-trait is arguably the most important male fitness determinant, it is probably easier to understand when broken down and dissected as one courtship element at a time. Indeed, most of these courtship elements have been subject to detailed investigation in one taxon or another, and in many instances specific genes have been identified that affect courtship and copulation (Hall 1994; Mackay et al. 2005).

6.4 The Genes

6.4.1 Single Genes

A number of studies have now uncovered single genes of large effect that influence courtship and mating in *Drosophila*. As noted above (see also Chapter 1 and 5), these genes typically have many pleiotropic effects and are not 'behaviour genes' *per se*, but typically affect normal behaviour via their effects on the development of the nervous system (Sokolowski 2001). One such gene is *dunce*, a mutation affecting learning and memory by disrupting the cyclic adenosine monophosphate (AMP) signal transduction cascade. The wild-type gene is expressed in many tissues but predominantly in the fly brain (in the mushroom bodies), and while *dunce* has many effects, it also impacts on female mating behaviour – females carrying this mutation remate more quickly than usual, seeming to have forgotten that they have already mated (Greenspan and Ferveur 2000; Kubli 2003). Other interesting examples of mutations with major effects on mating behaviour include *he's not interested* (*hni*), where males with the mutation do not orientate towards females (the first step in the behavioural cascade leading to mating) and *dissatisfaction (dsf)* where mutant males will court other males (Emmons and Lipton 2003). In fact, mutations have been identified that affect more or less every element of sexual behaviour in *D. melanogaster* (Yamamoto et al. 1997; Emmons and Lipton 2003; Gleason 2005; Anholt and Mackay 2015), and some of the developmental and neural pathways that lead from genes to sexual behaviours are now being carefully characterized (Asahina 2018).

Clearly, the sex of an animal determines its sexual behaviour, and in *Drosophila*, sexual differentiation of the neural circuits underlying sexual behaviours is dependent on the action of two transcription factors, *fruitless* (*fru*) and *doublesex* (*dsx*). Selective expression of these factors defines cell type-specific developmental programmes that govern neural connectivity and lay the foundations through which sex-specific sexual behaviours are genetically fixed (Dalton et al. 2013; Neville et al. 2014; Clough et al. 2014; Meissner et al. 2016). Historically, understanding of the way in which *fru* and *dsx* regulate courtship came from behavioural analyses of individuals carrying mutations at the respective loci (Ito et al. 1996; Ryner et al. 1996; Villella and Hall 1996; Villella et al. 1997). How these genes function in specifying sexual behaviour was inferred by the temporal and spatial patterns of *fru* and *dsx* expression in the nervous system, and this in turn facilitated the identification of candidate cellular components of this circuit (Cachero et al. 2010; Rideout et al. 2010; Robinett et al. 2010).

Expression of *fru* and *dsx* in sensory neurons, interneurons, and motor neurons suggests that they are organized into circuit elements capable of receiving, processing, and transferring the information that controls sexual behaviour (Pavlou and Goodwin 2013). Restricting the activity of all, or some, of these neurons has profound effects on male and female courtship behaviours (Stockinger et al. 2005; Billeter et al. 2006; Yapici et al. 2008; Häsemeyer et al. 2009; Rideout et al. 2010; Kohatsu et al. 2011; von Philipsborn et al. 2011; Pan et al. 2011, 2012; Rezával et al. 2012, 2014, 2016; Feng et al. 2014; Zhou et al. 2014, 2015; Bussell et al. 2014; Tran et al. 2014; Koganezawa et al. 2016; Pavlou et al. 2016; Chen et al. 2017). A central tenet in understanding sexual behaviour is that sexually dimorphic behaviours arise from anatomical and functional differences in neural circuits. In some cases, the sex differences are qualitative such that particular neurons are unique to one sex; in others, a quantitative sex difference may represent a dimorphism in the same cell or the molecular characteristics of shared neurons (Kimura et al. 2005; Billeter et al. 2006; Rideout et al. 2007; Cachero et al. 2010; Yu et al. 2010; Rideout et al. 2010; Kohl et al. 2013). For example, males and females may detect many of the same external signals but process them differently to produce distinct behavioural responses (Kohl et al. 2013).

Recent work has focused on identifying the neural and molecular components of sex-specific neural circuits, and mapping functional connectivity, to define causal relationships between circuit activity and sexual behaviour, in particular how pertinent sensory cues are differentially detected and integrated in the male and female brain (Asahina 2018), and how sexual behaviours are socially modifiable both on short and longer timescales (Villella and Hall 2008; Griffith and Ejima 2009; Dankert et al. 2009; Kim et al. 2012; Keleman et al. 2012; Coen et al. 2014; Lebreton et al. 2014; Inagaki et al. 2014; Lin et al. 2016) (see discussion below: Social environment and mating behaviour).

Questions of how decisions about different sexual behaviours are implemented, and how the associated internal states influence these decisions (e.g. experienced versus naïve), are starting to be elucidated (Ellendersen and von Philipsborn 2017). These internal states clearly affect many behaviours (see below), but *how* they promote arousal, stimulate social interactions, encode reward and/or control the progression from the seeking of mates to actual mating is only beginning to be unravelled. Furthermore, given the complex networks and pathways that ultimately lead to behaviour, epistatic effects seem likely, and genetic background–single gene interactions have been documented for some important behaviours that affect sexual success. For example, the retrotransposon upregulation of a cytochrome P450 gene, *Cyp6g1*, confers resistance to DDT and also decreases male sexual competitiveness and aggression (Rostant et al. 2015, 2017). However, these behavioural effects depend on the genetic background in which the resistance allele occurs (Smith et al. 2011). Similarly, single gene effects can be environmentally dependent (Kaun et al. 2007). So even in instances where single genes have large phenotypic effects, these can be mediated by other genes and/or the environment.

6.4.2 Many Additive Genes

Since behaviour is largely underpinned by complex genetic architecture (Anholt and Mackay 2004) (see Chapter 5), the single gene impacts documented above are snapshots of the total genetic effects and their interactions that contribute to fly behavioural phenotypes. Furthermore, mutations at single genes can affect hundreds of co-regulated loci, even when phenotypic effects are small (Ranz et al. 2003). To capture all this complexity, a range of experimental (and statistical) approaches have been employed – pedigrees, isogenic lines, selection experiments, and hybridization – and inevitably, studies find genetic variation for the sexual behaviours/phenotypes investigated (Gromko 1987; Welbergen and van Diijken 1992; Ritchie and Kyriacou 1996; Hoffmann 1999; Hine et al. 2002; Gleason et al. 2002; Huttunen and Aspi 2003; Ritchie et al. 2005; Taylor et al. 2007; Sharma et al. 2010; Debelle et al. 2014; Gaertner et al. 2015). Nonetheless, behaviours have been generally found to have heritabilities (the additive genetic contribution to phenotypic variation) that are relatively low, with *Drosophila* mating behaviours tending to have heritabilities of less than 30%, with most estimates less than about 20% (compared with morphology at about 50%) (Roff and Mousseau 1987). Part of the reason for this may be that because behaviours are often only measured once, the within-individual variability is included in the environmental variance, inflating the phenotypic variance (i.e. the denominator in the heritability estimate is falsely inflated) (for further explanation see Hoffmann 1999). Repeated behavioural measurement can remove this problem, such that behavioural heritabilities increase from effectively 0 to >50% (Hoffmann 1999). In any case, polygenic behavioural variation seems to be spread broadly across the genome (Gleason 2005), with only the X chromosome being gene poor for at least some behaviours (Turner and Miller 2012).

While these statistical-genetics approaches have been perfected over decades of plant and animal breeding, they cannot identify specific genes – that is not what they were designed for – and until we can identify all the genes involved in sexual behaviours and how they interact with each other and the environment, we cannot have a complete understanding of sexual phenotypes. However, when these tried and tested designs are combined with molecular techniques, powerful advances can be made (Lynch 2007).

6.5 The Environments and the Interactions

With heritabilities generally less than 30% (Roff and Mousseau 1987), much behavioural variation must be attributable to non-additive effects, be they interactions or environmental, and behaviour seems especially prone to non-additive genetic effects (Meffert et al. 2002; Gleason 2005). It is no surprise, then, that many recent interesting advances in the study of *Drosophila* sex are coming about through studying the statistical interactions that occur between genes and environment, and particularly genes and social environments, because, after all, sex is a social activity (Bleakley et al. 2010).

6.5.1 Social Environments

Arguably the most important environment is the one provided by other individuals, the social environment, and because these individuals have their own genomes, the genetic component of the social environment can also evolve (Moore et al. 1997; Schneider et al. 2017) (see also Chapters 4 and 7). That is, unlike the abiotic environment, the social environment can also evolve (see Chapters 1, 4, 5, and 7). Thus social interactions can have both direct effects on behavioural phenotypes and indirect genetic effects (IGEs) via the evolving social environment (Moore et al. 1997). IGEs have probably been underappreciated for animals like *Drosophila* that are not thought of as classically 'social', but we are starting to discover the complexity and consequences of fly social interactions (Sokolowski 2010). For example, *Drosophila* often form aggregations that are not linked to resource availability (Simon et al. 2012), they can form social networks (Schneider et al. 2017) and engage in social learning (Foucaud et al. 2013), and gene–social environment correlations have been documented (Saltz 2011). Social contact affects *Drosophila* sleep patterns and neuronal plasticity (Donlea et al. 2009, 2014), improves memory (Chabaud et al. 2009), alters their ageing patterns (Ruan and Wu 2008; Zajitschek et al. 2013; Leech et al. 2017; reviewed in Flatt 2011) and affects gene expression (Carney 2007; Ellis and Carney 2011; Fedorka et al. 2011; Immonen and Ritchie 2012; Mohorianu et al. 2017).

Sexual reproduction in internal fertilizers like flies inevitably involves some social interaction and if there is multiple mating, as with most *Drosophila*, mating decisions (e.g. how much to invest in current versus future reproduction) are needed about the value of a current mating and the probability of a future opportunity (Parker et al. 1996, 1997). This is where social information can be exploited and a range of models and experimental data have shown when this information is beneficial for males, particularly with copulatory behaviour (Parker and Pizzari 2010; Wedell et al. 2002; Bretman et al. 2011).

Social context can also drive *Drosophila* morphological plasticity (e.g. males develop larger accessory sex glands in the presence of other males) (Bretman et al. 2016), which exemplifies just how pervasive the effects of social environments can be on the expression of phenotypes.

One way in which social environments affect behaviour is through learning, and male *D. melanogaster* can learn to reduce costly courtship directed towards unreceptive, recently mated females (Dukas 2004; Ejima et al. 2007). Furthermore, mutation of genes like *rutabaga*, *amnesiac* or *dunce* disrupts learning and/or memory and therefore disturbs male learning about who to court as well as the normal remating inhibition that occurs in females after copulation (McGuire et al. 2005). RNA interference of the epigenetic modifier *euchromatin histone methyltransferase* (*EHMT*) likewise disrupts courtship memory (Kramer et al. 2011), and although epigenomic states can have long-lasting and even transgenerational effects, remodelling can occur within hours (Kangaspeska et al. 2008). This flexibility may increase the environmental sensitivity of the genome by facilitating changes in gene expression. Indeed, the transcriptome is sensitive to social learning, with expression of genes with predicted and known functions in nervous system development, chromatin biology, translation, cytoskeletal dynamics, and transcriptional regulation all altered by a male's previous courtship experience (Winbush et al. 2012). Furthermore, because Winbush et al. (2012) identified many differentially expressed transcript isoforms, it seems that sensitivity to the social environment is not just conveyed by binary switching (on/off), but also through alternative splicing (Winbush et al. 2012). This implies that integration of environmental cues can occur at various cellular levels and we will need to take a much broader view than the transcriptome to understand these phenomena.

The effects of social environments are especially complex because in addition to genetic variation for behaviour, and an environment which evolves, there can be genetic variation for the social environment preferred, potentially leading to social environment (niche) construction (Saltz et al. 2016), and this generates all kinds of feedback loops (Moore et al. 1997) (see also Chapters 1 and 4). For example, fighting success is one determinant of mating success for male *D. melanogaster*, and aggression and its outcomes depend on the genotypes of the interacting males (Saltz 2013). Furthermore, genotypes vary in the groups they prefer and in how they interact within groups (Saltz 2017), so group size and composition influence social experience and behaviour, and these both influence group size and composition, and all have a genetic component.

In most *Drosophila* species so far investigated, males with prior exposure to rival males increase copulation duration (Bretman et al. 2009; Mazzi et al. 2009; Lizé et al. 2012; Price et al. 2012). In *D. melanogaster* this response increases male fitness, at least in the short term, by increasing offspring production and reducing female willingness to remate (Bretman et al. 2009). Interestingly, increased copulation duration is not associated with a matching increase in ejaculate size or number of sperm transferred to the female (Gilchrist and Partridge 2000; Manier et al. 2010); that is, longer copulations need not mean more sperm/ejaculate, and neither is it a form of mate guarding (Bretman et al. 2009). However, exposure to rivals increases seminal fluid proteins in an ejaculate (Wigby et al. 2009) and the number and quality of sperm (Garbaczewska et al. 2013; Moatt et al. 2014), but sustained exposure to rivals also reduces lifespan and accelerates age-specific fitness declines (Bretman et al. 2013).

Responding appropriately to social environment is therefore critical for fitness and males flexibly alter their behaviours to match social environments (Bretman et al. 2012). This takes time, approximately 24 hours to build up a response when faced with a rival (Bretman et al. 2010), and has a finite retention duration (~15 hours after the rival is removed) (Rouse and Bretman 2016), although response speeds depend on the sensory cues received (Rouse and Bretman 2016). Males acquire environmental information using any combination of sound, touch or smell cues in order to respond adaptively (Bretman et al. 2011), but it is not clear how each of these determines male responses. Olfactory-defective *Orco* (formerly *Or83b*), an odorant binding co-receptor required for the majority of odours detectable by *Drosophila* (Larsson et al. 2004), or hearing-defective *inactive* (Gong et al. 2004) mutations do not alone erase behavioural response to rivals (Bretman et al. 2011). However, they do ablate the wild-type increase in sperm transferred (Garbaczewska et al. 2013). Although the 'rival response' is best understood in *D. melanogaster*, it is also mirrored in *Drosophila pseudoobscura*. Again, exposure to rivals causes males to increase copulation duration. The *D. pseudoobscura* ejaculate contains both apyrene (non-fertilizing) and eupyrene (fertilizing) sperm, yet exposure to rivals only increases the transfer of eupyrene sperm (Price et al. 2012). There also appears to be involvement of multiple sensory cues in determining this response (Maguire et al. 2015).

Again, researchers have looked to the transcriptome for clues about the underlying mechanisms linking the social environment to mating behaviours. Short-term (~20 minutes) exposure of males to females or rival males can alter spermatogenic transcription and odour perception gene expression (Carney 2007; Ellis and Carney 2011). Male exposure to rivals for up to 72 hours also generates differential expression in two of three seminal fluid protein genes, but not in four testis genes (Fedorka et al. 2011). This finding was somewhat echoed in a recent RNAseq study of differential expression at 2, 26, and 50 hours exposure to a rival, in which the response was characterized by early upregulation of a large number of sensory genes in the head/thorax, which then decayed over time (Mohorianu et al. 2017). This was associated with a wave of differential expression of abdominal genes encoding the structural components of the male ejaculate – specifically in seminal fluid, but not sperm, genes (Mohorianu et al. 2017). These genes are clear candidates for the mechanistic basis of the behavioural response to rivals, but other gene categories were also differentially expressed, including those implicated in immunity, stress responses, cognition, and lifespan.

However, we need to be cautious about inferring function from differential expression. The learning- and memory-associated genes *dunce* and *Nf1* were differentially expressed in response to rivals (Mohorianu et al. 2017), but the use of transgenic flies found these genes to be unnecessary to produce the extended mating duration (Rouse et al. 2018). This genetic dissection of the neuronal processes involved showed that the response uses a type of long-term, anaesthesia-sensitive memory, requires the genes *amnesiac* and *rutabaga*, and neural transmission in the γ lobes of the Mushroom bodies (brain regions associated with olfactory learning) (Rouse et al. 2018). Using this type of memory seems consistent with the temporal dynamics of the response, in that it allows behavioural outputs to persist through transient changes in cues (i.e. another male being present or absent) but to be overwritten if the environmental change persists.

In addition to responding to rivals, male responses can also depend on the females present. In many *Drosophila*, cuticular hydrocarbons (CHC) are an important sexual signal, but it has recently become apparent that males can plastically alter this signalling (Ingleby et al. 2014). For example, CHC signals are an important element of courtship in *Drosophila serrata* (Chenoweth and Blows 2005) and males alter their CHC signal based on their assessment of female quality, which has a genetic basis (Petfield et al. 2005). Thus the genes of the females influence the courtship phenotype of the males (i.e. = an IGE), and similar effects have been suggested by female genotype-dependent changes in male courtship behaviours seen in *D. melanogaster* (Casares et al. 1993).

In comparison to males, responses in female mating behaviour to social environments have been rather overlooked, possibly because of a lack of testable theoretical predictions. Nevertheless, social context can alter female mate choice, willingness to mate and oviposition behaviour. Females previously courted by small males were subsequently more likely to copulate than those previously courted by large males (Dukas 2005), suggesting that social experience alters female mate acceptance thresholds. When allowed to mate in groups of 12 (six males, six females), females mate more frequently in groups composed of males from different genetic strains (Billeter et al. 2012). Intriguingly, these changes in reproductive behaviour driven by the social context were dependent on female olfaction, not male–male interactions, as *Orco* mutant females did not increase mating frequency according to group composition. Females also show a preference for familiar mates, and again olfactory cues are important here as *Orco* mutants do not display this preference (Tan et al. 2013).

Once mated, naïve females can exhibit social learning in terms of oviposition site choice, changing their preference in the presence of mated females (Sarin and Dukas 2009), and aggregated oviposition can be induced in response to pheromones released by both sexes (Wertheim et al. 2002). As with males, the female transcriptome is sensitive to social cues. Being exposed to a courtship song in isolation from any other male traits resulted in relatively modest differences in gene expression, but included genes that function in sexual signalling, including *Orco*, and genes involved in immunity and stress response (Immonen and Ritchie 2012).

Evidence for the social effects on the expression of phenotypes is therefore accruing (see also Chapter 5) and modern technologies (see Chapter 11) are making it ever more possible to interrogate the interaction between the social environment and individual genes and genomes. There are clearly multiple challenges here, from comparing transcriptomes across studies and platforms when gene expression is so variable and transient, to assessing the actual functional significance of these changes, and determining which genes are causal in these responses and which are consequences of social contact.

6.5.2 Abiotic Environments

As with many phenotypes, *Drosophila* mating behaviours are influenced by abiotic factors and at times there are very strong genotype-by-environment (GxE) interactions. For example, GxEs in both female mate preference and choosiness are generated by cold shock (versus no cold shock) in *D. melanogaster* isolines, with heritabilities substantially reduced under the more stressful shock conditions (Narraway et al. 2010), and GxE across temperature environments also occurs for preference (but not choosiness) in *D. simulans* (Ingleby et al. 2013). Similarly, diet-dependent GxE for male courtship song has been documented in *Drosophila mojavensis* (Etges et al. 2007). In contrast, Delcourt et al. (2010) found little evidence that environmental quality (food variation) altered female preference functions in *D. serrata*. However, there is strong evidence in many *Drosophila* that abiotic environments alter the way males sexually signal (Ingleby et al. 2014), and nutritional environments, and fly perception of them, can also alter reproductive behaviour (Zajitschek et al. 2013; Gorter et al. 2016). As one example, female responses to male ejaculatory proteins depend on female nutrition (Fricke et al. 2010).

Finally, social environments can interact with abiotic environments and with male genotype to influence mating success. In *D. melanogaster* experiments with lines selected for increased male territoriality (versus controls), male mating success depended on food availability, population size and female mating status (Hoffmann 1994). For example, males selected for more territorial behaviour had high mating success when food was concentrated in a small area, populations were relatively dense and females were non-virgins, and did less well when there was no food resource to defend and females were virgins or when territorial males were relatively common even when food was concentrated (Hoffmann and Cacoyianni 1989). This particular study shows how inferences can fundamentally change depending on how experimental systems are interrogated, reflecting the complexity of genotype (G)–environment (E)–phenotype (P) links and feedbacks.

6.6 Conclusions

Behaviours, their causes, and their implementation pathways are complicated as the preceding discussion clearly shows. And even for those phenotypes that are well characterized, the full interplay between genes and environment, plus their interactions, have not been completely described, let alone the feedbacks between G, P, and E. However, *Drosophila* with its wealth of genetic resources, mass of genomic data and extensive shared databases arguably provides our best chance of establishing clear G–E–P links, especially if we employ the powerful experimental designs and analyses developed by statistical geneticists together with the powerful tools of molecular genetics. Even relatively simple approaches that employ experimental evolution, followed by phenotyping and sequencing, can be extremely powerful (Turner and Miller 2012; Immonen et al. 2014). This is why, despite or perhaps due to more than 100 years of research, *Drosophila* still have much to teach us and perhaps why they will enable us to generate the first clear mapping of G to P across environments, which remains a major aim of modern biology.

Acknowledgements

We thank the three referees for their comments that helped us clarify this chapter.

References

Anholt, R.H.R. and Mackay, T.F.C. (2004). Quantitative genetic analyses of complex behaviours in *Drosophila*. *Nature Reviews Genetics* 5: 838–849.

Anholt, R.H.R. and Mackay, T.F.C. (2015). Dissecting the genetic architecture of behaviour in *Drosophila melanogaster*. *Current Opinions in Behavioral Sciences* 2: 1–7.

Asahina, K. (2018). Sex differences in *Drosophila* behavior: qualitative and quantitative dimorphism. *Current Opinion in Physiology* 6: https://doi.org/10.1016/j.cophys.2018.04 .004.

Billeter, J.-C., Villella, A., Allendorfer, J.B. et al. (2006). Isoform-specific control of male neuronal differentiation and behavior in *Drosophila* by the *fruitless* gene. *Current Biology* 16: 1063–1076.

Billeter, J.-C., Jagadeesh, S., Stepek, N. et al. (2012). *Drosophila melanogaster* females change mating behaviour and offspring production based on social context. *Proceedings of the Royal Society B: Biological Sciences* 279: 2417–2425.

Bleakley, B.H., Wolf, J.B., and Moore, A.J. (2010). The quantitative genetics of social behaviour. In: *Social Behaviour: Genes, Ecology and Evolution* (ed. T. Szekely, A.J. Moore and J. Komdeur). Cambridge: Cambridge University Press.

Bretman, A., Fricke, C., and Chapman, T. (2009). Plastic responses of male *Drosophila melanogaster* to the level of sperm competition increase male reproductive fitness. *Proceedings of the Royal Society B: Biological Sciences* 276: 1705–1711.

Bretman, A., Fricke, C., Hetherington, P. et al. (2010). Exposure to rivals and plastic responses to sperm competition in *Drosophila melanogaster*. *Behavioral Ecology* 21: 317–321.

Bretman, A., Gage, M.J.G., and Chapman, T. (2011). Quick-change artists: male plastic behavioural responses to rivals. *Trends in Ecology & Evolution* 26: 467–473.

Bretman, A., Westmancoat, J., Gage, M.J.G., and Chapman, T. (2012). Individual plastic responses by males to rivals reveal mismatches between behaviour and fitness outcomes. *Proceedings of the Royal Society B: Biological Sciences* 279: 2868–2876.

Bretman, A., Westmancoat, J., Gage, M.J.G., and Chapman, T. (2013). Costs and benefits of lifetime exposure to mating rivals in male *Drosophila melanogaster*. *Evolution* 67: 2413–2422.

Bretman, A., Fricke, C., Westmancoat, J., and Chapman, T. (2016). Effect of competitive cues on reproductive morphology and behavioral plasticity in male fruitflies. *Behavioral Ecology* 27: 452–461.

Bussell, J.J., Yapici, N., Zhang, S.X. et al. (2014). *Abdominal-B* neurons control *Drosophila* virgin female receptivity. *Current Biology* 24: 1584–1595.

Cachero, S., Ostrovsky, A.D., Yu, J.Y. et al. (2010). Sexual dimorphism in the fly brain. *Current Biology* 20 (18): 1589–1601.

Carney, G.E. (2007). A rapid genome-wide response to *Drosophila melanogaster* social interactions. *BMC Genomics* 8: 288.

Casares, P., Carracedo, M.C., San Miguel, E. et al. (1993). Male mating speed in *Drosophila melanogaster*: differences in genetic architecture and relative performance according to female genotype. *Behavioural Genetics* 23: 349–358.

Chabaud, M.A., Isabel, G., Kaiser, L., and Preat, T. (2009). Social facilitation of long-lasting memory retrieval in *Drosophila*. *Current Biology* 19: 1654–1659.

Chen, D., Sitaraman, D., Chen, N. et al. (2017). Genetic and neuronal mechanisms governing the sex-specific interaction between sleep and sexual behaviors in *Drosophila*. *Nature Communications* 8: 154.

Chenoweth, S.F. and Blows, M.W. (2005). Contrasting mutual sexual selection on homologous signal traits in *Drosophila serrata*. *American Naturalist* 165: 281–289.

Clough, E., Jimenez, E., Kim, Y. et al. (2014). Sex- and tissue-specific functions of *Drosophila* Doublesex transcription factor target genes. *Developmental Cell* 31 (6): 761–773.

Coen, P., Clemens, J., Weinstein, A.J. et al. (2014). Dynamic sensory cues shape song structure in *Drosophila*. *Nature* 507: 233–237.

Dalton, J.E., Fear, J.M., Knott, S. et al. (2013). Male-specific fruitless isoforms have different regulatory roles conferred by distinct zinc finger DNA binding domains. *BMC Genomics* 14: 659.

Dankert, H., Wang, L., Hoopfer, E.D. et al. (2009). Automated monitoring and analysis of social behavior in *Drosophila*. *Nature Methods* 6: 297–303.

Debelle, A., Ritchie, M.G., and Snook, R.R. (2014). Evolution of divergent female mating preference in response to experimental sexual selection. *Evolution* 68: 2524–2533.

Delcourt, M., Blows, M.W., and Rundle, H.D. (2010). Quantitative genetics of female mate preferences in an ancestral and a novel environment. *Evolution* 64: 2758–2766.

Donlea, J.M., Ramanan, N., and Shaw, P.J. (2009). Use-dependent plasticity in clock neurons regulates sleep need in *Drosophila*. *Science* 324: 105–108.

Donlea, J.M., Ramanan, N., Silverman, N., and Shaw, P.J. (2014). Genetic rescue of functional senescence in synaptic and behavioral plasticity. *Sleep* 37: 1427–1437.

Drosophila 12 Genomes Consortium (2007). Evolution of genes and genomes on the *Drosophila* phylogeny. *Nature* 450: 203–218.

Dukas, R. (2004). Male fruit flies learn to avoid interspecific courtship. *Behavioral Ecology* 15: 695–698.

Dukas, R. (2005). Learning affects mate choice in female fruit flies. *Behavioral Ecology* 16: 800–804.

Ejima, A., Smith, B.P.C., Lucas, C. et al. (2007). Generalization of courtship learning in *Drosophila melanogaster* is mediated by *cis*-vaccenyl acetate. *Current Biology* 17: 599–605.

Ellendersen, B.E. and von Philipsborn, A.C. (2017). Neuronal modulation of *D. melanogaster* sexual behaviour. *Current Opinion in Insect Science* 24: 21–28.

Ellis, L.L. and Carney, G.E. (2011). Socially-responsive gene expression in male *Drosophila melanogaster* is influenced by the sex of the interacting partner. *Genetics* 187: 157–169.

Emmons, S.W. and Lipton, J. (2003). Genetic basis of male sexual behaviour. *Journal of Neurobiology* 54: 93–110.

Etges, W.J., Cardoso de Oliveira, C., Gragg, E. et al. (2007). Genetics of incipient speciation in *Drosophila mojavensis*. I. Male courtship song, mating success and genotype X environment interactions. *Evolution* 61: 1106–1119.

Fedorka, K.M., Winterhalter, W.E., and Ware, B. (2011). Perceived sperm competition intensity influences seminal fluid protein production prior to courtship and mating. *Evolution* 65: 584–590.

Feng, K., Palfreyman, M.T., Häsemeyer, M. et al. (2014). Ascending SAG neurons control sexual receptivity of *Drosophila* females. *Neuron* 83: 135–148.

Flatt, T. (2011). Survival costs of reproduction in *Drosophila*. *Experimental Gerontology* 46: 369–375.

Foucaud, J., Philippe, A.-S., Moreno, C., and Mery, F. (2013). A genetic polymorphism affecting reliance on personal versus public information in a spatial learning task in *Drosophila melanogaster*. *Proceedings of the Royal Society B: Biological Sciences* 280: 20130588.

Fricke, C., Bretman, A., and Chapman, T. (2010). Female nutritional status determines the magnitude and sign of responses to a male ejaculate signal in *Drosophila melanogaster*. *Journal of Evolutionary Biology* 23: 157–165.

Gaertner, B.E., Ruedi, E.A., McCoy, L.J. et al. (2015). Heritable variation in courtship patterns in *Drosophila melanogaster*. G3: genes. *Genomes, Genetics* 5: 531–539.

Garbaczewska, M., Billeter, J.-C., and Levine, J.D. (2013). *Drosophila melanogaster* males increase the number of sperm in their ejaculate when perceiving rival males. *Journal of Insect Physiology* 59: 306–310.

Gilchrist, A.S. and Partridge, L. (2000). Why it is difficult to model sperm displacement in *Drosophila melanogaster*: the relation between sperm transfer and copulation duration. *Evolution* 54: 534–542.

Gleason, J.M. (2005). Mutations and natural genetic variation in the courtship song of *Drosophila*. *Behavioral Genetics* 35: 265–277.

Gleason, J.M., Nuzhdin, S.V., and Ritchie, M.G. (2002). Quantitative trait loci affecting a courtship signal in *Drosophila melanogaster*. *Heredity* 89: 1–6.

Gong, Z., Son, W., Chung, Y.D. et al. (2004). Two interdependent TRPV channel subunits, inactive and Nanchung, mediate hearing in *Drosophila*. *Journal of Neuroscience* 24: 9059–9066.

Gorter, J.A., Jagadeesh, S., Gahr, C. et al. (2016). The nutritional and hedonic value of food modulate sexual receptivity in *Drosophila melanogaster* females. *Scientific Reports* 6: 19441.

Greenspan, R.J. and Ferveur, J.-F. (2000). Courtship in *Drosophila*. *Annual Reviews in Genetics* 34: 205–232.

Griffith, L.C. and Ejima, A. (2009). Courtship learning in *Drosophila melanogaster*: diverse plasticity of a reproductive behavior. *Learning and Memory* 16: 743–750.

Gromko, M.H. (1987). Genetic constraint on the evolution of courtship behaviour in *Drosophila melanogaster*. *Heredity* 58: 435–441.

Guss, K.A., Nelson, C.E., Hudson, A. et al. (2001). Control of genetic regulatory network by a selector genes. *Science* 292: 1164–1167.

Hall, J.C. (1994). The mating of a fly. *Science* 264: 1702–1714.

Häsemeyer, M., Yapici, N., Heberlein, U., and Dickson, B.J. (2009). Sensory neurons in the *Drosophila* genital tract regulate female reproductive behavior. *Neuron* 61: 511–518.

Hine, E., Lachish, S., Higgie, M., and Blows, M.W. (2002). Positive genetic correlation between female preference and offspring fitness. *Proceedings of the Royal Society B: Biological Sciences* 269: 2215–2219.

Hoffmann, A.A. (1994). Genetic analysis of territoriality in *Drosophila melanogaster*. In: *Quantitative Genetic Studies of Behavioural Evolution* (ed. C.R.B. Boake). Chicago: University of Chicago Press.

Hoffmann, A.A. (1999). Is the heritability for courtship and mating speed in *Drosophila* (fruit fly) low? *Heredity* 82: 158–162.

Hoffmann, A.A. and Cacoyianni, Z. (1989). Selection for territoriality in *Drosophila melanogaster*. Correlated responses in mating success and other fitness components. *Animal Behaviour* 38: 23–49.

Houle, D. (2010). Numbering the hairs on our heads: the shared challenge and promise of phenomics. *Proceedings of the National Academy of Sciences USA* 107 (Suppl. 1): 1793–1799.

Hunt, J. and Hosken, D.J. (eds.) (2014). *Genotype by Environment Interactions and Sexual Selection*. Oxford: Wiley.

Huttunen, S. and Aspi, J. (2003). Complex inheritance of male courtship song characters in *Drosophila virilis*. *Behavioral Genetics* 33: 17–24.

Immonen, E. and Ritchie, M.G. (2012). The genomic response to courtship song stimulation in female *Drosophila melanogaster*. *Proceedings of the Royal Society B: Biological Sciences* 279: 1359–1365.

Immonen, E., Snook, R.R., and Ritchie, M.G. (2014). Mating system variation drives rapid evolution of the female transcriptome in *Drosophila pseudoobscura*. *Ecology and Evolution* 4: 2186–2201.

Inagaki, H.K., Jung, Y., Hoopfer, E.D. et al. (2014). Optogenetic control of *Drosophila* using a red-shifted channelrhodopsin reveals experience-dependent influences on courtship. *Nature Methods* 11: 325–332.

Ingleby, F.C., Hunt, J., and Hosken, D.J. (2013). Genotype-by-environment interactions for female mate choice of male cuticular hydrocarbons in *Drosophila simulans*. *PLoS One* 8: e67623.

Ingleby, F.C., Hunt, J., and Hosken, D.J. (2014). Sexual selection and genotype-by-environment interactions in *Drosophila* cuticular hydrocarbons. In: *Genotype by Environment Interactions and Sexual Selection* (ed. J. Hunt and D.J. Hosken). London: Wiley.

Ito, H., Fujitani, K., Usui, K. et al. (1996). Sexual orientation in *Drosophila* is altered by the *satori* mutation in the sex-determination gene *fruitless* that encodes a zinc finger protein with a BTB domain. *Proceedings of the National Academy of Sciences USA* 93: 9687–9692.

Jiang, J. and White-Cooper, H. (2003). Transcriptional activation in *Drosophila* spermatogenesis involves mutually dependent function of *aly* and a novel meiotic arrest gene *cookie monster*. *Development* 130: 563–573.

Kangaspeska, S., Stride, B., Metivier, R. et al. (2008). Transient cyclical methylation of promoter DNA. *Nature* 452: 112–115.

Kaun, K.R., Riedl, C.A.L., Chakaborty-Chatterjee, M. et al. Natural polymorphism in a cGMP-dependent protein kinase affects food intake and absorption in *Drosophila*. *Journal of Experimental Biology* 210: 3547–3558.

Keleman, K., Vrontou, E., Krüttner, S. et al. (2012). Dopamine neurons modulate pheromone responses in *Drosophila* courtship learning. *Nature* 489: 145–149.

Kent, C., Azanchi, R., Smith, B. et al. (2008). Social context influences chemical communication in *D. melanogaster* males. *Current Biology* 18: 1384–1389.

Kim, W.J., Jan, L.Y., and Jan, Y.N. (2012). Contribution of visual and circadian neural circuits to memory for prolonged mating induced by rivals. *Nature Neuroscience* 15: 876–883.

Kimura, K., Ote, M., Tazawa, T., and Yamamoto, D. (2005). Fruitless specifies sexually dimorphic neural circuitry in the *Drosophila* brain. *Nature* 438: 229–233.

Koganezawa, M., Kimura, K., and Yamamoto, D. (2016). The neural circuitry that functions as a switch for courtship versus aggression in *Drosophila* males. *Current Biology* 26: 1395–1403.

Kohatsu, S., Koganezawa, M., and Yamamoto, D. (2011). Female contact activates male-specific interneurons that trigger stereotypic courtship behavior in *Drosophila*. *Neuron* 69: 498–508.

Kohl, J., Ostrovsky, A., Frechter, S., and Jefferis, G. (2013). A bidirectional circuit switch reroutes pheromone signals in male and female brains. *Cell* 155: 1610–1623.

Kramer, J.M., Kochinke, K., Oortveld, M.A. et al. (2011). Epigenetic regulation of learning and memory by *Drosophila* EHMT/G9a. *Plos Biology* 9: e1000569.

Kubli, E. (2003). Sex-peptides: seminal peptides of the *Drosophila* male. *Cellular and Molecular Life Science* 60: 1689–1704.

Larsson, M.C., Domingos, A.I., Jones, W.D. et al. (2004). Or83b encodes a broadly expressed odorant receptor essential for *Drosophila* olfaction. *Neuron* 43: 703–714.

Lasbleiz, C., Ferveur, J.-F., and Everaert, C. (2006). Courtship behaviour of *Drosophila melanogaster* revisited. *Animal Behaviour* 72: 1001–1012.

Lebreton, S., Grabe, V., Omondi, A.B. et al. (2014). Love makes smell blind: mating suppresses pheromone attraction in *Drosophila* females via Or65a olfactory neurons. *Scientific Reports* 4: 7119.

Leech, T., Sait, S.M., and Bretman, A. (2017). Sex-specific effects of social isolation on ageing in *Drosophila melanogaster*. *Journal of Insect Physiology* 102: 12–17.

Lin, H.H., Cao, D.S., Sethi, S. et al. (2016). Hormonal modulation of pheromone detection enhances male courtship success. *Neuron* 90: 1272–1285.

Lizé, A., Doff, R.J., Smaller, E.A. et al. (2012). Perception of male-male competition influences *Drosophila* copulation behaviour even in species where females rarely remate. *Biology Letters* 8: 35–38.

Lynch, M. (2007). *The Origins of Genome Architecture*. London: Sinauer Associates.

Mackay, T.F.C., Heinsohn, S.L., Lyman, R.F. et al. (2005). Genetics and genomics of *Drosophila* mating behaviour. *Proceedings of the National Academy of Science USA* 102 (Suppl 1): 6622–6629.

Maguire, C.P., Lize, A., and Price, T.A.R. (2015). Assessment of rival males through the use of multiple sensory cues in the fruitfly *Drosophila pseudoobscura*. *PLoS One* 10: e0123058.

Manier, M.K., Belote, J.M., Berben, D. et al. (2010). Resolving mechanisms of competitive fertilization success in *Drosophila melanogaster*. *Science* 328: 354–357.

Markow, T.A. and O'Grady, P.M. (2005). Evolutionary genetics of reproductive behavior in *Drosophila*: connecting the dots. *Annual Reviews in Genetics* 39: 263–291.

Mazzi, D., Kesaniemi, J., Hoikkala, A., and Clapert, K. (2009). Sexual conflict over the duration of copulation in *Drosophila montana*: why is longer better? *BMC Evolutionary Biology* 9: 132.

McGuire, S.E., Deshazewr, M., and David, R.L. (2005). Thirty years of olfactory learning and memory research in *Drosophila melanogaster*. *Progress in Neurobiology* 76: 328–347.

Meffert, L.M., Hicks, S.K., and Regan, J.L. (2002). Nonadditive genetic effects in animal behaviour. *American Naturalist* 160 (S6): S198–S213.

Meissner, G.W., Luo, S.D., Dias, B.G. et al. (2016). Sex-specific regulation of Lgr3 in *Drosophila* neurons. *Proceedings of the National Academy of Sciences USA* 113 (9): E1256–E1265.

Miller, G.T. and Pitnick, S. (2002). Sperm-female coevolution in *Drosophila*. *Science* 298: 1230–1233.

Moatt, J.P., Dytham, C., and Thom, M.D.F. (2014). Sperm production responds to perceived sperm competition risk in male *Drosophila melanogaster*. *Physiology and Behavior* 131: 111–114.

Mohorianu, I., Bretman, A., Smith, D. et al. (2017). Genomic responses to socio-sexual environment in male *Drosophila melanogaster* exposed to conspecific rivals. *RNA* 23: 1048–1059.

Moore, A.J., Brodie, E.D., and Wolf, J.B. (1997). Interacting phenotypes and the evolutionary process: I. Direct and indirect genetic effects of social interactions. *Evolution* 51: 1352–1362.

Narraway, C., Hunt, J., Wedell, N., and Hosken, D.J. (2010). Genotype by environment interactions for female preference. *Journal of Evolutionary Biology* 23: 2550–2557.

Neville, M., Nojima, T., Ashley, E. et al. (2014). Male-specific fruitless isoforms target neurodevelopmental genes to specify a sexually dimorphic nervous system. *Current Biology* 24: 229–241.

Okada, K., Blount, J.D., Sharma, M.D. et al. (2011). Male attractiveness, fertility and susceptibility to oxidative stress are influenced by inbreeding in *Drosophila simulans*. *Journal of Evolutionary Biology* 24: 363–371.

Pan, Y., Robinett, C.C., and Baker, B.S. (2011). Turning males on: activation of male courtship behavior in *Drosophila melanogaster*. *PLoS One* 6: e21144.

Pan, Y., Meissner, G.W., and Baker, B.S. (2012). Joint control of drosophila male courtship behavior by motion cues and activation of male-specific P1 neurons. *Proceedings of the National Academy of Sciences USA* 109: 10065–10070.

Parker, G.A. and Pizzari, T. (2010). Sperm competition and ejaculate economics. *Biological Reviews* 85: 897–934.

Parker, G.A., Ball, M.A., Stockley, P., and Gage, M.J.G. (1996). Sperm competition games: individual assessment of sperm competition intensity by group spawners. *Proceedings of the Royal Society B: Biological Sciences* 263: 1291–1297.

Parker, G.A., Ball, M.A., Stockley, P., and Gage, M.J.G. (1997). Sperm competition games: a prospective analysis of risk assessment. *Proceedings of the Royal Society B: Biological Sciences* 264: 1793–1802.

Pavlou, H.J. and Goodwin, S.F. (2013). Courtship behavior in *Drosophila melanogaster*: towards a 'courtship connectome'. *Current Opinion in Neurobiology* 23: 76–83.

Pavlou, H.J., Lin, A.C., Neville, M.C. et al. (2016). Neural circuitry coordinating male copulation. *eLife* 5: e20713.

Petfield, D., Chenoweth, S.F., Rundle, H.D., and Blows, M.W. (2005). Genetic variance in female condition predicts indirect genetic variance in male sexual display traits. *Proceedings of the National Academy of Sciences USA* 102: 6045–6050.

von Philipsborn, A.C., Liu, T., Yu, J.Y. et al. (2011). Neuronal control of *Drosophila* courtship song. *Neuron* 69: 509–522.

Price, T.A.R., Lizé, A., Marcello, M., and Bretman, A. (2012). Experience of mating rivals causes males to modulate sperm transfer, in the fly *Drosophila pseudoobscura*. *Journal of Insect Physiology* 58: 1669–1675.

Prokop, Z.M. and Drobniak, S.M. (2016). Genetic variation in male attractiveness: it is time to see the forest for the trees. *Evolution* 70: 913–921.

Ranz, J.M., Castillo-Davis, C.I., Meiklejohn, C.D., and Hartl, D.L. (2003). Sex-dependent gene expression and evolution of the *Drosophila* transcriptome. *Science* 300: 1742–1745.

Rezával, C., Pavlou, H.J., Dornan, A.J. et al. (2012). Neural circuitry underlying *Drosophila* female postmating behavioral responses. *Current Biology* 22: 1155–1165.

Rezával, C., Nojima, T., Neville, M.C. et al. (2014). Sexually dimorphic octopaminergic neurons modulate female postmating behaviors in *Drosophila*. *Current Biology* 24: 725–730.

Rezával, C., Pattnaik, S., Pavlou, H.J. et al. (2016). Activation of latent courtship circuitry in the brain of *Drosophila* females induces male-like behaviors. *Current Biology* 26: 2508–2515.

Rideout, E.J., Billeter, J.-C., and Goodwin, S.J. (2007). The sex-determination genes *fruitless* and *doublesex* specify a neural substrate required for courtship song. *Current Biology* 17: 1473–1478.

Rideout, E.J., Dornan, A.J., Neville, M.C. et al. (2010). Control of sexual differentiation and behavior by the *doublesex* gene in *Drosophila melanogaster*. *Nature Neuroscience* 13: 458–466.

Ritchie, M.G. and Kyriacou, C.P. (1996). Selection for courtship signal in *Drosophila melanogaster*. *Animal Behaviour* 52: 603–611.

Ritchie, M.G., Saarikettu, M., and Hoikkala, A. (2005). Variation, but no covariance, in female preference functions and male song in natural populations of *Drosophila montana*. *Animal Behaviour* 70: 849–854.

Robinett, C.C., Vaughan, A.G., Knapp, J., and Baker, B.S. (2010). Sex and the single cell. ii. there is a time and place for sex. *PLoS Biology* 8 (5): e1000365.

Roff, D.A. and Mousseau, T.A. (1987). Quantitative genetics and fitness: lessons from *Drosophila*. *Heredity* 28: 103–118.

Rostant, W.G., Kay, C., Wedell, N., and Hosken, D.J. (2015). Sexual conflict maintains variation at an insecticide resistance locus. *BMC Biology* 13 (1): 34.

Rostant, W.G., Bowyer, J., Coupland, J. et al. (2017). Pleiotropic effects of DDT resistance on male size and behaviour. *Behavior Genetics* 47: 449–459.

Rouse, J. and Bretman, A. (2016). Exposure time and sensory cues affect the speed of male plastic behavioural responses to sperm competition. *Animal Behaviour* 122: 1–8.

Rouse, J., Watkinson, K., and Bretman, A. (2018). Flexible memory controls sperm competition responses in male *Drosophila melanogaster*. *Proceedings of the Royal Society B: Biological Sciences* 285: ii.

Ruan, H. and Wu, C.-H. (2008). Social interaction-mediated extension of *Drosophila* Cu/Zn superoxide dismutase mutants. *Proceedings of the National Academy of Sciences USA* 105: 7506–7510.

Ryner, L.C., Goodwin, S.F., Castrillon, D.H. et al. (1996). Control of male sexual behavior and sexual orientation in *Drosophila* by the *fruitless* gene. *Cell* 87 (6): 1079–1089.

Saltz, J.B. (2011). Natural genetic variation in social environment choice: context dependent gene-environment correlation in *Drosophila melanogaster*. *Evolution* 65: 2325–2334.

Saltz, J.B. (2013). Genetic composition of social groups influences male aggressive behaviour and fitness in natural genotypes of *Drosophila melanogaster*. *Proceedings of the Royal Society B: Biological Sciences* 280: 20131926.

Saltz, J.B. (2017). Genetic variation in social environment construction influences the development of aggressive behaviour in *Drosophila melanogaster*. *Heredity* 118: 340–347.

Saltz, J.B. and Foley, B.R. (2011). Natural genetic variation in social niche construction: social effects of aggression drive disruptive sexual selection in Drosophila melanogaster. *American Naturalist* 177: 645–654.

Saltz, J.B., Geiger, A.P., Anderson, R. et al. (2016). What, if anything, is a social niche. *Evolutionary Ecology* 30: 349–364.

Sarin, S. and Dukas, R. (2009). Social learning about egg-laying substrates in fruitflies. *Proceedings of the Royal Society B: Biological Sciences* 276: 4323–4328.

Schneider, J., Atallah, J., and Levine, J.D. (2017). Social structure and indirect genetic effects: genetics of social behaviour. *Biological Reviews* 92: 1027–1038.

Sharma, M.D., Tregenza, T., and Hosken, D.J. (2010). Female mate preference in *Drosophila simulans*: evolution and costs. *Journal of Evolutionary Biology* 23: 1672–1679.

Shevtsova, E., Hansson, C., Janzen, D.H., and Kjærandsen, J.S. (2011). Stable structural color patterns displayed on transparent insect wings. *Proceedings of the National Academy of Science USA* 108: 668–673.

Simon, A.F., Chou, M.-T., Salazar, E.D. et al. (2012). A simple assay to study social behaviour in *Drosophila*: measurement of social space. *Genes, Brain and Behavior* 11: 243–252.

Smith, D.T., Hosken, D.J., Rostant, W.G. et al. (2011). DDT resistance, epistasis and male fitness in flies. *Journal of Evolutionary Biology* 24: 1351–1362.

Snook, R.R. and Hosken, D.J. (2004). Sperm death and dumping in *Drosophila*. *Nature* 428: 939–941.

Snook, R.R., Robertons, A., Crudgington, H.S., and Ritchie, M.S. (2005). Experimental manipulation of sexual selecxtion and the evolution of courtship song in *Drosophila pseudoobscura*. *Behavioral Genetics* 35: 245–255.

Sokolowski, M.B. (2001). *Drosophila*: genetics meets behaviour. *Nature Reviews Genetics* 2: 879–890.

Sokolowski, M.B. (2010). Social interactions in "simple" model systems. *Neuron* 65: 780–794.

Speith, H. (1974). Courtship behaviour in *Drosophila*. *Annual Reviews in Entomology* 19: 385–405.

Stockinger, P., Kvitsiani, D., Rotkopf, S. et al. (2005). Neural circuitry that governs *Drosophila* male courtship behavior. *Cell* 121: 795–807.

Tan, C.K.W., Løvlie, H., Greenway, E. et al. (2013). Sex-specific responses to sexual familiarity, and the role of olfaction in *Drosophila*. *Proceedings of the Royal Society B: Biological Sciences* 280: 20131691.

Taylor, M.L., Wedell, N., and Hosken, D.J. (2007). The heritability of attractiveness. *Current Biology* 17: R959–R960.

Taylor, M.L., Wedell, N., and Hosken, D.J. (2008a). Sexual selection and female fitness in *Drosophila simulans*. *Behavioral Ecology and Sociobiology* 62: 721–728.

Taylor, M.L., Wigmore, C., Hodgson, D.J. et al. (2008b). Multiple mating increases female fitness in *Drosophila simulans*. *Animal Behaviour* 76: 963–970.

Tran, D.H., Meissner, G.W., French, R.L., and Baker, B.S. (2014). A small subset of fruitless subesophageal neurons modulate early courtship in *Drosophila*. *PLoS One* 9: e95472.

Turner, T.L. and Miller, P.M. (2012). Investigating natural variation in drosophila courtship song by the evolve and resequence approach. *Genetics* 191: 633–642.

Villella, A. and Hall, J.C. (1996). Courtship anomalies caused by *doublesex* mutations in *Drosophila melanogaster*. *Genetics* 143: 331–344.

Villella, A. and Hall, J.C. (2008). Neurogenetics of courtship and mating in *Drosophila*. *Advances in Genetics* 62: 67–184.

Villella, A., Gailey, D.A., Berwald, B. et al. (1997). Extended reproductive roles of the *fruitless* gene in *Drosophila melanogaster* revealed by behavioral analysis of new *fru* mutants. *Genetics* 147: 1107–1130.

Wedell, N., Gage, M.J.G., and Parker, G.A. (2002). Sperm competition, male prudence and sperm limited females. *Trends in Ecology & Evolution* 17: 313–320.

Welbergen, P. and van Diijken, F.R. (1992). Asymmetric response to directional selection for licking behaviour of *Drosophila melanogaster* males. *Behavioral Genetics* 22: 113–124.

Wertheim, B., Dicke, M., and Vet, L.E.M. (2002). Behavioural plasticity in support of a benefit for aggregation pheromone use in *Drosophila melanogaster*. *Entomologia Experimentalis et Applicata* 103: 61–71.

Wigby, S., Sirot, L.K., Linklater, J.R. et al. (2009). Seminal fluid protein allocation and male reproductive success. *Current Biology* 19: 751–757.

Winbush, A., Reed, D., Chang, P.L. et al. (2012). Identification of gene expression changes associated with long-term memory of courtship rejection in *Drosophila* males. G3. *Genes, Genomic, Genetics* 2: 1437–1445.

Yamamoto, D., Jallon, J.M., and Komatsu, A. (1997). Genetic dissection of sexual behaviour in *Drosophila melanogaster*. *Annual Reviews in Entomology* 42: 551–585.

Yapici, N., Kim, Y.J., Ribeiro, C., and Dickson, B.J. (2008). A receptor that mediates the post-mating switch in drosophila reproductive behaviour. *Nature* 451 (7174): 33–37.

Yu, J.Y., Kanai, M.I., Demir, E. et al. (2010). Cellular organization of the neural circuit that drives *Drosophila* courtship behavior. *Current Biology* 20: 1602–1614.

Zajitschek, F., Zajitschek, S.R.K., Friberg, U., and Maklakov, A.A. (2013). Interactive effects of sex, social environment, dietary restriction, and methionine on survival and reproduction in fruit flies. *Age* 35: 1193–1204.

Zhou, S., Campbell, T.G., Stone, E.A. et al. (2012). Phenotypic plasticity of the *Drosophila* transcriptome. *PLoS Genetics* 8: e1002593.

Zhou, C., Pan, Y., Robinett, C.C. et al. (2014). Central brain neurons expressing *doublesex* regulate female receptivity in *Drosophila*. *Neuron* 83: 149–163.

Zhou, C., Franconville, R., Vaughan, A.G. et al. (2015). Central neural circuitry mediating courtship song perception in male *Drosophila*. *Elife* 4: e08477.

7

Nature and Nurture in Parental Care

Nick J. Royle[1] and Allen J. Moore[2]

[1] *Centre for Ecology & Conservation, University of Exeter, Penryn Campus, Penryn, TR10 9EZ, UK*
[2] *Department of Genetics, University of Georgia, Athens, 30602-7503, USA*

Parenting is both nurture and nature. The word 'nurture' is a Middle English word derived from the Old French *noureture* (nourishment), which is itself based on Latin *nutrire*, meaning to feed or cherish, and has links to the word 'nursery' (Oxford English Dictionary – OED). Modern usage of the word 'nurture' makes the link with parental care even more obvious. The OED defines nurture as to 'care for and protect (someone or something) while they are growing' (verb) and 'the action or process of nurturing someone or something' (noun). This definition aptly describes the popular conception of parental care. And yet, parental care is an evolutionary adaptation and to have evolved it must therefore have an underlying genetic basis. Thus, parental care and its causes and evolution are a highly appropriate context in which to parse nature-nurture.

The origins of the general usage of the terms *nature* and *nurture* can be traced back to Greek philosophy (in Plato's Meno, for example) and appear in the writings of Shakespeare, among others (in *The Tempest* Caliban, the wild-natured son of the witch Sycorax, is described as 'a born devil, on whose nature nurture can never stick'; Darwin 1927). However, the modern use of nature (genes)-nurture (environment) can be traced to Francis Galton, the English polymath and cousin of Charles Darwin, who worked on inheritance and conceived the idea of eugenics, and coined the term 'nature versus nurture' in the nineteenth century (Galton 1869). Galton held the view that traits such as intelligence were determined by nature, whereas others opposed this genetic determinism and favoured the *tabula rasa* or blank slate view, popularized by the philosopher John Locke in the late seventeenth century, that human behavioural traits are determined by the nurture (environment) experienced during growth and development. The debate has raged ever since, fuelled by political and sociological considerations (for example, due to concerns about the potential implications of eugenics) as much as by the science (Lewontin 1993). It is becoming increasingly evident, however, that nature versus nurture is a false dichotomy and that the relationship between genes and environment in determining the expression of traits, including parental care, is both rich and complex.

Parental care is taxonomically widespread and covers a wide diversity of behaviours and associated traits including offspring provisioning (feeding) and defence, and non-behavioural traits such as gamete provisioning, gestation and nests (Clutton-Brock

Genes and Behaviour: Beyond Nature-Nurture, First Edition.
Edited by David J. Hosken, John Hunt and Nina Wedell.

1991; Royle et al. 2012a). As a result, it can be difficult to define exactly what parental care constitutes, at least in terms of the specific behaviours that make up the behavioural category of 'care' (Royle et al. 2016). Smiseth et al. (2012) operationally defined parental care as 'any parental trait that increases the fitness of a parent's offspring, and is likely to have originated and/or is currently maintained for this function'. Whether non-behavioural traits such as gamete provisioning and behavioural traits such as resource defence can be considered to be a form of parental care or not may be difficult to determine, but for other traits such as feeding of offspring, the distinction is much clearer. In this chapter, we will therefore focus our attention on behavioural parental care traits such as offspring provisioning (Figure 7.1).

Despite the diversity across taxa in both the forms of care (from defence of eggs to extensive care of dependent offspring) and which sex provides care (female only, male only, both), there are some broadly universal defining features of parental behaviours: it is altruistic because it is costly to parents (it uses resources parents could otherwise allocate to their survival and future reproduction) but beneficial to offspring; it is responsive to the environment; it involves repeated, reciprocal interactions between parents and

Figure 7.1 Examples of postnatal parental care behaviours. Provisioning of dependent offspring in (a) *Nicrophorus orbicollis* burying beetles, (b) wattled starlings *Creatophora cinerea*, (c) humans *Homo sapiens* and (d) wild boar *Sus scrofa*. (e) An example of an affiliative behaviour (carrying young) in chacma baboons *Papio ursinus*. Source: Photo credits: (a) Jena Johnson, (b,e) Nick Royle, (c,d) pixabay.com.

offspring; and it is multivariate (Royle et al. 2012a, 2014, 2016). Parental care is multivariate because it can encompass many, related traits – parental care in mice, for example, involves lactation, licking and grooming and retrieval of errant young in addition to other behaviours such as nest making – and involves competing demands (e.g. more time spent foraging to provision offspring means less time for offspring defence), all of which fall under the general category of parental care. Moreover, the allocation of costly parental resources to these competing demands is responsive to environmental factors such as temperature and rainfall, food availability, risk of predation, pathogens and, perhaps most importantly, the social environment provided by other individuals, especially family members (mothers, fathers, offspring). The social environment is particularly important to the evolution of parental care because the development of phenotypes is shaped by the repeated, reciprocal behavioural interactions and transfer of resources and information between parents and offspring (Cheverud and Moore 1994; Royle et al. 2012b); parent and offspring behaviours are both the target and agent of selection.

The costs of parental care and imperfect relatedness among family members in sexual organisms mean that evolutionary conflicts of interest are expected over the provision of parental care (e.g. the providers of care – parents – do not share all their genes in common with the recipients – offspring) (Trivers 1972; Parker et al. 2002). In addition, parental care will be under antagonistic selection across life-stages because of parent–offspring conflict (Trivers 1974). Parental care is beneficial for the recipient and costly for the caregiver, so selection favours receiving care when an offspring but disfavours providing care as a parent (Kölliker et al. 2005, 2012). If parental care is to occur, these constraints must be surmounted and conflicts resolved so selection is expected to lead to the coadaptation of favourable parent–offspring combinations (Kölliker et al. 2012). Parental care is therefore fundamentally about interactions: between parents and offspring, between genes and environment. The distinction between nature and nurture is particularly blurred in the context of parental care because the environment contains genes, that is, other individuals that are not passive recipients of care but interact with one another to determine how parental care is expressed and evolves. Parental care is not so much nature versus nurture, nor even nature via nurture, but nature intertwined with nurture. In this chapter, we will review the evidence for the heritability, environmental sensitivity (phenotypic plasticity) and context dependence (gene-by-environment interaction; GxE) of parental behaviours.

7.1 Genetics Underlying Parental Care

Perhaps surprisingly, there are fewer studies of the genetic basis of parenting – that is, the behaviour of parents directed towards offspring – than might be imagined. There is a solid body of literature on the effects of parenting ('parental effects', most commonly maternal effects but also sometimes paternal effects) (Mousseau and Fox 1998; Reinhold 2002; Uller 2012; Kilner et al. 2015), and how the differences that affect offspring might reflect genetic differences amongst parents, but less on the parenting behaviour *per se*. Of course, we can infer that genetic maternal effects are likely to reflect genetic variation underlying parenting (Agrawal et al. 2001; Lock et al. 2004, 2007; Kilner et al. 2015), but this is by no means assured. 'Maternal effects' is a broad category that is most often defined by variation in offspring responses than in identified traits in the mother

(Cheverud and Moore 1994). In addition, 'parenting' and 'parental care' are just as broad and ambiguous and often used differently in biology and psychology. Nevertheless, there are patterns that can be discerned.

7.1.1 Quantitative Genetic Studies

Quantitative genetics is an approach that attributes phenotypic effects to components of variation. Causal studies benefit from more direct approaches, identifying specific genes or genetic contributions, but require genotypic information. Thus, the simplest and oldest studies that examine genetic components of parenting are quantitative genetic; those that require little information other than a well-described and measurable phenotype and some knowledge of genetic relatedness amongst individuals (Cheverud and Moore 1994). Such quantitative genetic studies tend to yield low heritability estimates (Kendler and Baker 2007) (see also Kasper et al. 2017 for alloparental care heritability). One reason for this that has been suggested is that parenting also reflects offspring not just parental characteristics (Kölliker et al. 2005; Kendler and Baker 2007). Moreover, parenting is a behavioural category but the traits that constitute parenting may vary. It is important that heritability and other measures of parenting are calculated on traits that are defined carefully in order to be able to compare amongst multiple organisms and not use composite traits or categories. Given that it is difficult to know the extent to which behavioural categories reflect homology, showing that homologous genetic underpinnings exist may help our understanding of the heritability of parenting traits. In addition, the trait(s) must be measured in a way to allow a quantification of variation and covariation in the traits defined as contributing to parenting. Nevertheless, several systems with well-defined adaptive parental behaviour have been studied in a quantitative genetic context.

One of the very first studies of quantitative genetic aspects of parenting involved nest building in mice. Carol Lynch showed that inbred lines of mice, *Mus musculus*, varied in nest building, reflecting genetic variation (Lynch and Hegmann 1972) and that there were GxE interactions with responses to changes in temperature in the amount of material used in nests, depending on strain (Lynch and Hegmann 1973). Artificial selection on a heterogeneous stock as the starting population showed that this genetic variation could lead to a response to selection for nest building (Lynch 1980).

Other studies have adopted a more classic quantitative genetic approach, using breeding designs with known relationships to quantify genetic variation and covariation. Walling et al. (2008) examined quantitative genetic architecture associated with parenting in burying beetles, *Nicrophorus vespilloides*, and found differences in the pattern of variances and covariances among males and females even though the parenting behaviour expressed was indistinguishable between the sexes. In Savannah sparrows, *Passerculus sandwichensis*, Freeman-Gallant and Rothstein (1999) provided evidence for heritability of provisioning by fathers and their sons, and Dor and Lotem (2009) also found that provisioning was heritable in house sparrows, *Passer domesticus*. Other studies have inferred heritable variation from repeatable differences amongst individuals (*P. domesticus*) (Nakagawa et al. 2007), or by consistent variation amongst populations (red-winged blackbirds *Agelaius phoeniceus*) (Westneat et al. 2011). Pedigrees are another way to determine the relatedness information required to calculate quantitative genetic parameters. Studying parenting in a pedigreed wild population may help increase the number of explicit quantitative genetics studies (Wilson et al. 2010). However, such analyses can be complicated as several environmental effects, including social environmental effects, need to be considered simultaneously when examining genetic influences on parental care (Adams et al. 2015).

7.1.2 Molecular Genetic Studies

With the rise of inexpensive genome sequencing and accessible bioinformatics tools (see Chapter 11), we are starting to see more studies that go beyond a phenotypic approach and begin to identify the genetic basis of parenting (Calisi and MacManes 2015). A very early example is the work of Amy Toth (Toth et al. 2007), who used next-generation sequencing to identify genes differentially expressed in wasps caring for offspring versus reproducing or otherwise working. However, while the 'omic' revolution has facilitated studies of non-model organisms (and most species with parenting are non-model), there are earlier studies that identified single gene influences on parenting.

Of these, the most well known is oxytocin and its receptor, along with the related vasopressin and its receptor in vertebrates, which influences social attachment and social tolerance in organisms as diverse as roundworms *Caenorhabditis elegans* (Garrison et al. 2012) and humans (Bakermans-Kranenburg and Ijzendoorn 2008; Tombeau Cost et al. 2017). The role of oxytocin in social attachment of mothers and offspring is well known and studied in vertebrates (Insel and Young 2001). The related neuropeptide vasopressin (oxytocin and vasopressin reflect a gene duplication early in the vertebrate lineage) also influences social interactions and affects pair bonding in males and female prairie voles, *Microtus ochrogaster*, for example (Lim and Young 2004; Hammock and Young 2005). Vasopressin is also involved in the differences in mouse lines selected for high and low nest building (Bult et al. 1992).

Other neuropeptides are also important in parenting behaviours. Galanin, for example, is a neuropeptide that influences a switch from pup-directed aggression to parenting in male mice, *M. musculus* (Wu et al. 2014). Expression of neuropeptide F and its receptor, the insect homologue of neuropeptide Y in vertebrates that influences feeding behaviour, is also associated with parenting in insects (Ament et al. 2011; Cunningham et al. 2016). The studies cited here adopt a genetic approach but of course these molecules are neurotransmitters and hormones. The role of these molecules in affecting parenting behaviour is discussed in more detail below in the section on environmental influences.

Despite the remarkable progress in single gene studies (also see Chapters 5 and 6), it would be misleading to assume from these studies that parenting typically reflects single gene influences. There have been a few studies that seek to link the expected multifactorial genetic influences from quantitative genetics with specific molecular underpinnings. Peripato and Cheverud (2002) and Peripato et al. (2002) took advantage of differences in parenting ability of two inbred strains of mice and mapped a quantitative trait locus (QTL) that influenced the propensity of mothers to care for their pups. Although clearly a lack of parental behaviour is not adaptive as many of the pups die (Peripato et al. 2002), it does provide an early insight into molecular mechanisms. One of the values of this approach is that it not only identifies genetic regions and potentially genes, it also makes it possible to look at gene interactions such as epistasis, which appears to be an important influence on parenting behaviour in the mice they studied (Peripato et al. 2002).

All genetic studies depend on there being heritable variants in the population. The approach by Cheverud and Peripato and colleagues depends on differences in inbred lines, and one criticism here is that inbred lines may provide information on mechanisms but not on natural variants. A different approach was adopted by Hoekstra and her colleagues in a recent study of another mammal, two sister species of *Peromyscus* mice (the promiscuous deer mouse *P. maniculatus* and the monogamous oldfield mouse *P. polionotus*) (Bendesky et al. 2017). These were derived from natural populations, but differed particularly in the extent to which fathers contributed to care (*P. polionotus* > *P. maniculatus*). Happily for the researchers, these species will interbreed and that led them to an extensive examination of molecular underpinnings of parental care involving nest building, licking of pups, huddling over pups and pup retrieval. All but nest building are highly intercorrelated. Bendesky et al. (2017) eliminated potential environmental and social stimuli for care, which indicated that it is heritable, and identified 18 QTLs, 12 with sex-specific effects. In accordance with single-gene studies, vasopressin was a top candidate gene and was confirmed to have a causal role in nest-building behaviour by pharmacological manipulation.

Two take-home messages arose from this work. First, the behaviour of males and females, while appearing to be similar, had somewhat different genetic architecture. This molecular work recapitulates the finding of Walling et al. (2008) in burying beetles that male and female parenting may follow different evolutionary trajectories even when the behaviour performed appears to be the same. This further highlights the difficulty in knowing if the phenotypic measures we have are indeed homologous behaviours. Genetic information helps provide some answers. Second, the genetic architecture parallels neuronal organization. Again, this is encouraging for those studies that may be limited in the extent that neurological studies are possible, especially in wild animals.

A different approach can be adopted if there is genomic information available. A sequenced genome allows researchers to examine gene expression and test hypotheses for the evolution of behavioural complexity. In the context of parenting, the ovarian ground plan hypothesis (co-opting genes related to reproduction for parenting) (West-Eberhard 1996) and behavioural heterochrony (changes in the timing of gene expression and maternal effects) (Linksvayer and Wade 2005) have both been proposed to explain the evolution of parent–offspring interactions (see also Roy-Zokan et al. 2015 and Cunningham et al. 2017).

Generating a sequenced genome is becoming increasingly accessible; for example, Moore and colleagues sequenced, assembled and annotated the genome of the burying beetle, *N. vespilloides*, specifically to study the molecular basis of parenting (Cunningham et al. 2015, 2017). This beetle has extraordinarily complex and yet variable parental care. Females feed begging offspring by regurgitating predigested carrion. Males can also participate in care, but this is not obligatory (Smiseth et al. 2005).

One relatively accessible way to identify multiple genetic factors is to examine differential expression of genes associated with parenting using RNA-seq (Calisi and MacManes 2015). This directly links genes to behaviour (Rehan and Toth 2015). Such a transcriptomic study showed that uniparental male and female care have similar genetic influences, but while the genes expressed during care by females when the male is present overlap with those expressed during uniparental care, biparental males were indistinguishable from non-caring individuals of either sex (Parker et al. 2015). With the genome in hand, additional questions could be asked. The transcriptomic study suggested candidate genes, and follow-up research confirmed that vitellogenin expression in the head is associated with parenting in both males and female (Roy-Zokan et al. 2015), as it is in wasp mothers (Toth et al. 2007, 2010). Specific neuropeptides, particularly those associated with feeding circuits, are associated with a shift to parenting in *N. vespilloides* (Cunningham et al. 2016, 2017).

In addition to examining the genes that differentially expressed during a shift in behavioural states from non-parenting to parenting, it is also possible to ask if the same genes are involved in variation within parenting (Benowitz et al. 2017). This begins to reconcile studies based on gross differences (populations, species, inbred lines) or changes in state (e.g. comparing caring to non-caring) and the subtle variation within a population in a behaviour state typically addressed by quantitative geneticists looking at continuous variation. The answers are not yet in but the initial indications are that there may be little overlap in the genes involved in a transition to care from a non-caring state and the genes associated with continuous variation in the level of care provided when caring, supporting the view that the processes involved in origins of care are likely to be very different from those involved in the maintenance of care (Royle et al. 2016).

7.2 Parental Care is Environmentally Sensitive

Parenting is variable both within and amongst species, and while much of the variation is evolved and therefore must reflect genetic influences, environments also play a strong role. Populations vary as well as species. Human parenting, for example, is characterized by mothers as the primary caregivers, with alloparenting also common, beneficial and universal across cultures, but care by fathers much more variable (Sear 2015). A universal characteristic of parental behaviour, not just in humans but perhaps in all organisms with parental care, is the flexibility of care. More specifically, parental behaviours are responsive to environmental changes; that is, they are phenotypically plastic (Royle et al. 2014). If such behavioural plasticity improves fitness then it is adaptive (Gomez-Mestre and Jovani 2013), although plasticity does not have to be adaptive to have implications for the evolution of traits (Ghalambor et al. 2015). Resource limitations, costs of parental care and the multivariate demands of parenting can heighten this environmental sensitivity via the effects of trade-offs. If, for example, parents spend more time on nest defence because of an increase in the threat to offspring from nest predators, this may lead to less time spent on provisioning young.

Environmental factors that parents respond to include abiotic factors such as rainfall and temperature in addition to biotic factors, both non-social (food availability, the threat of predators and infection by pathogens) and social (the environment provided by other conspecifics, particularly other family members). These environmental effects can be complex. Abiotic environmental factors can have both direct and indirect effects on parental care. Vincze et al. (2013) showed that ambient temperature was strongly associated with the duration and share of parental care in two closely related species of biparental plovers. The total amount of time spent incubating eggs increased at very high temperatures, with the effect largely a result of an increase in the prevalence of male care across 10 geographically distinct populations of Kentish (*Charadrius alexandrines*) and snowy plovers (*Charadrius nivosus*).

More commonly, abiotic factors such as temperature will have indirect effects on parental behaviour, via non-social biotic factors such as food availability and, perhaps most importantly, the social environment, which acts as a sieve through which other environmental factors are filtered (Figure 7.2). For example, the amount of food that an individual red-winged blackbird (*Agelaius phoenicius*) nestling receives from its mother (the primary caregiver) (Figure 7.2a) depends on food availability in the wider environment, in addition to social environmental factors such as red-winged blackbird population density, both of which, in addition to the amount of help she will receive from her partner (red-winged blackbirds are polygynous so male care is highly variable) (Searcy and Yasukawa 1995; Beletsky and Orians 1997), will affect the amount of competition the mother will experience in accessing the food (Figure 7.2b). These factors will contribute to determining the rate of provisioning to the nest by the mother, which will also be influenced by other social environmental factors including nestling demand, which is likely to exceed supply because red-winged blackbirds, like many other birds, produce more offspring than they can typically rear as a hedge against uncertainty in the wider environment (Forbes et al. 1997; Forbes 2010). This increases competition among siblings for access to the limited resources provided by parents (Hall et al. 2010). The probability of success in competition for the limited supply of food depends primarily on the size of nestlings, with larger chicks enjoying an advantage over their smaller nest-mates. Such size asymmetries result from staggered patterns of hatching and mean that the fate of first-hatched nestlings is largely independent of the availability of resources in the wider environment. In contrast, the fate of last-hatched nestlings is strongly contingent on food availability – mortality is almost inevitable when food is anything other than abundant (Forbes et al. 1997; Forbes 2010). As a result, it is social environmental factors that actually generate variation among offspring in fitness prospects, not wider environmental factors such as food availability *per se* (see Figure 7.2b).

Parents respond to changes in care provided by their partners (Harrison et al. 2009), offspring respond to changes in behaviour by their siblings (Mock and Parker 1997; Roulin and Dreiss 2012) and these interactions among siblings affect how offspring interact with their parents and their parents interact with them (Mock and Parker 1997; Kilner and Hinde 2012). The studies showcased in the above reviews show that the social environment provided by family members can induce changes in behaviour that vary with the form of parental care, sex and the species involved and that it is dynamic and characterized by reciprocal interactions among parents and offspring that affect the outcome of parental care (Royle et al. 2014).

7.2.1 Hormonal Mediation of Parental Care

The mechanistic link between environment and genes is often hormonally mediated. The octopaminergic and serotenergic systems are often implicated in parenting (Bakermans-Kranenburg and Ijzendoorn 2008; Dulac et al. 2014; Tombeau Cost et al. 2017). Such systems can be studied at the molecular genetic level, as has been done with burying beetles, *N. vespilloides* (Cunningham et al. 2014, 2017). There has been increasing recent work to elucidate the hormonal and neural mechanisms that underlie the behavioural response to environmental changes involved in parental care,

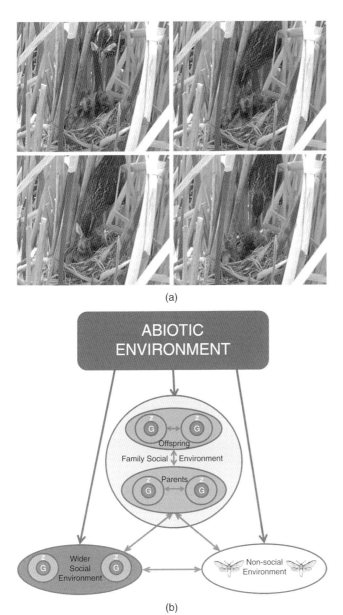

(a)

(b)

Figure 7.2 (a) Stills from video footage showing a female red-winged blackbird *Agelaius phoenicius* feeding begging young. Photo credit: Nick Royle. (b) Schematic showing relationship between abiotic environmental factors (e.g. rainfall, temperature), non-social environment (e.g. food supply), wider social environment (i.e. other individuals in the population) and family social environment on focal individuals (parents and offspring). Individuals are represented by circles with z = phenotype and G = genotype or the total genetic effects (i.e. including dominance, epistasis, etc.). The amount of food offspring receive from parents depends upon the outcome of interactions with siblings and parents (the family social environment), which depends on parental interactions with other individuals in the population (the wider social environment) and the availability of resources (the non-social environment), all of which are ultimately dependent on the abiotic environment. See main text for further details.

particularly in mammals (Dulac et al. 2014; Rilling and Young 2014). For example, the switch from potential aggression towards young in male mice to affiliative, parental behaviour is triggered by mating and continues until the offspring are weaned (vom Saal 1985; Wu et al. 2014). It has been hypothesized that these dramatic shifts in behaviour are a result of time-dependent synaptic or transcriptional changes triggered by mating and chemical cues released by the female (vom Saal 1985; Mennella and Moltz 1988). Such switches in behaviour associated with mating are not just restricted to rodents or mammals in general, but have been found in other taxa with biparental care such as cichlid fish, a variety of species of birds and some invertebrates (Elwood 1994; Oldekop et al. 2007), indicating a degree of universality.

A particularly notable example of female manipulation of males occurs in *N. vespilloides*, where females use an antiaphrodisiac pheromone, methyl geranate, to communicate their hormonal status to their partners (Engel et al. 2016). This signals that the female is (temporarily) infertile and stops males trying to mate with them while they are providing parental care (mating is costly to females) (Head et al. 2014). With the cooling of male ardour, females increase the probability that males will engage more in care and less in mating, promoting the coevolution of male and female parental care (Royle 2016).

Parenting behaviour of the main caregiver (usually females) (Kokko and Jennions 2012) is also highly sensitive to environmental factors such as stress, which can affect hormone levels (Dulac et al. 2014). For example, virgin female (and male) rats will exhibit parental care behaviours such as licking and grooming if sensitized through repeated exposure to pups (Rosenblatt 1967) and administering virgin females with an appropriate cycle of the hormones involved in reproduction (i.e. oestrogen, progesterone, oxytocin, and prolactin) elicits parental care behaviour (Moltz et al. 1970). These hormones act on a suite of neural systems to trigger parental care (Dulac et al. 2014). In mammals, the ovaries secrete oestrogen and progesterone during pregnancy before a drop in progesterone signals that parturition is imminent, which sensitizes particular regions of the brain to oxytocin and prolactin through an increase in the density of appropriate receptors (Rilling and Young 2014).

In particular, the medial preoptic area (MPOA) has been identified as a key area of the vertebrate brain involved in the control of parental care behaviours in rodents, and probably other mammals, as it suppresses the parallel system of neural circuits involved in the inhibition of parental care associated with another region of the brain, the medial amygdala (Dulac et al. 2014; Rilling and Young 2014). The MPOA is activated in parents because of interactions with young (Dulac et al. 2014); ablation of the subset of MPOA neurons that express the neuropeptide galanin leads to a decline in care in parent mice and increased aggression towards pups in virgin females whilst optogenetic activation of these neurons in virgin male mice stimulated pup grooming and other parental care behaviours (Wu et al. 2014). Oxytocin acts in the MPOA to activate the mesolimbic dopaminergic ventral tegmental area (VTA), stimulating dopamine production and activating dopamine receptors that inhibit amygdala-based agonism towards young (Rilling and Young 2014). In humans, activation of parents' mesolimbic dopamine system occurs in response to crying by infants, which also stimulates activation of other brain areas such as the anterior insula, which is involved in emotional empathy, and the prefrontal cortex, which is associated with the regulation of negative reactions to infant crying (Rilling 2013; Mascaro et al. 2014).

In addition to dopamine, serotonin is also known to be important in the regulation of parental care (Dulac et al. 2014), but it is oxytocin, a highly conserved neuropeptide, that is perhaps of most widespread importance in the regulation of social tolerance and affiliation across taxa (Dulac et al. 2014). It facilitates maternal responsiveness (rodents) and bond forming and maintenance (*C. elegans*, sheep, and humans), and may have a more generalized role in other social contexts (Rilling and Young 2014). For example, in monogamous Mandarin voles, *M. mandarinus*, pair bonding in adulthood is affected by the quality of parenting experienced during development, with poor parenting disrupting pair-bonding behaviour (Yu et al. 2013). In contrast, pharmacological stimulation of oxytocin neurons has the opposite effect (Barrett et al. 2013).

Such downstream effects of variation in the quality of parenting experienced by offspring during development are widespread and are an example, more broadly, of early life history effects, for which there is considerable evidence across taxa (Monaghan 2008; Hayward et al. 2013). In humans, for example, orphans reared without parental care in early life suffered long-term changes to brain development, including a larger amygdala and reduced connectivity between the amygdala and medial prefrontal cortex (Marshall et al. 2004; Nelson et al. 2009), that adversely affected intellectual and emotional development. This was (slightly) offset by improvements in brain development if the children subsequently received (high-quality) parenting, especially if this occurred before the age of two (Smyke et al. 2010; Vanderwert et al. 2010; Marshall 2014). Similar results have been found in rodents, showing a link between lack of parenting received and increased anxiety and socially mediated stress in later life (Tottenham et al. 2012), and in fish, where past exposure of mothers to predators influences subsequent male care and offspring development (McGhee et al. 2012). Early life exposure may also influence future generations (Burton and Metcalfe 2014).

Studies demonstrating extreme adverse effects of little or no care in taxa with obligate parental care such as mammals illustrate the importance of care for normal development and the interdependence of parental and offspring behaviours; the social environment provided by parents has a profound effect on offspring development (Royle et al. 2012b). These effects are not, however, limited to extreme social environments (e.g. no care); any variation in parental care can potentially affect the development of offspring phenotypes (Royle et al. 2012a). Moreover, effects of variation in parental care can also depend upon the genotypes of offspring and, conversely, offspring behavioural phenotypes can also depend on the genotypes of parents (Ashbrook et al. 2015). These genotype by (family) environment interactions (GxFE) (Kölliker et al. 2012) are essential components of the evolution and expression of parental care behaviours.

7.3 Gene by Family Environment Interactions

7.3.1 GxFE Studies

Numerous recent studies have provided evidence of GxFE effects associated with polymorphisms of genes associated with hormonal and/or neurological pathways implicated in parental care behaviours in humans. For example, Micali et al. (2017) found that the prevalence of eating disorders (binge eating) in adult women was affected by a GxFE with oxytocin receptor gene polymorphisms interacting with the quality of maternal care received during development, with the rs2254298 genotype and poor maternal care more likely to lead to binge eating in adulthood. Other studies have demonstrated links with parental care and brain-derived neurotrophic factor (BDNF) gene polymorphisms (val66met) which affect neuron growth and development. Suzuki et al. (2011, 2012) showed that the BDNF polymorphism significantly interacts with the quality of parental care received during development to influence personality traits and interpersonal sensitivity in adulthood (with met/met genotypes being more responsive to variation in care than val66/val66 genotypes). Such effects may be dependent on which sex is providing care. So, for example, Ibarra et al. (2014) provide evidence to show that high levels of paternal care are associated with a reduction in the probability of mental health issues such as depression in adulthood, whereas high levels of maternal care are associated with an increase in paranoia and obsessive-compulsive behaviours. The expression of these outcomes was dependent on the genotype of offspring, with met carriers more sensitive to variation in parental care than val/val genotypes (Ibarra et al. 2014).

Brain-derived neurotrophic factor has also been shown to be important in rodents. Tabbaa et al. (2017) found that prairie vole, *M. ochrogaster*, offspring reared without fathers (the species is monogamous and exhibits biparental care) received less licking and grooming than offspring reared under biparental care and at adulthood were more socially affiliative and had elevated gene and protein expression of BDNF (although did not differ in measures of anxiety). Evidence for GxFE involving parental care on the expression of stress and anxiety can also be found in studies of rodents. For example, inbred strains of mice with different binding levels of the benzodiazepine (BZ/GABA) receptor, which is involved in the regulation of stress, were used to show, using cross-fostering, that strain differences in BZ/GABA were a function of variation in parental care levels (Caldji et al. 2004).

Whilst most of the above GxFE studies focused on parental care as the FE, with the response variable being anxiety or some other stress-related phenotype, in a smaller number of studies parental care has been the response variable or, sometimes, both the FE and the response variable. In most cases there is evidence for GxFE effects. In long-tailed tits, *Aegithalos caudatus*, which are co-operative breeders, individuals (parents and helpers) were found to be consistent in their provisioning at a given nest but adjusted effort in relation to who was in their group, which differed across years and social contexts (Adams et al. 2015). In humans, higher levels of risk to health at birth were associated with less sensitive parenting, but only when carrying a specific polymorphism of the dopamine D4 receptor gene (seven repeat allele), with the greatest sensitivity shown when offspring scored low on the risk index (Fortuna et al. 2011). This indicated that plasticity of parenting in response to a change in the environment experienced by offspring (health risk) was dependent on genetic variation at the dopamine D4 receptor gene. Mileva-Seitz et al. (2013) investigated polymorphisms in the oxytocin peptide gene (OXT) and the oxytocin receptor gene (OXTR) in humans. They found that OXT interacted with the quality of care mothers experienced themselves in early life to predict variation in their own parenting behaviour at six months post birth, but OXTR was not related to parenting behaviour. Similarly, in an earlier study Mileva-Seitz et al. (2011) reported that there were interactive effects of maternal genotype at a serotonin transporter-linked polymorphism with early-life experience being recapitulated in their own maternal behaviour (at six months post birth) for some genotypes but not others. In contrast to the above studies, Kelley et al. (2013) found that in prairie voles

M. ochrogaster, variations in a vasopressin gene (polymorphisms in avpr1a gene) were not associated with variation in paternal care behaviours, but prior experience of parental care (whether they had reared an earlier litter) was, with fathers that were on their second litter less likely to lick pups but quicker at retrieving pups that had strayed from the nest.

7.3.2   Heritability of the Social Environment and IGEs

Although all the above studies provide examples of GxFE interactions, one potentially important component of the interaction is often not explicitly accounted for: that the family environment also contains genes (Cheverud and Moore 1994; Moore et al. 1997; Wolf et al. 1998) (see Chapter 4). This is a key consideration not only if we want to understand the evolution of traits involved in parental care, but also if we want to understand the health and welfare implications of GxFE interactions in humans and other animals because when individuals interact with one another, genes expressed in one individual can affect the expression of genes in another individual (indirect genetic effect (IGE)) (Moore et al. 1997) (Figure 7.3). This means that particular offspring genotypes, for example, may be more sensitive to particular genotypes in parents than others, with consequences for the expression of behaviours such as anxiety or their own parental care behaviour when adult. Although IGEs may arise from any form of social interaction

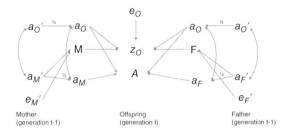

Mother
(generation t-1)

Offspring
(generation t)

Father
(generation t-1)

Figure 7.3 Summary of influences of mothers and fathers on offspring. Illustrated here are an offspring phenotype (z_O) and causal sources of variation. In this example, we assume biparental care. Arrows indicate direction of causality; double-headed arrows reflect correlated traits with joint effects. Given Mendelian laws of heredity, mothers pass on genetic influences on the offspring trait, which also influence the 'breeding value' (A, the sum of the average effects of all of the alleles that an individual carries; this can be considered the sum of its additive genetic influences regardless of the trait influenced. See Riska et al. 1985; Cheverud and Moore 1994). The mother also contributes parental care or a maternal environment M. This is an environmental effect for the offspring, but the expression of this trait in the parent may be influenced by the parent's genes which is symbolized by a_M'. The prime is used to indicate that these effects are expressed in the parental generation (t-1 from the offspring). This maternal genetic effect also contributes to the offspring breeding value even though the genes are expressed in a different generation. This separation of the expression of the genes and their effects has been termed 'indirect genetic effects' to distinguish them from the direct effects of genes on traits (Riska et al. 1985; Moore et al. 1997). There are also general environmental effects e_M', contributing to the maternal effect. Under biparental conditions, the father can also contribute care that influences the offspring trait so that there can be paternal effects (P) that are influenced by paternal genes (a_P') and paternal environment (e_P'). These paternal effect genes also contribute to the offspring trait. The father also contributes ½ of the genes carried by the offspring a_O that directly influence the offspring trait. Finally, there environmental contributions specific to the offspring, e_O, that also contribute to the offspring phenotype A full path analysis relating biparental care to its effects on offspring fitness, as well as other relationships, is given in Lynch (1987) (see also Riska et al. 1985; Arnold, 1994; Cheverud and Moore, 1994).

(Wolf et al. 1998; Wilson 2014) there are at least three reasons why they are expected to be particularly important in the context of parental care (Kölliker et al. 2012).

1. IGEs are likely to be particularly effective when social interactions involve repeated, reciprocal interactions such as occur when parents feed soliciting offspring (Cheverud and Moore 1994). These effects will be enhanced by the stability of group genotypic structure and composition (typically one or both parents plus their genetic offspring).
2. Social interactions during parental care can take place across three social dimensions (parent-offspring, offspring-offspring, parent-parent) (Parker et al. 2002) and involve a great diversity of traits that have evolved as adaptations to family life. As a result, they are likely to have fitness consequences for other family members.
3. Families are different from other kin-group structures in that the constituents are not different categories of related individuals (e.g. helper versus parent) but are two life-stages of individuals that overlap across generations (parents and offspring) that compete with kin in one stage (offspring) but are then altruistic towards kin in another stage (parents) (Kölliker et al. 2012). Families can be complicated, with co-operation amongst relatives and conflict between parents and offspring. How, then, are alleles at loci that influence offspring and parental traits associated and inherited across generations?

Parental care usually influences offspring fitness indirectly via offspring traits that convert care into offspring fitness (Kölliker et al. 2012). These are known as offspring performance traits (Cheverud and Moore 1994). A typical example would be growth rate in the context of variation in parental provisioning. Evolutionary genetic models of parental care show that parental care evolution is facilitated by a positive genetic covariance between parental traits and offspring performance traits (Cheverud and Moore 1994). For example, offspring with faster growth (because of high parental provisioning rates) may also provide food at high rates when adults (Agrawal et al. 2001). This pattern is not universal (Kölliker et al. 2005). In *N. vespilloides* burying beetles, there are substantial IGEs associated with parental care and total heritabilities for offspring performance traits are low for both maternal (Lock et al. 2004) and paternal (Head et al. 2012) (uniparental) care because of negative direct-indirect genetic covariance between parents and offspring. This means that offspring performance traits such as growth rate may be evolutionarily constrained, which maintains variation in these traits (Head et al. 2012).

Indirect genetic effects may not be explicitly discussed often, but especially in family studies may be identified experimentally. For example, in humans, parental feelings (positive or negative) towards offspring and thus their interactions with their children have higher heritability than parenting strategies towards their children (Oliver et al. 2014). More specifically, negative parenting strategies and negative feelings have higher genetic influence than positives of either. Although Oliver et al. discuss control and feelings as environments, these are identified in their work as heritable traits with 'genetic influences' on the offspring. Despite the undeniable logic of IGEs, rooted in the reciprocity of interacting phenotypes (and therefore genotypes), studies that explicitly account for the genetic covariance between parental and offspring traits involved in care have been largely ignored in favour of genetic relatedness (Kölliker et al. 2012).

Very recently, Kong et al. (2018) used genome-wide association studies (GWAS), associating variation in genetic markers across the chromosome to variation in a trait, to examine how genetic variation in parental educational attainment influences offspring educational attainment. There are a number of notable aspects to this study (see also Perspective by Koellinger and Harden 2018). First, it involves humans and therefore expands our molecular quantitative genetic studies of IGEs to a familiar organism. Second, the authors cleverly examine only the effect of the non-transmitted alleles (remember that a given parent only transmits one of the two alleles found in the offspring) and so are truly examining how genetic variation in the environment experienced by offspring matters. Finally, the combination of quantitative genetics and molecular genetics provides a nice integration. Given humans were studied and we have excellent genomic resources for this model organism, over 600 000 single nucleotide polymorphisms (SNPs) – the markers used to identify variation in specific regions of the chromosome – were used so there is good power. Unfortunately, the authors termed the effect studied 'genetic nurturing effects', when in fact it is simply IGEs or parental effects they examined. However, this study does only examine the environmental portion of the parental effect, further clarifying why separating genetic and environmental effects matters and how an environment for one individual may be influenced by genes expressed in another individual (Moore et al. 1997). Truly, there is no nurture without nature.

7.3.3 Coadaptation and Correlational Selection (Social Epistasis as a Special Case of GxFE)

In the offspring life-stage, individuals are under selection to adapt to the care provided by parents. In the parental life-stage, there will be adaptation in traits that influence offspring traits (Cheverud and Moore 1994). Thus, parenting crosses life history stages both within and between individuals and therefore coadaptation is predicted. Such coadaptation can be modelled by exploring how patterns of genetic variance and covariance of parental and offspring traits are expected to evolve (Kölliker 2005; Kölliker et al. 2012). For coadaptation to occur requires correlational selection, which favours combinations of parent-offspring traits rather than individual traits isolated from selection on other traits. Correlational selection is expected when there are reciprocal parent and offspring traits that affect fitness (see Chapter 4). If these traits are heritable, such that similar phenotypic effects/fitness can arise as a result of different combinations of parent and offspring genotypes (Kölliker et al. 2012), then 'social epistasis' (Linksvayer 2007) will occur. Social epistasis therefore represents a subcategory of GxFE when the FE is heritable (Kölliker et al. 2012).

Coadaptation models predict that correlational selection will favour the physical linkage of loci or their expression, or pleiotropic effects of genes involved in parental care. Evidence for this comes from mice, *M. musculus*, in which the Peg3 gene affects both the amount of milk let down by mothers and the suckling efficiency of offspring (Curley et al. 2004), and in rhesus macaques, *Macaca mulatta*, where a mu-opioid receptor gene affects both how attentive mothers are and how demanding of attention offspring are (Barr et al. 2008; Higham et al. 2011). However, parental care is multivariate with coadaptation expected for each combination of parent-offspring traits as well as across different parent-offspring traits and across different stages of care (Lock et al. 2007; Gomez and Kölliker 2013) and there are likely to be many different genes involved. As a result, quantitative genetic studies such as those involving cross-fostering are often the most appropriate method for assessing coadaptation as the explicit variance partitioning approach allows researchers to control for confounding variables that are hidden in other phenotypic approaches.

Cross-fostering of offspring and demonstrating that 'mismatched' fostered offspring fare worse than 'matched' genetic offspring provides evidence for coadaptation. A study by Hager and Johnstone (2003) using crosses of inbred strains of mice that were provided with cross-fostered offspring from their own or a different strain showed that mismatched offspring (i.e. those with a maternal line background different from their own strain) received less milk than offspring from matched parent-offspring strains. Whilst this study provided evidence for social epistasis, it is not clear whether it provides evidence for coadaptation due to, for example, potential issues associated with inbreeding depression and because crosses between different populations (lines) or species test for coadaptation at the level of differences in mean traits that reflect local adaptation, genetic drift or speciation rather than whether correlational selection on parent-offspring traits shapes the genetic architecture of parental care (Kölliker et al. 2012). Coadaptation is therefore best studied by focusing on families within populations.

A good example of this comes from a study by Hinde et al. (2010) on domestic canaries, *Serinus canaria*. They used a repeated measures design where mothers reared a brood of their own offspring and a brood of foster offspring consecutively. The difference in the begging behaviour of chicks in cross-fostered broods compared to broods of their own young was used as a measure of the mismatch as a result of coadaptation and compared to the growth of chicks (a measure of offspring performance). As predicted by coadaptation theory, chick growth decreased in proportion to the level of the mismatch between parents and offspring (the difference in begging between foster and genetic young). This provided evidence for covariance between parental provisioning and offspring begging as a result of correlational selection, either as a result of a maternal effect (the authors suggest steroid hormones in the eggs) (Hinde et al. 2009, 2010) or through genetic covariance.

More recent studies on coadaptation in canaries have failed to provide support for a role of steroid hormones in the eggs as a source of correlational selection (Estramil et al. 2017) or for genetic covariance (Estramil et al. 2014), as even though both begging (at the offspring stage) and provisioning (at the parental stage) were weakly heritable, there was no significant covariance between the two. Similarly, Dor and Lotem (2010) showed that food delivery rate of house sparrows, *P. domesticus*, was repeatable and heritable across four generations of cross-fostered broods, but experimentally induced changes in begging did not lead to a repeatable or heritable parental response, nor was the parental response correlated with nestling begging rate.

In contrast, other studies across a range of species have provided evidence for both positive (e.g. great tits, *Parus major*, Kölliker et al. 2000; burying beetles, *N. vespilloides*, Lock et al. 2004; mice, *M. musculus*, Curley et al. 2004; blue tits, *Cyanistes caeruleus*, Lucass et al. 2016a) and negative (e.g. burrower bugs, *Sehirus cinctus*, Agrawal et al. 2001; rhesus macaques, *M. mulatta*, Maestripieri 2004) parent-offspring covariances. Most of the studies have used cross-fostering, making it difficult to distinguish the source of the covariance (environmental maternal effect versus genetic covariance), which matters if you are interested in evolutionary potential of traits. Furthermore, differences across studies and species may reflect life history variation (Kölliker et al. 2005) and whether it is parents or offspring that primarily control the allocation of resources (Royle et al. 2002; Lucass et al. 2016b; Takata et al. 2016) as well as differences in experimental protocol.

In addition, variation across studies may also arise because the focus has been on mean parental and offspring behaviours (i.e. 'static' traits) and not the changes in behaviour (i.e. the reaction norms) (Smiseth et al. 2008). Parental supply and offspring demand are not static traits, but are the outcome of coevolved plastic response rules (behavioural reaction norms – BRNs; Dobler and Kölliker 2009). A recent study by Lucass et al. (2016c) conducted a large-scale cross-fostering experiment over three seasons in wild blue tits, *C. caeruleus*, focusing on both static traits (parental provisioning and offspring begging behaviours respectively) and BRNs (offspring begging as a function of food deprivation and parental provisioning as a function of offspring demand). They found no evidence for parent-offspring coadaptation in static traits but evidence for coadaptation in BRNs, for fathers at least. Fathers that were more responsive to offspring demand sired offspring that were less responsive in begging behaviour to a change in hunger (Lucass et al. 2016c).

This result is unusual because it is fathers and not mothers that are coadapted with offspring behaviours. More typically (e.g. in *N. vespilloides* burying beetles), coadaptation occurs between mothers (Lock et al. 2004) but not fathers (Head et al. 2012) and their offspring, presumably because females are typically the main caregivers (Kokko and Jennions 2012) and this magnifies all the key factors that drive coadaptation (see earlier). However, there is increasing evidence that individuals that are less involved in providing care (typically males) can have indirect impacts on the parenting behaviour of the main caregiver. Rodent fathers, for example, can have transgenerational, epigenetic effects on offspring phenotype (e.g. response to stress) even when they do not provide any care, via altered DNA methylation patterns in sperm (Champagne 2016). Furthermore, in mice, *M. musculus*, females that mated with males that had been exposed to chronic social stress showed reduced parenting behaviour towards offspring, which had reduced growth rates compared to the offspring of males that had experienced environmental enrichment (Mashoodh et al. 2012). In addition, paternally expressed (imprinted) genes can also affect levels of maternal care (Champagne 2016). Such changes in gene expression because of environmental experience may be an important source of (epigenetic) variation on which selection can act (Uller 2012), and illustrate how closely genes and environment are intertwined in parental care (Champagne 2016).

7.4 Summary and Conclusion

Traits that contribute to the category of behaviour we call 'parenting' are often genetically influenced. There are two consequences of this; first, parenting can evolve if there is a genetic basis and second, the proximate causes of variation are open to investigation and manipulation. For example, evidence from behavioural (Hopwood et al. 2015), quantitative genetic (Walling et al. 2008), transcriptomic (Parker et al. 2015) and molecular genetic studies shows that in species with biparental care, male behaviours may be superficially like female care behaviours but have different motivations, genetic architecture and neuronal circuitry. Consideration of genetics allows an explicit assessment of what parental care is composed of and whether the contributing behaviours that are similar from a phenotypic perspective are homologous across species or between the sexes within species. Does provisioning by a dung beetle involve the same genetic underpinnings as provisioning by a red-winged blackbird, for example? This improves our understanding of the origin, maintenance and evolution of parenting and its role in development and health. In addition, there is strengthening evidence for GxFE and coadaptation of traits involved in parental care. From these studies a picture is emerging that shows that parental and offspring traits cannot be considered in isolation from one another and that parental care traits are shaped by both parents and offspring.

There is still a large amount that we do not know and the data currently available are very taxonomically restricted (molecular studies are almost exclusively on rodents and humans, although with an increasing number on insects) and focused on a limited number of behaviours (e.g. licking of offspring and nest building in rodents) with unknown importance in terms of fitness. In addition, in species where we know a lot about the likely fitness consequences associated with variation in parental investment, we usually know little about the genetics and vice versa (burying beetles are a notable exception). Future studies would benefit from taking the multifaceted molecular genetics approach employed by Bendesky et al. (2017) but set within a coadaptation framework (i.e. consideration of both parent and offspring behaviours within populations) using traits that have a clear link to fitness variation. This is a tall order perhaps but one worth nurturing.

References

Adams, M.J., Robinson, M.R., Mannarelli, M.E., and Hatchwell, B.J. (2015). Social genetic and social environment effects on parental and helper care in a cooperatively breeding bird. *Proceedings of the Royal Society B: Biological Sciences* 282: 20150689.

Agrawal, A.F., Brodie, E.D. III, and Brown, J. (2001). Parent-offspring coadaptation and the dual genetic control of maternal care. *Science* 292: 1710–1712.

Ament, S.A., Velarde, R.A., Kolodkin, M.H. et al. (2011). Neuropeptide Y-like signaling and nutritionally mediated gene expression and behaviour in the honey bee. *Insect Molecular Biology* 20: 335–345.

Arnold, S.J. (1994). Multivariate inheritance and evolution: a review of concepts. In: *Quantitative Studies of Behavioral Evolution* (ed. C.R.B. Boake). Chicago: Chicago University Press.

Ashbrook, D.G., Gini, B., and Hager, R. (2015). Genetic variation in offspring indirectly influences the quality of maternal behaviour in mice. *eLife* 4: e11814.

Bakermans-Kranenburg, M.J. and Ijzendoorn, M.H. (2008). Oxytocin receptor (OXTR) and serotonin transporter (5-HTT) genes associated with observed parenting. *Social Cognitive and Affective Neuroscience* 3: 128–134.

Barr, C.S., Schwandt, M.L., Lindell, S.G. et al. (2008). Variation at the mu-opioid receptor gene (OPRM1) influences attachment behaviour in infant primates. *Proceedings of the National Academy of Sciences USA* 105: 5277–5288.

Barrett, C.E., Modi, M.E., Zhang, B.C. et al. (2013). Neonatal melanocortin receptor agonist treatment reduces play fighting and promotes adult attachment in prairie voles in a sex-dependent manner. *Neuropsychopharmacology* 85: 357–366.

Beletsky, L.D. and Orians, G.H. (1997). *Red-Winged Blackbirds: Decision-Making and Reproductive Success*. Chicago: Chicago University Press.

Bendesky, A., Kwon, Y.-M., Lassance, J.-M. et al. (2017). The genetic basis of parental care evolution in monogamous mice. *Nature* 544: 434–439.

Benowitz, K.M., McKinney, E.C., Cunningham, C.B., and Moore, A.J. (2017). Relating quantitative variation within a behaviour to variation in transcription. *Evolution* 71: 1999–2009.

Bult, A., van der Zee, E.A., Compaan, J.C., and Lynch, C.B. (1992). Differences in the number of arginine-vasopressin-immunoreactive neurons exist in the suprachiasmatic nuclei of house mice selected for differences in nest-building behavior. *Brain Research* 578: 335–338.

Burton, T. and Metcalfe, N.B. (2014). Can environmental conditions experienced in early life influence future generations? *Proceedings of the Royal Society B: Biological Sciences* 281: 20140311.

Caldji, C., Diorio, J., Anisman, H., and Meaney, M.J. (2004). Maternal behavior regulates benzodiazepine/GABA(A) receptor subunit expression in brain regions associated with fear in BALB/c and C57BL/6 mice. *Neuropsychopharmacology* 29: 1344–1352.

Calisi, R.M. and MacManes, M.D. (2015). RNAseq-ing a more integrative understanding of animal behavior. *Current Opinion in Behavioral Science* 6: 65–68.

Champagne, F.A. (2016). Epigenetic legacy of parental experiences: dynamic and interactive pathways to inheritance. *Development and Psychopathology* 28: 1219–1228.

Cheverud, J.M. and Moore, A.J. (1994). Quantitative genetics and the role of the environment provided by relatives in behavioral evolution. In: *Quantitative Studies of Behavioral Evolution* (ed. C.R.B. Boake). Chicago: Chicago University Press.

Clutton-Brock, T.H. (1991). *The Evolution of Parental Care*. Princeton: Princeton University Press.

Cunningham, C.B., Douthit, M.K., and Moore, A.J. (2014). Octopaminergic gene expression and flexible social behaviour in the subsocial beetle, *Nicrophorus vespilloides*. *Insect Molecular Biology* 23: 391–404.

Cunningham, C.B., Ji, L., Wiberg, R.A.W. et al. (2015). The genome and methylome of a beetle with complex social behavior, *Nicrophorus vespilloides* (Coleoptera: Silphidae). *Genome Biology and Evolution* 7: 3383–3396.

Cunningham, C.B., VanDenHeuvel, K., Khana, D.B. et al. (2016). The role of neuropeptide F in a transition to parental care. *Biology Letters* 12: 20160158.

Cunningham, C.B., Badgett, M., Meagher, R.B. et al. (2017). Ethological principles predict the neuropeptides co-opted to influence parenting. *Nature Communications* 8: 14225.

Curley, J.P., Barton, S., Surani, A., and Keverne, E.B. (2004). Coadaptation in mother and infant regulated by a paternally expressed imprinted gene. *Proceedings of the Royal Society B: Biological Sciences* 271: 1303–1309.

Darwin, L. (1927). Nature and nurture in Shakespeare's plays and elsewhere. *Eugenics Review* 19: 181–191.

Dobler, R. and Kölliker, M. (2009). Behavioural attainability of evolutionarily stable strategies in repeated interactions. *Animal Behaviour* 77: 1427–1434.

Dor, R. and Lotem, A. (2009). Heritability of nestling begging intensity in the house sparrow (*Passer domesticus*). *Evolution* 63: 738–748.

Dor, R. and Lotem, A. (2010). Parental effort and response to nestling begging in the house sparrow: repeatability, heritability and parent-offspring co-evolution. *Journal of Evolutionary Biology* 23: 1605–1612.

Dulac, C., O'Connell, L.A., and Wu, Z. (2014). Neural control of maternal and paternal behaviors. *Science* 345: 765–770.

Elwood, R.W. (1994). A switch in time saves mine. *Behavioral Processes* 33: 15–24.

Engel, K.C., Stöki, J., Schweizer, R. et al. (2016). A hormone-related female anti-aphrodisiac signals temporary infertility and causes sexual abstinence to synchronize parental care. *Nature Communications* 7: 11035.

Estramil, N., Eens, M., and Müller, W. (2014). On the coadaptation of offspring begging and parental supply-a within-individual approach across life stages. *Behavioral Ecology and Sociobiology* 68: 1481–1491.

Estramil, N., Groothuis, T.G.G., Eens, M. et al. (2017). Coadaptation of offspring begging and parental provisioning: a role for prenatal maternal effects? *Hormones and Behavior* 87: 129–136.

Forbes, S. (2010). Family structure and variation in reproductive success in blackbirds. *Behavioral Ecology and Sociobiology* 64: 475–483.

Forbes, L.S., Thornton, S., Glassey, B. et al. (1997). Why parent birds play favourites. *Nature* 390: 351–352.

Fortuna, K., van Ijzendoorn, M.H., Mankuta, D. et al. (2011). Differential genetic susceptibility to child risk at birth in predicting observed maternal behavior. *PLoS One* 6: e19765.

Freeman-Gallant, C.R. and Rothstein, M.D. (1999). Apparent heritability of parental care in savannah sparrows. *The Auk* 116: 1132–1136.

Galton, F. (1869). *Hereditary Genius – An Inquiry into its Laws and Consequences*. London: Macmillan.

Garrison, J.L., Macosko, E.Z., Bernstein, S. et al. (2012). Oxytocin/vasopressin-related peptides have an ancient role in reproductive behavior. *Science* 338: 540–543.

Ghalambor, G.K., Hoke, K.L., Ruell, E.W. et al. (2015). Non-adaptive plasticity potentiates rapid adaptive evolution of gene expression in nature. *Nature* 525: 372–375.

Gomez, Y. and Kölliker, M. (2013). Maternal care, mother-offspring aggregation and age-dependent coadaptation in the European earwig. *Journal of Evolutionary Biology* 26: 1903–1911.

Gomez-Mestre, I. and Jovani, R. (2013). A heuristic model on the role of plasticity in adaptive evolution: plasticity increases adaptation, population viability and genetic variation. *Proceedings of the Royal Society B: Biological Sciences* 280: 20131869.

Hager, R. and Johnstone, R.A. (2003). The genetic basis of family conflict resolution in mice. *Nature* 421: 533–535.

Hall, M.E., Blount, J.D., Forbes, S., and Royle, N.J. (2010). Does oxidative stress mediate the trade-off between growth and self-maintenance in structured families? *Functional Ecology* 24: 365–373.

Hammock, E.A. and Young, L.J. (2005). Microsatellite instability generates diversity in brain and socio-behavioral traits. *Science* 308: 1630–1634.

Harrison, F., Barta, Z., Cuthill, I., and Székely, T. (2009). How is sexual conflict over parental care resolved? A meta-analysis. *Journal of Evolutionary Biology* 22: 1800–1812.

Hayward, A.D., Rickard, I.J., and Lummaa, V. (2013). Influence of early-life nutrition on mortality and reproductive success during a subsequent famine in a preindustrial population. *Proceedings of the National Academy of Sciences USA* 110: 13886–13891.

Head, M.L., Berry, L.K., Royle, N.J., and Moore, A.J. (2012). Paternal care: direct and indirect genetic effects of fathers on offspring performance. *Evolution* 66: 3570–3581.

Head, M.L., Hinde, C.A., Moore, A.J., and Royle, N.J. (2014). Correlated evolution in parental care in females but not males in response to selection on paternity assurance behaviour. *Ecology Letters* 17: 803–810.

Higham, J.P., Barr, C.S., Hoffman, C.L. et al. (2011). Mu-opioid receptor (OPRM1) variation, oxytocin levels and maternal attachment in free-ranging rhesus macaques *Macaca mulatta*. *Behavioral Neuroscience* 125: 131–136.

Hinde, C.A., Buchanan, K.L., and Kilner, R.M. (2009). Prenatal environmental effects match offspring begging to parental provisioning. *Proceedings of the Royal Society B: Biological Sciences* 276: 2787–2794.

Hinde, C.A., Johnstone, R.A., and Kilner, R.M. (2010). Parent-offspring conflict and coadaptation. *Science* 327: 1373–1376.

Hopwood, P.E., Moore, A.J., Tregenza, T., and Royle, N.J. (2015). Male burying beetles extend, not reduce, parental care duration when reproductive competition is high. *Journal of Evolutionary Biology* 28: 1394–1402.

Ibarra, P., Alemany, S., Fatjo-Vilas, M. et al. (2014). The BDNF-Val66Met polymorphism modulates parental rearing effects on adult psychiatric symptoms: a community twin-based study. *European Psychiatry* 29: 293–300.

Insel, T.R. and Young, L.J. (2001). The neurobiology of attachment. *Nature Reviews Neuroscience* 2: 129–132.

Kasper, C., Kölliker, M., Postma, E., and Taborsky, B. (2017). Consistent cooperation in a cichlid fish is caused by maternal and developmental effects rather than heritable genetic variation. *Proceedings of the Royal Society B: Biological Sciences* 284: 20170369.

Kelley, R.A., Castelli, F.R., Mabry, K.E., and Solomon, N.G. (2013). Effects of experience and avpr1a microsatellite length on parental care in male prairie voles (*Microtus ochrogaster*). *Behavioral Ecology and Sociobiology* 67: 985–992.

Kendler, K.S. and Baker, J.H. (2007). Genetic influences on measures of the environment: a systematic review. *Psychological Medicine* 37: 615–626.

Kilner, R.M. and Hinde, C.A. (2012). Parent-offspring conflict. In: *The Evolution of Parental Care* (ed. N.J. Royle, P.T. Smiseth and M. Kölliker). Oxford: Oxford University Press.

Kilner, R.M., Boncoraglio, G., Henshaw, J.M. et al. (2015). Parental effects alter the adaptive value of an adult behavioural trait. *eLife* 4: e07340.

Koellinger, P.D. and Harden, K.P. (2018). Using nature to understand nurture. *Science* 359: 386–387.

Kokko, H. and Jennions, M. (2012). Sex differences in parental care. In: *The Evolution of Parental Care* (ed. N.J. Royle, P.T. Smiseth and M. Kölliker). Oxford: Oxford University Press.

Kölliker, M. (2005). Ontogeny in the family. *Behavior Genetics* 35: 7–18.

Kölliker, M., Brinkhof, M.W.G., Heeb, P. et al. (2000). The quantitative genetic basis of offspring solicitation and parental response in a passerine bird with biparental care. *Proceedings of the Royal Society B: Biological Sciences* 267: 2127–2132.

Kölliker, M., Brodie, E.D. III, and Moore, A.J. (2005). The coadaptation of parental supply and offspring demand. *American Naturalist* 166: 506–516.

Kölliker, M., Royle, N.J., and Smiseth, P.T. (2012). Parent-offspring co-adaptation. In: *The Evolution of Parental Care* (ed. N.J. Royle, P.T. Smiseth and M. Kölliker). Oxford: Oxford University Press.

Kong, A., Thorleifsson, G., Frigge, M.L. et al. (2018). The effect of nurture: effects of parental genotypes. *Science* 359: 424–428.

Lewontin, R.C. (1993). *The Doctrine of DNA*. London: Penguin.

Lim, M.M. and Young, L.J. (2004). Vasopressin-dependent neural circuits underlying pair bond formation in the monogamous prairie vole. *Neuroscience* 125: 35–45.

Linksvayer, T.A. (2007). Ant species differences determined by epistasis between brood and worker genomes. *PLoS One* 2: e994.

Linksvayer, T.A. and Wade, M.J. (2005). The evolutionary origin and elaboration of sociality in the aculeate hymenoptera: maternal effects, sib-social effects and heterochrony. *Quarterly Review of Biology* 80: 317–336.

Lock, J.E., Smiseth, P.T., and Moore, A.J. (2004). Selection, inheritance and the evolution of parent-offspring interactions. *American Naturalist* 164: 13–24.

Lock, J.E., Smiseth, P.T., Moore, P.J., and Moore, A.J. (2007). Coadaptation of prenatal and postnatal maternal effects. *American Naturalist* 170: 709–718.

Lucass, C., Fresneau, N., Eens, M., and Müller, W. (2016a). Sex roles in nest keeping – how information asymmetry contributes to parent-offspring co-adaptation. *Ecology and Evolution* 6: 1825–1833.

Lucass, C., Stowe, M., Eens, M., and Müller, W. (2016b). Favored parent-offspring trait combinations? On the interplay of parental and offspring traits. *Behavioral Ecology* 27: 134–140.

Lucass, C., Korsten, P., Eens, M., and Müller, W. (2016c). Within-family parent-offspring co-adaptation in a wild bird: on static traits, behavioural reaction norms, and sex differences. *Functional Ecology* 30: 274–282.

Lynch, C.B. (1980). Response to divergent selection for nesting behavior in *Mus musculus*. *Genetics* 96: 757–765.

Lynch, M. (1987). Evolution of intrafamilial interactions. *Proceedings of the National Academy of Sciences USA* 84: 8507–8511.

Lynch, C.B. and Hegmann, J.P. (1972). Genetic differences influencing behavioral temperature regulation in small mammals. I. Nesting by *Mus musculus*. *Behavior Genetics* 2: 43–53.

Lynch, C.B. and Hegmann, J.P. (1973). Genetic differences influencing behavioral temperature regulation in small mammals. II. Genotype-environment interactions. *Behavior Genetics* 3: 145–154.

Maestripieri, D. (2004). Genetic aspects of mother-offspring conflict in rhesus macaques. *Behavioral Ecology and Sociobiology* 55: 381–387.

Marshall, E. (2014). An experiment in zero parenting. *Science* 345: 752–754.

Marshall, P.J., Fox, N.A., and the BEIP Core Group (2004). A comparison of the electroencephalogram between institutionalized and community children in Romania. *Journal of Cognitive Neuroscience* 16: 1327–1338.

Mascaro, J.S., Hackett, P.D., Gouzoules, H. et al. (2014). Behavioral and genetic correlates of the neural response to infant crying among human fathers. *Social Cognitive and Affective Neuroscience* 9: 1704–1712.

Mashoodh, R., Franks, B., Curley, J.P., and Champagne, F.A. (2012). Paternal social enrichment effects on maternal behavior and offspring growth. *Proceedings of the National Academy of Sciences USA* 109: 17232–17238.

McGhee, K.E., Pintor, L.M., Suhr, E.L., and Bell, A.M. (2012). Maternal exposure to predation risk decreases offspring antipredator behavior and survival in threespined stickleback. *Functional Ecology* 26 (4): 932–940.

Mennella, J.A. and Moltz, H. (1988). Infanticide in rats: male strategy and female counter-strategy. *Physiology Behavior* 42: 19–28.

Micali, N., Crous-Bou, M., Treasure, J., and Lawson, E.A. (2017). Association between oxytocin receptor genotype, maternal care, and eating disorder behaviours in a community sample of women. *European Eating Disorders Review* 25: 19–25.

Mileva-Seitz, V., Kennedy, J.L., Atkinson, L. et al. (2011). Serotonin transporter allelic variation in mothers predicts maternal sensitivity and attitudes towards 6-month-old infants. *Genes, Brain and Behavior* 10: 325–333.

Mileva-Seitz, V., Steiner, M., Atkinson, L. et al. (2013). Interaction between oxytocin genotypes and early experience predicts quality of mothering and postpartum mood. *PLoS One* 8: e61443.

Mock, D.W. and Parker, G.A. (1997). *The Evolution of Sibling Rivalry*. Oxford: Oxford University Press.

Moltz, H., Lubin, M., Leon, M., and Numan, M. (1970). Hormonal induction of maternal behaviour in the ovariectomized nulliparous rat. *Physiology Behavior* 5: 1373–1377.

Monaghan, P. (2008). Early growth conditions, phenotypic development and environmental change. *Philosophical Transactions of the Royal Society B: Biological Sciences* 363: 1635–1645.

Moore, A.J., Brodie, E.D. III, and Wolf, J.B. (1997). Interacting phenotypes and the evolutionary process: I. direct and indirect genetic effects of social interactions. *Evolution* 51: 1352–1362.

Mousseau, T.A. and Fox, C.W. (1998). *Maternal Effects as Adaptations*. New York: Oxford University Press.

Nakagawa, S., Gillespie, D.O.S., Hatchwell, B.J., and Burke, T. (2007). Predictable males and unpredictable females: sex difference in repeatability of parental care in a wild bird population. *Journal of Evolutionary Biology* 20: 1674–1681.

Nelson, C.A., Furtado, E.A., Fox, N.A., and Zeanah, C.H. (2009). The deprived human brain. *American Scientist* 97: 222–229.

Oldekop, J.A., Smiseth, P.T., Piggins, H.D., and Moore, A.J. (2007). Adaptive switch from infancticide to parental care: How do beetles time their behaviour? *Journal of Evolutionary Biology* 20: 1998–2004.

Oliver, B.R., Trzaskowski, M., and Plomin, R. (2014). Genetics of parenting: the power of the dark side. *Developmental Psychology* 50: 1233–1240.

Parker, G.A., Royle, N.J., and Hartley, I.R. (2002). Intrafamilial conflict and parental investment: a synthesis. *Philosophical Transactions of the Royal Society B: Biological Sciences* 357: 295–307.

Parker, D.J., Cunningham, C.B., Walling, C.A. et al. (2015). Transcriptomes of parents identify parenting strategies and sexual conflict in a subsocial beetle. *Nature Communications* 6: 8449.

Peripato, A.C. and Cheverud, J.M. (2002). Genetic influences on maternal care. *American Naturalist* 160: s173–s185.

Peripato, A.C., de Brito, R.A., Vaughn, T.T. et al. (2002). Quantitative trait loci for maternal performance for offspring survival in mice. *Genetics* 162: 1341–1353.

Rehan, S.M. and Toth, A.L. (2015). Climbing the social ladder: the molecular evolution of sociality. *Trends in Ecology & Evolution* 30: 426–433.

Reinhold, K. (2002). Maternal effects and the evolution of behavioral and morphological characters: a literature review indicates the importance of extended maternal care. *Journal of Heredity* 93: 400–405.

Rilling, J.K. (2013). The neural and hormonal bases of human parental care. *Neuropsychologia* 51: 731–747.

Rilling, J.K. and Young, L.J. (2014). The biology of mammalian parenting and its effect on offspring social development. *Science* 345: 771–776.

Riska, B., Rutledge, J.J., and Atchley, W.R. (1985). Covariance between direct and maternal genetic effects with a model of persistent environmental influences. *Genetical Research* 45: 287–297.

Rosenblatt, J.S. (1967). Nonhormonal basis of maternal behavior in the rat. *Science* 156: 1512–1514.

Roulin, A. and Dreiss, A.N. (2012). Sibling competition and cooperation over parental care. In: *The Evolution of Parental Care* (ed. N.J. Royle, P.T. Smiseth and M. Kölliker). Oxford: Oxford University Press.

Royle, N.J. (2016). Parental care: when the sex has to stop. *Current Biology* 26: R478–R480.

Royle, N.J., Hartley, I.R., and Parker, G.A. (2002). Begging for control: when are offspring solicitation behaviours honest? *Trends in Ecology & Evolution* 17: 434–440.

Royle, N.J., Smiseth, P.T., and Kölliker, M. (2012a). *The Evolution of Parental Care*. Oxford: Oxford University Press.

Royle, N.J., Smiseth, P.T., and Kölliker, M. (2012b). The evolution of parental care: summary, conclusions and implications. In: *The Evolution of Parental Care* (ed. N.J. Royle, P.T. Smiseth and M. Kölliker). Oxford: Oxford University Press.

Royle, N.J., Russell, A.F., and Wilson, A.J. (2014). The evolution of flexible parenting. *Science* 345: 776–781.

Royle, N.J., Alonzo, S.H., and Moore, A.J. (2016). Co-evolution, conflict and complexity: what have we learned about the evolution of parental care behaviors? *Current Opinion in Behavioral Sciences* 12: 30–36.

Roy-Zokan, E.M., Cunningham, C.B., Hebb, L.E. et al. (2015). Vitellogenin and vitellogenin receptor gene expression is associated with male and female parenting in a subsocial insect. *Proceedings of the Royal Society B: Biological Sciences* 282: 20150787.

Sear, R. (2015). Beyond the nuclear family: an evolutionary perspective on parenting. *Current Opinion in Psychology* 7: 98–103.

Searcy, W.A. and Yasukawa, Y. (1995). *Polygyny and Sexual Selection in Red-Winged Blackbirds*. Princeton: Princeton University Press.

Smiseth, P.T., Dawson, C., Varley, E., and Moore, A.J. (2005). How do caring parents respond to mate loss? Differential response by males and females. *Animal Behaviour* 69: 551–559.

Smiseth, P.T., Wright, J., and Kölliker, M. (2008). Parent-offspring conflict and co-adaptation: behavioural ecology meets quantitative genetics. *Proceedings of the Royal Society B: Biological Sciences* 275: 1823–1830.

Smiseth, P.T., Kölliker, M., and Royle, N.J. (2012). What is parental care? In: *The Evolution of Parental Care* (ed. N.J. Royle, P.T. Smiseth and M. Kölliker). Oxford: Oxford University Press.

Smyke, A.T., Zeanah, C.H., Fox, N.A. et al. (2010). Placement in foster care enhances quality of attachment among young institutionalized children. *Child Development* 81: 212–223.

Suzuki, A., Matsumoto, Y., Shibuya, N. et al. (2011). The brain-derived neurotrophic factor Val66Met polymorphism modulates the effects of parental rearing on personality in healthy subjects. *Genes, Brain and Behavior* 10: 385–391.

Suzuki, A., Matsumoto, Y., Shibuya, N. et al. (2012). Interaction effect between the BDNF Val66Met polymorphism and parental rearing for interpersonal sensitivity in healthy subjects. *Psychiatry Research* 200: 945–948.

Tabbaa, M., Lei, K., Liu, Y., and Wang, Z.X. (2017). Paternal deprivation affects social behaviors and neurochemical systems in the offspring of socially monogamous prairie voles. *Neuroscience* 343: 284–297.

Takata, M., Doi, H., Thomas, C.E., and Koyama, S. (2016). Control of parental investment changes plastically over time with residual reproductive value. *Journal of Evolutionary Biology* 29: 199–204.

Tombeau Cost, K., Unternaehrer, E., Plamondon, A. et al. (2017). Thinking and doing: the effects of dopamine and oxytocin genes and executive function on mothering behaviours. *Genes, Brain and Behavior* 16,: 285–295.

Toth, A.L., Varala, K., Newman, T.C. et al. (2007). Wasp gene expression supports an evolutionary link between maternal behavior and eusociality. *Science* 318: 441–444.

Toth, A.L., Varala, K., Henshaw, M.T. et al. (2010). Brain transcriptomic analysis in paper wasps identifies genes associated with behaviour across social lineages. *Proceedings of the Royal Society B: Biological Sciences* 277: 2139–2148.

Tottenham, N., Shapiro, M., Telzer, E.H., and Humphreys, K.L. (2012). Amygdala response to mother. *Developmental Science* 15: 307–319.

Trivers, R.L. (1972). Parental investment and sexual selection. In: *Sexual Selection and the Descent of Man* (ed. B.G. Campbell). Chicago: Aldine.

Trivers, R.L. (1974). Parent-offspring conflict. *American Zoologist* 14: 249–264.

Uller, T. (2012). Parental effects in development and evolution. In: *The Evolution of Parental Care* (ed. N.J. Royle, P.T. Smiseth and M. Kölliker). Oxford: Oxford University Press.

Vanderwert, R.E., Marshall, P.J., Nelson, C.A. et al. (2010). Timing of intervention affects brain electrical activity in children exposed to severe psychosocial neglect. *PLoS One* 5: 1–5.

Vincze, O., Székely, T., Kupper, C. et al. (2013). Local environment but not genetic differentiation influences biparental care in ten plover populations. *PLoS One* 8: e60998.

Vom Saal, F.S. (1985). Time-contingent change in infanticide and parental behavior induced by ejaculation in male mice. *Physiology Behavior* 34: 7–15.

Walling, C.A., Stamper, C.E., Smiseth, P.T., and Moore, A.J. (2008). Genetic architecture of sex differences in parental care. *Proceedings of the National Academy of Sciences USA* 105: 18430–18435.

West-Eberhard, M.J. (1996). Wasp societies as microcosms for the study of development and evolution. In: *Natural History and Evolution of Paper-Wasps* (ed. S. Turilazzi and M.J. West-Eberhard), 290–317. Oxford: Oxford University Press.

Westneat, D.F., Hatch, M.I., Wetzel, D.P., and Ensminger, A.L. (2011). Individual variation in parental care reaction norms: integration of personality and plasticity. *American Naturalist* 178: 652–667.

Wilson, A.J. (2014). Competition as a source of constraint on life history evolution in natural populations. *Heredity* 112: 70–78.

Wilson, A.J., Réale, D., Clements, M.N. et al. (2010). An ecologist's guide to the animal model. *Journal of Animal Ecology* 79: 13–26.

Wolf, J.B., Brodie, E.D. III, Cheverud, J.M. et al. (1998). Evolutionary consequences of indirect genetic effects. *Trends in Ecology & Evolution* 13: 64–69.

Wu, Z., Autry, A.E., Bergan, J.F. et al. (2014). Galanin neurons in the medial preoptic area govern parental behaviour. *Nature* 509: 325–330.

Yu, P., An, S., Tai, F. et al. (2013). Early social deprivation impairs pair bonding and alters serum corticosterone and the NAcc dopamine system in mandarin voles. *Psychoneuroendocrinology* 38: 3128–3138.

8

The Effect of Non-Self Genes on the Behaviour of Hosts

Nina Wedell

Centre for Ecology & Conservation, University of Exeter, Penryn Campus, Penryn, TR10 9FE, UK

Classic debates about nature-nurture have only really focused on self-genes, but it is increasingly apparent that 'non-self genes' (NSGs) can significantly influence the behaviour of 'hosts' (in some sense not dissimilar from social environmental effects: see Chapters 4 and 7). This chapter focuses the discussion on genomic parasites and symbionts, including microbes such as the gut microbiota and their impacts beyond 'normal' patterns of pathology. The primary goal is to describe how NSGs can influence the behaviour of their hosts in unexpected ways, both directly and indirectly. This is a rapidly expanding field that is still in its infancy, so while the discussion illustrates the breadth of the impact of NSGs on a variety of behaviours, it is not possible to cover all the exciting recent research spanning this large topic. Instead, a number of systems are highlighted that have been studied in greater depth, including, for example, what has been learned about the population level effects, both positive and negative, of the intracellular bacterium *Wolbachia* on the reproductive biology of its insect hosts. I will also highlight several outstanding examples of how endosymbionts, viruses, and various mobile genetic elements can alter host behaviour, often by targeting the nervous system (Moore 2002; Poulin 2011; Lefevre et al. 2009; Adamo 2012; Hughes et al. 2012; Eisthen and Theis 2015; Barber et al. 2017). In some cases, the underlying functional mechanisms of host manipulation and/or host influence are well understood (and I will touch on these), but in the majority of cases, the functional connections that give rise to NSG/host interactions are still not characterized.

This chapter focus primarily on invertebrates, but will include a few vertebrate examples to illustrate the general impact of NSGs on animal behaviour. NSGs, including microbes that have pathogenic effects on animal behaviour, have been reviewed extensively elsewhere (Moore 2002; Hughes et al. 2012; Barber et al. 2017), and will not be considered further in any detail. Instead, the focus is on microbes, viruses, and selfish genes that reside within the cells and genomes of animals that have been shown to alter the behaviour of the host either mutualistically or in a way that is not associated with 'disease pathology' that adversely affects health (i.e. 'sickness behaviour'; Hart 1988). The interactions and behaviour between microbes themselves will also not be covered.

Genes and Behaviour: Beyond Nature-Nurture, First Edition.
Edited by David J. Hosken, John Hunt and Nina Wedell.
© 2019 John Wiley & Sons Ltd. Published 2019 by John Wiley & Sons Ltd.

8.1   What are Non-self Genes (NSGs)?

I define 'genes within', or NSGs, as genes that do not originate from the host's own genome. This definition is not clear-cut since many mobile genetic elements with bacterial, viral, plasmid, or sometimes even eukaryotic origin become incorporated into the host genome and can make up a large part of it, especially in invertebrates (Drezen et al. 2017; Dunning Hotopp et al. 2007). For example, in bdelloid rotifers ~8–9% of the genome derives from horizontal gene transfer from other organisms (Boschetti et al. 2012). Most organisms harbour a variety of selfish genetic elements that manipulate the host's genome in ways to ensure that they are transmitted at a higher rate relative to the rest of the genome and therefore selfishly increase in frequency (Burt and Trivers 2006). Such selfish genetic elements include driving chromosomes (including sex-chromosomes that can give rise to sex-ratio distortion), homing endonucleases, transposable elements (TEs), and a variety of postsegregation distorters such as maternal effect lethals, and endosymbionts. While these selfish genetic elements can be suppressed, inactivated, and/or sometimes domesticated to take on a beneficial role for the host organism, they persist because of their selfish vertical transmission. In invertebrates such as aphids (Douglas 1998), endosymbiotic bacteria are commonly found inside the host's cells or residing in specialized structures. They can also sometimes be found in specialized structures of the gut, although the majority of gut bacteria are free living (Dillon and Dillon 2004).

Endosymbionts housed in specific organs or organelles are frequently vertically transmitted from mother to offspring, and hence share part of their evolutionary history with their hosts. This association can lead to cospeciation between hosts and their mutualistic endosymbionts (Moran et al. 1993; Bandi et al. 1995). In contrast, the evolutionary history of vertically transmitted parasitic endosymbionts and viruses is characterized by horizontal host switching, potentially promoting their long-term persistence (O'Neill et al. 1992; Weinert et al. 2009; Longdon et al. 2011). Some organisms become entirely reliant on their endosymbionts for successful reproduction. Females of the parasitic wasp *Asobara tabida* are unable to develop ovaries in the absence of their obligate endosymbiotic *Wolbachia* bacteria (Dedeine et al. 2001).

We expect the evolutionary dynamics of NSG on behaviour to be qualitatively different depending upon the mode of inheritance of symbionts, with vertically transmitted NSGs exerting stronger impacts on host behaviour than horizontally transmitted NSGs. While to date there has been little systematic examination of this possibility, a recent comparative study of >100 host–symbiont associations showed that host dependence is strongly associated with vertical transmission (Fisher et al. 2017). The potential impact of NSGs will be discussed more generally on host behaviour including both 'free' living, horizontally, and vertically transmitted NSG, although endosymbionts that are predominantly vertically transmitted can occasionally also be horizontally transmitted, so clear-cut distinctions can become blurred.

8.2   Indirect Effects of NSGs

Non-self genes (like other indirect genetic effects: see Chapters 4 and 7) can affect their host in a variety of ways. These effects can be indirect, whereby the action of a selfish gene or microbe has a knock-on effect on the subsequent behaviour of the host population, but one that is not due to the direct action of the NSG. For example, some populations of the butterfly *Hypolimnas bolina* harbour male-killing endosymbionts (*Wolbachia*) that cause population-level female-biased sex ratios. In high-prevalence populations, males are rare – 1 male per 100 females – but the frequency varies between populations (Dyson and Hurst 2004). Variation in male killer prevalence in turn affects female mating behaviour, with females evolving to be less choosy, and male ejaculate size is negatively correlated with the frequency of male killers (Charlat et al. 2007). In high-frequency populations, females run the risk of not obtaining sufficient sperm to fertilize all their eggs due to a shortage of males. This sperm shortage is exacerbated by males in these populations suffering ejaculate depletion due to high mating rate, further promoting increased female mating to obtain more sperm (Charlat et al. 2007). In African *Acraea encedon* butterflies, in populations harbouring male-killing endosymbionts, there is even evidence of sex role reversal with females adopting lekking behaviour to advertise their presence to the rare males and thereby increase their mating success (Jiggins et al. 2000). In contrast, in populations without male killers, lekking is only performed by males, illustrating the impact of sex ratio distorters in indirectly shaping female mating behaviour.

Similarly, sex ratio distorters in flies promote female mating strategies that affect male ejaculate evolution. Some populations of *Drosophila pseudoobscura* flies harbour a selfish gene (an X-linked meiotic driving chromosome) that kills Y-linked sperm (because they do not pass on the selfish gene) resulting in population-level, female-biased sex ratios (Price et al. 2014). Male flies that carry the sperm killer suffer reduced paternity in sperm competition due to low sperm number (Price et al. 2008a). This in turn favours multiple mating by females as a strategy that effectively biases paternity against sex ratio distorting males. Experiments have shown that females in populations that are at risk of sex ratio distorting males rapidly evolve increased remating frequency to promote sperm competition, demonstrating the potency of sex ratio distorters to alter female mating behaviour (Price et al. 2008b). As a consequence of increased female remating, male ejaculates evolve in response to the higher risk of sperm competition in these populations (Price et al. 2010). This male-female coevolution occurs even when the sex ratio distorter is present at low frequency (~5%; Price et al. 2010). These are examples where the indirect impact of NSGs (in this case sex ratio distorters) can shape the entire mating system of a population, even when present at low frequency. There are additional examples of such indirect effects of NSGs on sexual selection and mating systems (see Price and Wedell 2008; Wedell 2013).

8.3 Direct Effects of NSGs

Many NSGs directly affect host behaviour. They achieve this through a variety of sophisticated manipulations by targeting neurological pathways, including neural peptides and neurotransmitters (e.g. octopamine, serotonin), neurotransmitter receptors, or by invading specific brain regions that regulate behaviour. This is also the case for gut bacteria, and there is growing evidence that gut microbiota can communicate directly with the host nervous system (Cryan and Dinan 2012). For example, bacteria produce a variety of neurotransmitters and neuromodulators such as GABA, noradrenaline, and dopamine (Lyte 2011; Forsythe and Kunze 2013). Their impacts on behaviour are well documented in vertebrates where specific bacterial neurotransmitters can affect anxiety levels (Forsythe and Kunze 2013), and influence behaviours ranging from cognitive performance to sleep (Cryan and Clark 2016). Since neurotransmitter-producing bacteria are also common in insects (e.g. *Lactobacillus* and *Bifidobacterium* spp.), it seems likely that they influence the behaviour of their insect hosts in similar ways (Engel et al. 2012). As a consequence, it has been suggested that NSGs can even shape the architecture and development of the host's central nervous system (Rohrscheib and Brownlie 2013).

In insects and arthropods, the brain includes the antennal lobes, that receive input from the olfactory sensory neurons, and mushroom bodies, that play a central role in sensory learning and memory, and microbes have been identified that specifically target these brain regions (Temple and Richard 2015; Strunov et al. 2017). In vertebrates, viruses such as rabies, herpes, and measles enter neurons through cell surface receptors, and once inside the neuron, use synapses to spread from cell to cell (Mothes et al. 2010; Kramer and Enquist 2013). Similarly, the endosymbiont *Wolbachia* has been recorded in the central nervous system of flies, butterflies, mosquitoes, springtails, and terrestrial isopod hosts (Strunov et al. 2017). Gut microbiota can also directly affect the development and function of the nervous system by influencing neurogenesis, neurotransmitter signalling and neurodevelopment and thereby also influence the behaviour of animals (Diaz Heijtz et al. 2011; Ezenewa et al. 2012; Archie and Tung 2015; Voung et al. 2017).

8.4 Host Responses

Not surprisingly, hosts mount their own defences against manipulation by NSGs and have evolved a variety of counter-measures to reduce potential fitness costs. Some animals, including insects, engage in self-medication. Ingestion of specific nutrients can enhance host immunity (Mason et al. 2014), or reduce the cost of infection (Povey et al. 2014).

Prophylactic behaviours also occur at the group level in social animals, referred to as 'social immunity' (Cremer et al. 2007; Ezenwa et al. 2016; Schmid-Hempel 2017). Honeybees engage in group-level thermoregulation generating a 'social fever' against the fungus *Ascosphaera apis*, and behaviours such as grooming and disposal of infected nest-mates in social insect colonies can result in reduced pathogen transmission (Starks et al. 2000; Wong et al. 2015). There is even evidence that the host can make use of their endosymbionts to reduce the risk of additional infection. Fungus-growing ants employ a specific behaviour whereby they use antibiotics produced by actinomycetous bacteria housed in specialized structures (infrabuccal pockets) to kill spores of a virulent parasite (*Escovopsis*) attacking their fungal gardens (Little et al. 2006). There is also evidence that an initially pathogenic NSG can take on a beneficial role for the host if circumstances change and the interests of host and bacteria align.

Many insects, nematodes, and arachnids harbour maternally transmitted endosymbionts, with more than >50% of all insects infected with *Wolbachia* (Hilgenboecker et al. 2008; Weinert et al. 2015) and that is just one of many infective agents. However, since endosymbionts can only be passed on via females to eggs, the interest of females and the endosymbiont are likely to align over time. For example, in *Drosophila simulans*, the

Riverside strain of *Wolbachia* has rapidly spread in populations throughout California. Interestingly, this spread in frequency has been accompanied by a shift from parasitic to mutualistic associations with the host. From initially causing a 20–15% loss in fecundity to females, *Wolbachia* has become associated with a 10% fecundity advantage after only 20 years of host–microbe coevolution (Weeks et al. 2007). Hence there is evidence of rapid coevolution between the NSG and the host in this species.

Coevolutionary interactions between host and endosymbionts can result in beneficial effects to both the host and the NSG (Werren et al. 2008; Correa and Ballard 2016). A variety of mutualistic endosymbiotic microbes aid their hosts by providing nutrients or defence against pathogens. Bumblebees and honeybees harbour distinct bacterial communities in their guts that are not shared with related solitary bee species. These microbiota protect bee hosts against a natural trypanosomatid gut parasite, and hence provide an additional benefit of group living to these social insects (Koch and Schmid-Hempel 2011).

It has been suggested that one overlooked benefit of group living and sociality is that this serves to facilitate the transmission of beneficial microbes (Lombardo 2008). Microbes that manipulate their hosts to act altruistically in certain situations may be favoured by selection and may therefore also play a role in the evolution of co-operation. Co-operation could favour the microbes as they can be transferred horizontally between hosts during social interactions. Altruistic behaviour could also be favoured by bacteria that are vertically transmitted between mother and offspring, as helping behaviour will increase host survival and reproduction and hence transmission of the microbes (Lewin-Epstein et al. 2017). However, to date this hypothesis has not been examined experimentally. The flip side of social interactions is that they can of course potentially lead to increased risk of disease transmission, and both processes (transmission of pathogenic and protective microbes) are likely to have shaped social behaviour of hosts.

Micro-organisms were the first life forms on our planet and therefore have a long history of associating with later emerging multicellular life forms. Thus the association with microbes (e.g. bacteria, archaea, fungi, protozoa, viruses) is as old as multicellularity itself, providing plenty of opportunity for coevolution between microbes and host. Microbial and foreign genes can make up a large proportion of the host's body, with the number of bacteria being similar to the number of human cells (Sender et al. 2016). So it comes as no surprise that microbes have shaped host evolution, and may directly influence the nervous system of their hosts to alter their behaviour (Archie and Tung 2015; Eisthen and Theis 2015). Here I highlight some of the ways in which NSGs can affect specific host behaviours. It is worth noting that it is not always clear whether the NSG or the host has the upper hand.

8.5 Odour is a Key Signal

The way an individual smells can have a dramatic impact on its social interactions and thus its social environment. The microbiota can directly affect the chemical signals used in social communication and therefore influence a variety of the host's social behaviour, including sexual signalling (Natsch 2017; Wong et al. 2015). Odour-based signals are also key components in most animal communication. Odour is used as a cue to signal individual identity (e.g. species, sex, age, reproductive status), including relatedness, in both vertebrates and non-vertebrates (van Zweden and d'Ettorre 2010; Lize et al. 2014; Ingleby 2015; Lihoreau et al. 2016). Since these are traits known to shape mate preferences, odour also plays a large role in mate choice. Microbial products are used to signal the presence of food sources or pathogens that in turn affect foraging and avoidance behaviour, but the response can be modulated by NSGs.

Wolbachia infections are associated with changes in responses to olfactory cues (Peng and Wang 2009; Rohrscheib et al. 2015). In *Drosophila simulans* flies, the *w*Ri strain of *Wolbachia* increases the responsiveness of flies to food cues, whereas the *Wolbachia* strains *w*Mel and *w*MelPop in *D. melanogaster* instead result in a slight reduction in responsiveness to food (Peng et al. 2009).

There is a growing realization that microbes can also regulate behaviours between individuals in a social context, and that microbe-based chemical communication commonly occurs between species, as discussed below. Furthermore, these interactions are not restricted to communication between animals. Plants also release volatiles when being consumed by herbivorous insects that in turn can attract parasitoid wasps that attack the herbivores and thereby provide some protection to the plant (Moraes et al. 1998). Plant pathogens have taken this signalling one step further. The bacterial pathogen *Candidatu*s modifies the odours released by its citrus tree plant host to attract its vector, the psyllid *Diaphorina citri*, and thereby facilitate its own proliferation (Martini et al. 2014).

8.6 Kin Recognition

Bacteria can contribute to the costs and benefits of sociality by shaping olfactory cues used in social interactions such as kin recognition and mate choice. In humans, the bacterial composition of armpit odours functions as a reliable individual recognition cue as it shows stability over time and conveys distinctive odour profiles (Penn et al. 2007; Natsch et al. 2010). Bacterial armpit odours can reflect an individual's genotype and be used to establish kinship. For instance, humans are able to match the scent of monozygotic twins even if they do not live together (Roberts et al. 2005), and mothers can recognize odours of their offspring, but not their stepchildren (Weisfeld et al. 2003). A large and sometimes conflicting body of data suggests that humans can identify specific genotypes based on odour cues correlated with major histocompatibility complex (MHC) haplotype diversity (Havilek and Roverst 2009), but to what extent this is due to armpit odours as opposed to other body odours is not clear. Similar findings have also been shown in a range of mammals where gut microbiota shape the odour cues used in kin recognition (Archie and Tung 2015). Similarly, social insects use cuticular hydrocarbons odours to identify kin (van Zweden and d'Ettorre 2010). The same seems true also in other non-social insect species (Lihoreau et al. 2016), and the gut microbiota appear to play a role in shaping odour cues such as cuticular hydrocarbons in *Drosophila*, just as food eaten during development has been shown to later mask kin recognition cues in *D. melanogaster* (Lize et al. 2014). There is therefore scope for a complex interaction between diet, the gut microbiota, and resulting individual odours (also see below).

8.7 Mate Choice and Reproductive Behaviour

Bacterial communities can be internal (gut biota), but also present on the skin, cuticle, fur, and feathers. In birds, feather-degrading bacteria can affect plumage coloration and therefore influence mate choice. This has been shown in house finches where females prefer redder males that have fewer feather-degrading bacteria than dull males (Shawkey et al. 2009), whereas in bluebirds, males with experimentally elevated bacterial loads have brighter feathers that are more preferred by females (Gunderson et al. 2009). Currently, the precise mechanisms underlying these different results are unclear, but it is possible the former is a case of parasite-mediated sexual selection, with birds of higher quality having lower bacterial loads. In the latter case, brighter males may spend less time preening their feathers and socially dominant males may pay a health cost for dominance, and/or higher bacterial loads are in fact beneficial, but bright birds are better at acquiring these bacteria than their dull male counterparts (Archie and Theis 2011).

Internal gut microbiota can also affect odours of animals that directly regulate their sexual behaviours (Sharon et al. 2010; Damodaram et al. 2016). In *D. melanogaster* flies, the diet determines gut bacteria, and this in turn affects major cuticular hydrocarbons (Sharon et al. 2010) that are known to be subject to sexual selection and used in mate choice in insects (Ingleby 2015).

Evidence that endosymbionts such as *Wolbachia* are present in the central nervous system of certain insects and terrestrial isopod hosts suggests that they could act to affect mate preferences (Strunov et al. 2017). In *D. melanogaster*, the *Wolbachia* strain *w*Mel is found throughout the insect brain (Albertson et al. 2013), whereas in *D. paulistorum* a different *Wolbachia* strain (*w*Pau) is restricted to brain areas that are associated with processing olfactory and auditory information (Strunov et al. 2017). In the *D. paulistorum* species complex, different *Wolbachia* strains cause bidirectional reproductive incompatibility resulting in embryo mortality and hybrid male sterility between semi-species crosses. Intriguingly, in mate choice assays, female mate preferences are dependent on their own *Wolbachia* variant, with females preferring to mate with males that carry the same compatible *Wolbachia* strain as themselves. This mate preference disappears after partial depletion of *Wolbachia* (Miller et al. 2010) (Figure 8.1). It is therefore possible that the restriction of *Wolbachia* to brain areas involved in processing cues relating to sexual behaviour may have evolved to reduce any fitness costs of unrestricted *Wolbachia* presence in the brain (Strunov et al. 2017).

The difference in localization of *Wolbachia* in brain tissue between fly species has been proposed to be the outcome of the age of the association and therefore the potential for coevolution between the host and the endosymbiont. The *Wolbachia* strain *w*Pau is considered an obligate mutualist that has had a long association with its host and clearing of *Wolbachia* in *D. paulistorum* results in lethality (Miller et al. 2010), whereas in *D. melanogaster* the *w*Mel strain has recently replaced the more ancestral *Wolbachia* strain *w*MelCS globally (Riegler et al. 2005).

In other *Drosophila* species, findings are mixed for the role of *Wolbachia*-mediated mate preferences. In *D. melanogaster* flies, *Wolbachia* has been shown to have no influence on mate choice in some studies (Champion de Crespigny and Wedell 2007), whereas other studies have shown that it can influence mate choice in some populations but not others (Markov et al. 2009, Arburthnott et al. 2016). This difference indicates the existence of a genotype by *Wolbachia* interaction affecting mate preference, and

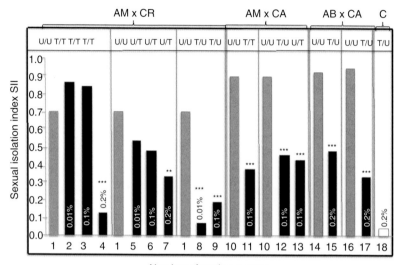

Figure 8.1 *Wolbachia* affect mate choice in *Drosophila paulistorum* semi-species. Removal of species-specific *Wolbachia* infection removes assortative mating preferences. Mating preferences in combination between untreated and treated heterogametic pairs. The y-axis represents sexual isolation index (SII): number of mating assays (1–18) is shown on the x-axis. Grey bars indicate untreated controls; black bars indicate assays with antibiotic-treated flies. Two-tailed P values were calculated by comparing SII of untreated and treated pairs of mating choice assays by Fisher's exact test. Significance indicated by one, two, or three asterisks (P <0.05; P <0.01; P <0.001 respectively). AB, Andean-Brazilian; AM, Amazonian; CA, Centroamerican; OR, Oriniocan; T, treated; U, untreated. Source: Redrawn after Miller et al. (2010) with permission of http://journals.plos.org/plospathogens.

is similar to other findings showing that the impact of *Wolbachia* on behaviour can vary depending upon *Wolbachia* strain and host genotype (e.g. food preferences; Peng et al. 2009). However, it is also possible that the different findings could be due to the presence of other agents such as viruses that may also influence mate preferences (Ritschof et al. 2013). In other insects such as the jewel wasp *Nasonia* sp., *Wolbachia* has been shown to decrease mate discrimination (Chafee 2011). In the terrestrial isopod *Armadillidium vulgare*, feminizing *Wolbachia* influences mate attraction by altering the cuticular compounds and therefore female odours. Males prefer *Wolbachia*-free females that have different odour profiles, and this preference will result in increased fitness through the production of both sons and daughters (Richard 2017).

It remains to be seen to what extent the differential impact of *Wolbachia* on host mate choice influences the success of NSGs as opposed to affecting host fitness, and whether there is evidence for coevolution between endosymbionts and hosts analogous to what has been found for female fecundity (Weeks et al. 2007). Sex-specific effects seem likely for maternally inherited endosymbionts that cannot be passed on through males.

The impact of other NSGs on mate preferences, such as segregation distorters that are associated with severe fitness costs to females of mating with male carriers, is not clear (Price and Wedell 2008). In mice populations that carry an autosomal segregation distorter (the *t*-complex), heterozygous females discriminate against heterozygous males using odour cues (Lenington 1991). This is advantageous because homozygosity at the *t*-locus is lethal. In populations of stalk-eyed flies that harbour a sex ratio distorter, females prefer males with longer eye-stalks, as this is associated with carrying a suppressor of sex ratio distortion. Females that preferentially mate with males with large eye-stalks enjoy the production of both sons and daughters, and sons in turn enjoy a higher mating success as they will also have large eye-stalks and a mating advantage (Wilkinson et al. 1998). In contrast, in *D. pseudoobscura* flies, there is no evidence of sex ratio distortion affecting female mate preferences. Females are just as likely to mate with males whether they carry the sex ratio distorter or not (Price et al. 2012).

In general, the modest evidence of mate preferences based on selfish genetic elements (Price and Wedell 2008; Wedell 2013) may be due to a lack of linkage between the selfish gene and the female preference gene due to recombination (Nicholls and Butlin 1998). Mate choice appears to only be present when there are reliable cues (e.g. odour profiles in mice; Lenington 1991) or a linkage between the preference alleles and the selfish genes (e.g. stalk-eyed flies; Johns et al. 2005).

Non-self genes can also affect overall mating activity. When a virus is transmitted sexually, the host may exhibit increased sexual activity, which can increase virus transmission (Knell and Webberley 2004). Female *Helicoverpa zea* moths infected with a parasitic virus (Hz-2V) have increased pheromone production and calling frequencies. Virus-infected females are therefore more attractive to males than uninfected females, resulting in increased mating activity and higher reproductive success. This seems to be due to virus-infected females producing six- to sevenfold more pheromone than uninfected females (Burand et al. 2005). The Hz-2V virus is both vertically transmitted by infected females to eggs, and horizontally transmitted during mating (Hamm et al. 1996). Since infected females are more attractive to males, the virus will spread.

Similarly, *Wolbachia* infection appears to be associated with differences in male mating activity in flies. In *D. melanogaster* and *D. simulans*, infected males mate at a higher rate than uninfected males (Champion de Crespigny et al. 2006). This could be a strategy whereby males try to increase their reproductive success by increasing the likelihood of mating with infected and reproductively compatible females, since crosses with uninfected females result in poor offspring production caused by *Wolbachia*-induced reproductive incompatibility. Mating at higher rates also restores male reproductive compatibility with uninfected females by depleting his *Wolbachia*-modified sperm (Awrahman et al. 2014), and taken together, results in higher male reproductive success.

Other NSGs such as TEs can also affect reproductive behaviour by influencing levels of gene expression and at times this can have a remarkably wide-reaching effect. In *D. melanogaster*, males can carry a TE in the promoter region of a P450 detoxification gene (*Cyp6g1*) that confers DDT resistance by upregulation. This TE appears to have a widespread pleiotropic effect on male behaviour. In the *Canton-S* genetic background, DDT-resistant males have reduced mating success compared to their susceptible counterparts (Smith et al. 2011). This reduction appears to be largely due to reduced effectiveness in courting females, as a higher proportion of initiated courtship attempts are aborted by resistant males. Resistant males are also less aggressive, and it is likely that reduced courtship activity and lower level of aggression contribute to the reduced mating success of DDT-resistant males (Rostant et al. 2017) (Figure 8.2). In contrast, DDT-resistant females do not show lower reproductive success, but are instead almost twice as fecund as susceptible females (McCart et al. 2005).

It is as yet unknown why resistant males display such reduced mating success and why there is such a dramatic sex-specific effect of carrying the TE responsible for DDT

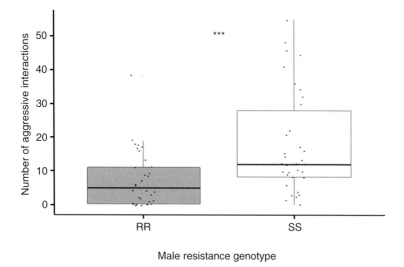

Figure 8.2 Example of a selfish gene affecting male aggressive behaviour. Counts of the total number of aggressive interactions in pairs of DDT-resistant and DDT-susceptible *D. melanogaster* males. DDT resistance is associated with upregulation of the cytochrome P450 gene *Cyp6g1*. Grey = DDT resistant; White = DDT susceptible; Asterisk represents significance of main effect of male genotype in GLMM: *** P <0.001. Source: Reproduced from Rostant et al. (2017).

resistance, but it is possible it may be mediated by sex-specific differences in metabolic activity of expressing the DDT-R allele. A similar finding has also been observed in a different population of *D. melanogaster* where males selected for increased mating success were found to have significantly reduced levels of *Cyp6g1* expression, although it is unclear if expression differences of *Cypg6g1* in successful and unsuccessful males were related to DDT resistance as this was not examined (Drnevich et al. 2004).

8.8 Aggressiveness

Aggressiveness can be directly influenced by NSGs. For example, one particular strain of *Wolbachia* (*w*MelPop) influences male aggression levels in *D. melanogaster* flies by reducing the initiation of aggressive encounters in infected males compared to uninfected controls (Rohrscheib et al. 2015). Aggression is critically important in mate competition, and is hence a key fitness-related male trait. Aggressive behaviour in many insect species is influenced by the neurotransmitter octopamine (Hoyer et al. 2008). Transcriptional analysis of the octopamine biosynthesis pathway in male flies revealed that two essential genes (tyrosine decarboxylase and tyramine hydroxylase) were significantly downregulated in *Wolbachia*-infected flies. Quantitative chemical analysis also showed that total octopamine levels were significantly reduced in adult heads (Rohrscheib et al. 2015). It is not clear whether low levels of octopamine are caused by *Wolbachia* directly modifying neural pathways and thereby influencing transcription and/or high infection densities of *w*MelPop resulting in consumption of host metabolites important to neurotransmitter production, or are simply due to brain damage. Similarly, when transferred to a novel mosquito vector host, *Wolbachia* has been shown to affect dopamine levels (Moreira et al. 2011), indicating that *Wolbachia* can directly affect a range of neurotransmitters with potential impacts on subsequent fitness-related behaviours such as aggression.

Microbiota can similarly affect level of aggression, and can even have sex-specific effects. In Siberian hamsters (*Phodopus sungorus*), a broad-spectrum antibiotic was used to explore the consequences of an altered gut flora on subsequent social behaviour. In males, antibiotic treatment reduced the level of aggression, but this effect was transient and aggression returned to normal levels following recovery. In females, on the other hand, antibiotic treatment rapidly reduced aggression levels, and these did not return to normal levels following recovery (Sylvia et al. 2017). This shows that even a modest treatment by antibiotics can result in a marked disruption of the hamsters' gut microbiota, which in turn can have a strong sex-specific effect on aggression. The different effect on aggression levels observed in males treated with antibiotics may be because males are under strong selection to maintain aggression via alternative mechanisms since this is a trait that is associated with high reproductive fitness in many rodent species. In social insects, a link has also been shown between gut microbiota and aggressiveness, with changes in both inter- and intraspecific aggression levels after antibiotic treatment (Wei et al. 2007). In termites, changes in aggression following alterations to the bacterial communities with antibiotic treatment appear to be mediated by disruption of faecal odour cues that are used in nest-mate recognition (Matsuura 2001).

The impact of the gut microbiota on aggression levels can also have implications for the establishment success of invasives. In Argentine ants (*Linepithema humile*), invasion success is related to level of aggression and survival that in turn may be influenced by gut microbiota. Antibiotic treatment slightly increased levels of interspecific aggression, but resulted in significant decrease in survival. This increased mortality may be due to ants engaging in aggressive interactions for longer periods of time (Lester et al. 2017). In general, the release from old pathogens and/or the spread of novel pathogens during interaction with new competitors during the invasion phase is an important component dictating the success of invasive species. For example, red fire ants (*Solenopsis invicta*) have fewer pathogenic infections in their introduced range compared to their home range (Yang et al. 2010). Invading harlequin ladybirds (*Harmonia axyridis*) gain a competitive edge by carrying parasitic microsporidia in their haemolymph that are fatal to native ladybird beetles (Vilcinskas et al. 2013). To what extent the gut microbiota affects this interaction remains to be quantified, but there are tantalizing data suggesting it may play a role.

8.9 Activity, Aggregation, and Dispersal

A well-known example of a NSG affecting the activity levels of its hosts is the impact of toxoplasmosis on exploratory behaviour in mice by removing their fear of cats. *Toxoplasma gondii* can only reproduce sexually in cats, and hence this behavioural alteration results in increased transmission of toxoplasmosis as infected mice are more likely to be eaten (Ingram et al. 2013). Similarly, other NSGs have been shown to influence host activity.

In mice, females that are heterozygous for the *t*-complex (see above) are less likely to perform exploratory behaviours and eat less food than homozygous wild-type females, and it has been suggested this may be an adaptive behaviour as heterozygous females live longer than wild-type females (Auclair et al. 2013). It is unclear whether this may be a side effect of lower metabolic rate associated with carrying the *t*-haplotype. Interestingly, there is no difference in activity level of males with respect to carrying the *t*-haplotype or not. In *D. melanogaster*, infection by *Drosophila* C-virus is associated with increased sleep in female flies, but has no detectable impact on males' sleep patterns. Similarly, *Wolbachia* has been shown to affect activity levels in *D. melanogaster*, with infected flies being more lethargic when awake than uninfected flies (Vale and Jardin 2015). Recently, *Wolbachia* has been shown to cause increased sleep in male and female *D. melanogaster* flies, probably by affecting dopamine-related genes as infected flies showed upregulation of *Pale* and *Ddc* (Bi et al. 2018).

In *Aedes triseriatus* mosquitos, gregarine endosymbionts affect larval behaviour in a way that has been documented to reduce the risk of predation. Infected larvae change their feeding behaviour, thrash less frequently and spend more time in refuges in the presence of predators than uninfected larvae (Soghigian et al. 2017). Overall, the effect of reduced predation on infected hosts means that gregarines may in fact be mutualistic in some circumstances. Predator detection appears to be based on olfactory cues, and it is possible that infected hosts are less susceptible to predation because altered feeding behaviour changes their smell and/or because lower activity levels of infected larvae reduce their ability to be detected by predators as a side effect.

Non-self genes can influence behaviours such as the propensity to aggregate. Locusts (*Locusta migratoria*) are well known for their ability to shift between gregarious and solitary behaviour based on insect density (Applebaum and Heifetz 1999). Aggregation behaviours are achieved through a combination of visual, tactile, and chemical cues, with faecal odour playing a large role (Pener and Yerushalmi 1998). Infection by the microsporidian parasite *Paranosema (Nosema) locustae* is known to inhibit aggregation of solitary locusts and to induce gregarious locusts to shift back to solitary behaviour (Figure 8.3). Recent work has shown that this behavioural alteration of locusts by microsporidia involves modulation of the host immune response that suppresses the growth of hindgut bacteria that produce aggregation pheromones (Tan et al. 2015), and that this in turn reduces the production of the neurotransmitter serotonin that is involved in regulating aggregation behaviour. The microsporidia also suppress production of dopamine, which has been implicated in maintaining gregariousness (Shi et al. 2014). Aggregation on food by *D. melanogaster* larvae is also correlated with the presence of chemical cues derived from the gut microbiota. Behavioural experiments show that these odour cues are important as long-distant attractants to guide flies to suitable food sources (Venu et al. 2014).

Just as NSGs affect aggregation, they can also affect dispersal behaviour. In the spider *Erigone atra*, the presence of maternally inherited *Rickettsia* endosymbionts has been found to correlate with the tendency for long-distance 'ballooning' behaviour (Goodacre et al. 2009).

8.10 Feeding

A large and rich literature exploring the varied impact that NSGs such as microbes have on host feeding behaviour has accumulated and only some of this will be touched on here.

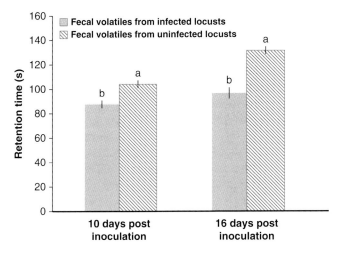

Figure 8.3 Example of NSG affecting aggregation behaviour in locusts. Healthy locusts placed in chambers containing scats from microsporidia-infected locusts are less likely to display swarming behaviours than locusts placed in chambers with scat from uninfected individuals. This is caused by microsporidia suppressing the growth of hindgut bacteria that produce the aggregation pheromones, and by suppressing production of a neurotransmitter that helps maintain gregariousness. Retention responses (time spent in faecal volatiles containing compartment; mean ± SE) of healthy locusts to faecal volatiles from microsporidia infected locusts for 10 or 16 days. Values with different letter are significantly different (P <0.05, Kruskal-Wallis). Source: Redrawn from Shi et al. (2014) with permission from PNAS.

The role of gut bacteria on host feeding and resulting fitness has been well studied (Wong et al. 2015) and this long-lasting association has resulted in a variety of mechanisms by which the microbiota can directly influence host behaviour. The gut is full of receptors that are directly connected with the neuroendocrine system, and several of these receptors are designed to detect microbial metabolites that in turn affect host feeding (Tan et al. 2014). Gut bacteria can affect diet intake by altering cravings by the host for specific food types and nutrients that are beneficial for the microbes themselves (Alcock et al. 2014; Ayres 2016). For example, the gut microbiota directly modifies the feeding preferences of *D. melanogaster* by influencing their chemosensory responses (Leitao-Goncalves et al. 2017; Wong et al. 2017). Flies prefer beneficial bacteria, but these preferences are modified by the gut microbiota, and early microbial exposure shape the specific preference of the adult fly. This indicates that maternal deposition of microbes on eggs, for example, can influence the feeding preference of their offspring (Funkhouser and Bordenstein 2013).

Symbiotic NSGs can also have profound positive fitness benefits to hosts by providing key nutrients missing from the host diet. This is evident in blood-feeding bed bugs where *Wolbachia* provide hosts with vitamin B that is essential for successful development but absent from the blood diet (Hosokawa et al. 2010). Similarly, in ants there is extensive evidence of the importance of the gut microbiota in providing key nutrients such as nitrogen to plant sap-feeding and insect honeydew-feeding ant species whose diet is low in nitrogen (Russell et al. 2017). The ability of endosymbionts to synthesize vital nutrients in turn has the potential to affect host feeding behaviour more generally as it means that hosts can utilize food sources even if they are lacking in some key nutritional constituents as these can instead be provided by the gut microbes. Coevolution between host dietary needs on the one hand and ability of the microbiome to synthesize particular nutrients on the other, has shaped host–symbiont associations, at times promoting obligate dependence typically resulting in highly reduced symbiont genome size (Moran and Sloan 2015; Wilson and Duncan 2015; Fisher et al. 2017).

Considering the intimate association between the gut microbiota and host in affecting host feeding, it is not surprising that microbes can have far-reaching effects on host health. In honeybees, the gut microbiota provides pathogen protection (Kwong et al. 2017) and nutritional benefits (Kwong and Moran 2016), as well as directly affecting bee growth and behaviour. Recent research shows that metabolites resulting from the gut bacteria's own metabolism of pollen ingested by the bees directly influences bee growth and hormone profile that in turn regulate the bees' hunger level and therefore foraging behaviour (Zheng et al. 2017). There is also a growing realization that the gut microbiota is a key regulator of the host immune system (Rosenberg and Zilber-Rosenberg 2016). For example, honeybees upregulate gene expression of antimicrobial peptides in the presence of gut microbes that appear to confer pathogen protection (Kwong et al. 2017). However, at times the gut microbiome can have negative impact on host feeding behaviour by resulting in overeating and obesity or intake of potentially harmful substances (Rosenberg and Zilber-Rosenberg 2016). Host–microbe associations clearly involve both positive and negative feedbacks, emphasizing that this relationship may not be intrinsically parasitic or mutualistic, but will depend on the fitness outcomes that are likely to vary depending on environmental conditions (Wong et al. 2015).

Non-self genes from viruses are also known to affect feeding behaviour. Rice-stripe-virus (RSV) is transovarially transmitted within its vector, the brown planthopper *Laodelphax striatellus*, where it causes serious disease. Infected planthoppers show significant fecundity decrease but also significantly shortened nymphal stage duration. The accelerated development is associated with downregulation of a gene in the juvenile hormone pathway (*JHAMT*) and upregulation of another gene (*CYP307A1*) in the ecdysteroid pathway. Infected individuals also differ in their feeding behaviour by feeding for longer, possibly to compensate for their lower fecundity (Wan et al. 2015). This suggests that NSGs may have the potential to shape life histories such as duration of juvenile stages, a key trait implicated in the evolution of social behaviour in the Hymenoptera (Macke et al. 2017). There is also evidence that TE can influence host metabolism with knock-on effects on nutrient assimilation. In mice, for example, TE-mediated chromatin variation in liver cells underlies differences in metabolic rate that translate into variation in how mice respond to a high-fat/high-sucrose diet (Du et al. 2016).

8.11 Learning and Memory

There is extensive evidence that parasites and pathogens can adversely affect the cognitive ability of hosts and therefore influence learning and memory. Bumblebees infected by gut parasites show reduced ability to use floral information during foraging (Gegear et al. 2006). It is therefore not surprising to find that NSGs also have the potential to influence host cognition. Many microbes and endosymbionts directly target the mushroom bodies of insect brains that influence sensory learning and memory. *Wolbachia* infection is known to adversely affect learning and memory in *A. vulgare* isopods. Infected individuals are less able to learn particular odour cues that aid orientation in a Y-test chamber, as they are less likely to learn and remember the correct direction with social reinforcement compared to uninfected individuals (Temple and Richard 2015). Reduced learning ability can be costly since terrestrial crustaceans such as isopods use associative learning in activities such as mating, egg laying, and foraging (Dukas 2008). Similarly, *Wolbachia* can impair learning and memory in *Trichogramma brassicae* parasitoid wasps. Infected wasps show reduced ability to learn to associate a novel odour with a reward (a host egg) (Farahani et al. 2017). Previous work has also shown that *Wolbachia* impairs decision making during patch exploration in this species, with infected wasps being less efficient in assessing the nutritional value of hosts showing a reduced ability to discriminate between unparasitized and parasitized hosts, and needed to forage more frequently than uninfected individuals (Farahani et al. 2015). Collectively, these effects favour transmission of *Wolbachia* by encouraging infected wasps to seek out new environments. However, it is not clear whether selection has favoured *Wolbachia* manipulation because of increased transmission or whether this effect is simply a by-product of infection.

In honeybees and bumblebees, individuals with experimentally challenged immune systems have poorer learning and memory (Mallon et al. 2003; Riddell and Mallon 2006). This may in part be due to the energetic stress imposed by activating the immune system, but also due to a compromised gut microbiota (Kwong and Moran 2016; Gomez-Moracho et al. 2017). Conversely, memory formation was found to be impaired in germ-free mice lacking a gut microbiota, and diets that induced changes in the microbiota also altered the memory in mice (Gareau et al. 2011).

These findings suggest that cognitive abilities and memory can be determined by an animal's microbial communities to a greater extent than previously realized. There is also evidence to suggest that TEs such as retrotransposons can directly affect cognition and potentially even brain development in mammals. For example, the ancient *Arc* gene (*Arg3.1*) is a neuron-specific gene derived from a retrotransposon that is required for learning and memory in mice (Plath et al. 2006). Similarly, disruption of the retrotransposon-derived *Zcchc16/Mart4/Sirh11* gene in mice results in abnormal behaviours related to cognition and memory (Irie et al. 2015). Taken together, these findings suggest that such NSGs have the potential to play a large role in shaping learning and memory in a variety of animals, even if they have now been integrated within the host genome (Naville et al. 2016).

8.12 Summary and Conclusion

It is well established (in at least some circles) that behaviour of animals is determined by their genes as well as their environment – see chapters throughout this book. However, we now realize that the genetic make-up of individuals also includes myriad 'foreign' genes including microbes, and a battery of mobile and selfish genetic elements. As discussed in this chapter, the effect of these NSGs can be profound and affect most aspects of an animal's behaviour, from feeding and activity levels to social and reproductive behaviour. In many cases the interests of the host and the NSG coincide and the relationship is largely mutually beneficial. However, this relationship can be fragile and at times the NSG has the upper hand and the impact on host behaviour can be associated with costs that in turn will modify the NSG–host association.

Many NSGs, particularly microbes and symbionts, have evolved remarkably sophisticated means to influence the behaviour of their host, often by invading the nervous system and hijacking signalling pathways that regulate a variety of host behaviours. It has been suggested that such NSGs may provide novel insights into how nervous systems work and regulate behaviour more generally as they have evolved to target specific neurological pathways to alter key behaviours (Adamo 2013). Similarly, many mobile genetic elements are able to regulate expression of genes in key regulatory gene networks with large impacts on behaviour. Seeing the ancient association between NSGs and hosts, it is perhaps not surprising that they can have such dramatic effects on animal behaviour.

Given the prevalence of NSGs and their far-reaching effects on a range of behaviours, it is clear we need to take them into consideration when studying the behaviour of animals in the future. Not surprisingly, there are several outstanding questions that need addressing. We still do not know how frequently NSGs influence behaviour and if certain types of behaviours are more likely to be under the influence of NSGs than others, and if so what type of NSGs is involved (e.g. endosymbiont, gut bacteria, mobile genetic elements). We also do not have a good understanding of whether the impact of NSGs on behaviour is predominantly indirect (i.e. consequences of population sex ratio bias) or if NSGs more commonly directly affect animal behaviour (i.e. mate preferences). Moreover, when NSGs directly influence the behaviour of animals, it is often unknown by what mechanism this is achieved. There are examples of NSGs such as *Wolbachia* directly invading insect brain regions and directly modulate host behaviour, but also plenty of examples where NSGs hijack regulatory pathways and key gene networks.

Once we have a better understanding of the prevalence and mechanisms whereby NSGs affect host behaviour, we can start to quantify the consequences to host and NSG fitness. This will allow us to formulate hypotheses to predict the evolutionary dynamics of NSGs and the potential for coevolution between NSGs and hosts. Given the prevalence and diversity of host–NSG associations, it is clear that we need to consider NSGs alongside other genetic influences for unravelling the genetic basis to animal behaviour.

References

Adamo, S.A. (2012). The strings of the puppet master: how parasites change host behaviour. In: *Parasitic Manipulation* (ed. D. Hughes and F. Thomas), 36–51. Oxford: Oxford University Press pp.

Adamo, S.A. (2013). Parasites: evolution's neurobiologists. *Journal of Experimental Biology* 216: 3–10.

Albertson, R., Tan, V., Leads, R.R. et al. (2013). Mapping *Wolbachia* distributions in the adult *Drosophila* brain. *Cell Microbiology* 15: 1527–1544.

Alcock, J., Maley, C.C., and Aktipis, C.A. (2014). Is eating behavior manipulated by the gastrointestinal microbiota? Evolutionary pressures and potential mechanisms. *Bioessays* 36: 940–949.

Applebaum, S.W. and Heifetz, Y. (1999). Density-dependent physiological phase in insects. *Annual Reviews of Entomology* 44: 317–341.

Arburthnott, D., Levin, T.C., and Promislow, D.E. (2016). The impacts of *Wolbachia* and the microbiome on mate choice in *Drosophila melanogaster*. *Journal of Evolutionary Biology* 29: 461–468.

Archie, E.A. and Theis, K.R. (2011). Animal behavior meets microbial ecology. *Animal Behaviour* 82: 425–436.

Archie, E.A. and Tung, J. (2015). Social behavior and the microbiome. *Current Opinions in Behavioral Sciences* 6: 28–34.

Auclair, Y., Konig, B., and Lindholm, A.K. (2013). A selfish genetic element influencing longevity correlates with reactive behavioural traits in female house mice. *PLoS One* 8: e67130.

Awrahman, Z.A., Champion de Crespigny, F.E., and Wedell, N. (2014). The impact of *Wolbachia*, male age and mating history on cytoplasmic incompatibility and sperm transfer in *Drosophila simulans*. *Journal of Evolutionary Biology* 27: 1–10.

Ayres, J.S. (2016). Cooperative microbes tolerance behaviors in host-microbiota mutualism. *Cell* 155: 1323–1331.

Bandi, C., Sironi, M., Damiani, G. et al. (1995). The establishment of intracellular symbiosis in an ancestor of cockroaches and termites. *Proceedings of the Royal Society B: Biological Sciences* 259: 293–299.

Barber, I., Mora, A.B., Payne, E.M. et al. (2017). Parasitism, personality and cognition in fish. *Behavioral Processes* 141: 205–219.

Bi, J., Sehgal, A., Williams, J.A., and Wang, Y.-F. (2018). *Wolbachia* affects sleep behavior in *Drosophila melanogaster*. *Journal of Insect Physiology* 107: 81–88.

Boschetti, C., Carr, A., Crisp, A. et al. (2012). Biochemical diversification through foreign gene expression in bdelloid rotifers. *PLoS Genetics* 8: e1003035.

Burand, J.P., Tan, W., Kim, W. et al. (2005). Altered mating behavior and pheromone production in female *Helicoverpa zea* moths infected with the insect virus Hz-2v. *Journal of Insect Science* 5: 6.

Burt, A. & Trivers, R. 2006. *Genes in Conflict* Harvard: Harvard University Press.

Chafee, M.E. (2011). Decoupling of host–phage coadaptations following transfer between insect species. *Genetics* 187: 203–215.

Champion de Crespigny, F.E. and Wedell, N. (2007). Mate preferences in *Wolbachia* infected *Drosophila*. *Behavioral Ecology and Sociobiology* 61: 1229–1235.

Champion de Crespigny, F.E., Pitt, T., and Wedell, N. (2006). *Wolbachia* increases male mating rate in *Drosophila*. *Journal of Evolutionary Biology* 19: 1964–1972.

Charlat, S., Reuter, M., Dyson, E.A. et al. (2007). Male-killing bacteria increase female mating frequency. *Current Biology* 17: 273–277.

Correa, C.C. and Ballard, J.W.O. (2016). *Wolbachia* associations with insects: winning or losing against a master manipulator. *Frontiers in Ecology and Evolution* 3: 153.

Cremer, S., Armitage, S.A.O., and Schmid-Hempel, P. (2007). Social immunity. *Current Biology* (16): R693–R702.

Cryan, J.F. and Clark, G. (eds.) (2016). *Gut Microbiome and Behavior*, International Review of Neurobiology, vol. 131. Cambridge: Academic Press.

Cryan, J.F. and Dinan, T.G. (2012). Mind-altering microorganisms: the impact of the gut microbiota on brain and behavior. *Nature Reviews Neuroscience* 13: 701–712.

Damodaram, K.J.P., Ayyasamy, A., and Kempraj, V. (2016). Commensal bacteria aid mate-selection in the fruit fly, *Bactrocera dorsalis*. *Microbial Ecology* 72: 725–729.

De Moraes, C., Lewis, W.J., Pare, P. et al. (1998). Herbivore-infested plants selectively attract parasitoids. *Nature* 393: 570–573.

Dedeine, F., Vavre, F., Fleury, F. et al. (2001). Removing symbiotic *Wolbachia* bacteria specifically inhibits oogenesis in a parasitic wasp. *Proceedings of the National Academy of Sciences USA* 98: 6247–6252.

Diaz Heijtz, R., Wang, S., Anuar, F. et al. (2011). Normal gut microbiota modulates brain development and behavior. *Proceedings of the National Academy of Sciences USA* 108: 3047–3052.

Dillon, R.J. and Dillon, V.M. (2004). The gut bacteria of insects: nonpathogenic interactions. *Annual Reviews of Entomology* 49: 71–92.

Douglas, A.E. (1998). Nutritional interactions in insect-microbial symbioses: aphids and their symbiotic bacteria *Buchnera*. *Annual Reviews of Entomology* 43,: 17–37.

Drezen, J.-M., Gauthier, J.G., Josse, T. et al. (2017). Foreign DNA acquisition by invertebrate genomes. *Journal of Invertebrate Pathology* 147: 157–168.

Drnevich, J.M., Reedy, M.M., Ruedi, E.A. et al. (2004). Quantitative evolutionary genomics: differential gene expression and male reproductive success in *Drosophila melanogaster*. *Proceedings of the Royal Society B: Biological Sciences* 271: 2267–2273.

Du, J., Leung, A., Trac, C. et al. (2016). Chromatin variation associated with liver metabolism is mediated by transposable elements. *Epigenetics & Chromatin* 9: 28.

Dukas, R. (2008). Evolutionary biology of insect learning. *Annual Reviews of Entomology* 53: 145–160.

Dunning Hotopp, J.C. et al. (2007). Widespread lateral gene transfer from intracellular bacteria to multicellular eukaryotes. *Science* 317: 1753–1756.

Dyson, E.A. and Hurst, G.D.D. (2004). Persistence of an extreme sex-ratio bias in a natural population. *Proceedings of the National Academy of Sciences USA* 101: 6520–6523.

Eisthen, H.L. and Theis, K.R. (2015). Animal-microbe interactions and the evolution of nervous systems. *Philosophical Transactions of the Royal Society B: Life Sciences* 371: 20150052.

Engel, P., Martinson, V.G., and Moran, N.A. (2012). Functional diversity within the simple gut microbiota of the honey bee. *Proceedings of the National Academy of Sciences USA* 109: 11002–11007.

Ezenewa, V.O., Gerardo, N.M., Inouye, D.W. et al. (2012). Animal behavior and the microbiome. *Science* 338: 198–199.

Ezenwa, V.O., Ghai, R.R., McKay, A.F., and Williams, A.E. (2016). Group living and pathogen infection revisited. *Current Opinions in Behavioral Science* 12: 66–72.

Farahani, H.K., Ashouri, A., Goldansaz, S.H. et al. (2017). Decrease of memory retention in a parasitic wasp: an effect of host manipulation by *Wolbachia*? *Insect Science* 24: 569–583.

Fisher, R.M., Henry, L.M., Cornwallis, C.K. et al. (2017). The evolution of host-symbiont dependence. *Nature Communications* 8: 15973.

Forsythe, P. and Kunze, W.A. (2013). Voices from within: gut microbes and the CNS. *Cellular Molecular Life Sciences* 70: 55–69.

Funkhouser, L.J. and Bordenstein, S.R. (2013). Mom knows best: the universality of maternal microbial transmission. *PLoS Biology* 11: e1001631.

Gareau, M.G., Wine, E., Rodrigues, D.M. et al. (2011). Bacterial infection causes stress-induced memory dysfunction in mice. *Gut* 60: 307–317.

Gegear, R.J., Otterstatter, M.C., and Thomson, D.J. (2006). Bumble-bee foragers infected by a gut parasite have an impaired ability to utilize floral information. *Proceedings of the Royal Society B: Biological Sciences* 273: 1073–1078.

Gomez-Moracho, T., Heeb, P., and Lihoreau, M. (2017). Effects of parasites and pathogens on bee cognition. *Ecological Entomology* 42: 51–64.

Goodacre, S.L., Martin, O.Y., Bonte, D. et al. (2009). Microbial modification of host long-distance dispersal capacity. *BMC Biology* 7: 32.

Gunderson, A.R., Forsyth, M.H., and Swaddle, J.P. (2009). Evidence that plumage bacteria influence feather colouration and body condition in eastern bluebirds *Sialia Sialis*. *Journal of Avian Biology* 40: 440–447.

Hamm, J.J., Carpenter, J.E., and Styer, E.L. (1996). Oviposition day effect on incidence of agonadal progeny of *Helicoverpa zea* (Lepidotera: Noctuidae) infected with a virus. *Annual Entomological Society of America* 89: 266–275.

Hart, B.L. (1988). Biological basis of the behavior of sick animals. *Neuroscience and Biobehavioral Review* 12: 123–137.

Havilek, J. and Roberts, S.C. (2009). MHC-correlated mate choice in humans: a review. *Psychoneuroendochrinology* 34: 497–512.

Hilgenboecker, K., Hammerstein, P., Schlattmann, P. et al. (2008). How many species are infected with *Wolbachia*? – a statistical analysis of current data. *FEMS Microbiology Letters* 281: 215–220.

Hosokawa, T., Koga, R., Kikuchi, Y. et al. (2010). *Wolbachia* as a bacteriocyte-associated nutritional mutualis. *Proceedings of the National Academy of Sciences USA* 107: 769–774.

Hoyer, S.C., Eckart, A., Herrel, A. et al. (2008). Octopamine in male aggression of *Drosophila*. *Current Biology* 18: 159–167.

Hughes, D.P., Brodeur, J., and Thomas, F. (eds.) (2012). *Host Manipulation by Parasites*. Oxford: Oxford University Press.

Ingleby, F. (2015). Insect cuticular hydrocarbons as dynamic traits in sexual communication. *Insects* 6: 732–742.

Ingram, W.M., Goodrich, L.M., Robey, E.A., and Eisen, M.B. (2013). Mice infected with low-virulence strains of *Toxoplasma gondii* lose their innate aversion to cat urine, even after extensive parasite clearance. *PLoS One* 8: e75246.

Irie, M., Yoshikawa, M., Ono, R. et al. (2015). Cognitive function related to the *Sirh11/Zcchc16* gene acquired from an LTR retrotransposon in eutherians. *PLoS Genetics* 11: e1005521.

Jiggins, F., Hurst, G.D.D., and Majerus, M.E.N. (2000). Sex-ratio distorting *Wolbachia* causes sex-role reversal in its butterfly host. *Proceedings of the Royal Society B: Biological Sciences* 267: 69–73.

Johns, P.M., Wolfenbarger, L.L., and Wilkinson, G.S. (2005). Genetic linkage between a sexually selected trait and X chromosome meiotic drive. *Proceedings of the Royal Society B: Biological Sciences* 272: 2097–2103.

Knell, R.J. and Webberley, K.M. (2004). Sexually transmitted diseases of insects: distribution, evolution, ecology and host behaviour. *Biological Reviews* 79: 557–581.

Koch, H. and Schmid-Hempel, P. (2011). Socially transmitted gut microbiota protect bumble bees against an intestinal parasite. *Proceedings of the National Academy of Sciences USA* 108: 19288–19292.

Kramer, T. and Enquist, L.W. (2013). Directional spread of alpha herpes viruses in the nervous system. *Viruses* 5: 678–707.

Kwong, W.K. and Moran, N.A. (2016). Gut microbial communities of social bees. *Nature Reviews Microbiology* 14: 374–384.

Kwong, W.K., Mancenido, A.L., and Moran, N.A. (2017). Immune system stimulation by the native gut microbiota of honey bees. *Royal Society Open Science* 4: 170003.

Lefèvre, T., Adamo, S.A., Biron, D.G. et al. (2009). Invasion of the body snatchers: the diversity and evolution of manipulative strategies in host-parasite interactions. *Advances in Parasitology* 68: 45–83.

Leitao-Goncalves, R., Carvalho-Santos, Z., Francisco, A.P. et al. (2017). Commensal bacteria and essential amino acids control food choice behavior and reproduction. *PLoS Biology* 15: e2000862.

Lenington, S. (1991). The t-complex: a story of genes, behavior, and populations. In: *Advances in the Study of Behavior*, vol. 20, 51–86.

Lester, P.J., Sebastien, A., Suarez, A.V. et al. (2017). Symbiotic bacterial communities in ants are modified by invasion pathway bottlenecks and alter host behavior. *Ecology* 98: 861–874.

Lewin-Epstein, O., Aharonov, R., and Hadany, L. (2017). Microbes can help explain the evolution of host altruism. *Nature Communications* 8: 14040.

Lihoreau, M., Rivault, C., and van Zweden, J.S. (2016). Kin discrimination increases with odordistance in the German cockroach. *Behavioral Ecology* 27: 1694–1701.

Little, A.E.F., Murakami, T., Mueller, U.G., and Currie, C.R. (2006). Defending against parasites: fungus-growing ants combine specialized behaviours and microbial symbionts to protect their fungus gardens. *Biology Letters* 2: 12–16.

Lize, A., McKay, R., and Lewis, Z. (2014). Kin recognition in *Drosophila*: the importance of ecology and gut microbiota. *ISME* 8: 469–477.

Lombardo, M.P. (2008). Access to mutualistic endosymbiotic microbes: an underappreciated benefit of group living. *Behavioral Ecology & Sociobiology* 62: 479–497.

Longdon, B., Wilfert, L., Osei-Poku, J. et al. (2011). Host-switching by a vertically transmitted rhabdovirus in *Drosophila*. *Biology Letters* 7.

Lyte, M. (2011). Probiotics function mechanistically as delivery vehicles for neuroactive compounds: microbial endocrinology in the design and use of probiotics. *BioEssays* 33: 574–581.

Macke, E., Tasiemski, A., Massol, F. et al. (2017). Life history and eco-evolutionary dynamics in light of the gut microbiota. *Oikos* 126: 508–531.

Mallon, E.B., Brockmann, A., and Schmid-Hempel, P. (2003). Immune response inhibits associative learning in insects. *Proceedings of the Royal Society B: Biological Sciences* 270: 2471–2473.

Markov, V., Lazebny, O.E., Goryacheva, I.I. et al. (2009). Symbiotic bacteria affect mating choice in *Drosophila melanogaster*. *Animal Behaviour* 77: 1011–1017.

Martini, X., Hoffmann, M., Coy, M.R. et al. (2015). Infection of an insect vector with a bacterial plant pathogen increases its propensity for dispersal. *PLoS ONE* 10 (6): e0129373.

Mason, P.A., Smilanich, A.M., and Singer, M.S. (2014). Reduced consumption of protein-rich foods follows immune challenge in a polyphagous caterpillar. *Journal of Experimental Biology* 217: 2250–2260.

Matsuura, K. (2001). Nestmate recognition mediated by intestinal bacteria in a termite, *Reticulitermes speratus*. *Oikos* 92: 20–26.

McCart, C., Buckling, A., and ffrench-Constant, R.H. (2005). DDT resistance in flies carries no cost. *Current Biology* 15: R587–R589.

Miller, W.J., Ehrman, L., and Schneider, D. (2010). Infectious speciation revisited: impact of symbiont-depletion on female fitness and mating behavior of *Drosophila paulistrorum*. *PLoS Pathogens* 6: e1001214.

Moore, J. (2002). *Parasites and the Behavior of Animals*. Oxford, UK: Oxford University Press.

Moran, N.A. and Sloan, D.B. (2015). The hologenome concept: helpful or hollow? *PLoS Biology* 13: e1002311.

Moran, N.A., Munson, M.A., Baumann, P., and Ishikawa, H. (1993). A molecular clock in endosymbiotic bacteria is calibrated using insect hosts. *Proceedings of the Royal Society B: Biological Sciences* 253: 167–171.

Moreira, L.A., Ye, Y.H., Turner, K. et al. (2011). The wMelPop strain of *Wolbachia* interferes with dopamine levels in *Aedes aegypti*. *Parasite Vectors* 4: 28.

Mothes, W., Sherer, N.M., Jin, J., and Zhong, P. (2010). Virus cell-to-cell transmission. *Journal of Virology* 84: 8360–8368.

Natsch, A. (2017). Biochemistry and genetics of human axilla odor. In: *Springer Handbook of Odor* (ed. A. Buettner), 123–124. Cham: Springer.

Natsch, A., Kuhn, F., and Tiercy, J.M. (2010). Lack of evidence for HLA-linked patterns of odorous carboxylic acids released from glutamine conjugates secreted in the human axilla. *Journal of Chemical Ecology* 36: 837–846.

Naville, M., Warren, I.A., Haftek-Terreau, Z. et al. (2016). Not so bad after all: retroviruses and long terminal repeat retrotransposons as a source of new genes in vertebrates. *Clinical Microbiology and Infection* 22: 312–323.

Nichols, R.A. and Butlin, R.K. (1998). Does runaway sexual selection work in finite populations. *Journal of Evolutionary Biology* 2: 299–313.

O'Neill, S.L., Giordano, R., Colbert, A.M. et al. (1992). 16S rRNA phylogenetic analysis of the bacterial endosymbionts associated with cytoplasmic incompatibility in insects. *Proceedings of the National Academy of Sciences USA* 89: 2699–2702.

Pener, M.P. and Yerushalmi, Y. (1998). The physiology of locust phase polymorphism: an update. *Journal of Insect Physiology* 44: 365–377.

Peng, Y. and Wang, Y. (2009). Infection of *Wolbachia* may improve the olfactory response of *Drosophila*. *Chinese Sciences Bulletin* 54: 1369–1375.

Peng, Y., Nielsen, J.E., Cunningham, J.P., and McGraw, E.A. (2009). *Wolbachia* infection alters olfactory-cued locomotion in *Drosophila* spp. *Applied Environmental Microbiology* 74: 3943–3948.

Penn, D., Oberzaucher, E., Grammer, K. et al. (2007). Individual and gender fingerprints in human body odour. *Journal of Royal Society Interface* 22: 331–340.

Plath, N., Ohana, O., Dammermann, B. et al. (2006). Arc/Arg3.1 is essential for the consolidation of synaptic plasticity and memories. *Neuron* 52: 437–444.

Poulin, R. (2010). Parasite manipulation of host behavior: an update and frequently asked questions. In: *Advances in the Study of Behavior*, vol. 41, 151–186.

Poulin, R. (2011). *Evoltionary Ecology of Parasites*, 2nd edn. Princeton: Princeton University Press.

Povey, S., Cotter, S.C., Simpson, S.J., and Wilson, K. (2014). Dynamics of macronutrient self-medication and illness-induced anorexia in virally infected insects. *Journal of Animal Ecology* 83: 245–255.

Price, T.A.R. and Wedell, N. (2008). Selfish genetic elements and sexual selection: their impact on male fertility. *Genetica* 132: 295–307.

Price, T.A.R., Bretman, A.J., Avent, T. et al. (2008a). Sex ratio distorter reduces sperm competitive ability in an insect. *Evolution* 62: 1644–1652.

Price, T.A.R., Hodgson, D.J., Lewis, Z. et al. (2008b). Selfish genetic elements promote polyandry in a fly. *Science* 322: 1241–1243.

Price, T.A.R., Lewis, Z., Smith, D.T. et al. (2010). Sex ratio drive promotes sexual conflict and sexual coevolution in the fly *Drosophila pseudoobscura*. *Evolution* 65: 1504–1509.

Price, T.A.R., Lewis, Z., Smith, D.T. et al. (2012). No evidence of mate discrimination against males carrying a sex ratio distorter in *Drosophila pseudoobscura*. *Behavioral Ecology and Sociobiology* 66: 561–568.

Price, T.A.R., Bretman, A., Gradilla, A.C. et al. (2014). Does polyandry control population sex ratio via regulation of a selfish gene? *Proceedings of the Royal Society B: Biological Sciences* 281: 20133259.

Richard, F.-J. (2017). Symbiotic bacteria influence the odor and mating preference of their host. *Frontiers in Ecology and Evolution* 5: 143.

Riddell, C.E. and Mallon, E.B. (2006). Insect psychoneuroimmunology: immune response reduces learning in protein starved bumblebees (*Bombus terrestris*). *Brain, Behavior and Immunity* 20: 135–138.

Riegler, M., Sidhu, M., Miller, W.J., and O'Neill, S.L. (2005). Evidence for global *Wolbachia* replacement in *Drosophila melanogaster*. *Current Biology* 15: 1428–1433.

Ritschof, C.C., Pattanaik, S., Johnson, L. et al. (2013). Sigma virus and male reproductive success in *Drosophila melanogaster*. *Behavioral Ecology and Sociobiology* 67: 529–540.

Roberts, S.C., Gosling, L.M., Spector, T.D. et al. (2005). Body odor similarity in noncohabiting twins. *Chemical Senses* 30: 651–656.

Rohrscheib, C. and Brownlie, J. (2013). Microorganisms that manipulate complex animal behaviours by affecting the host's nervous system. *Springer Science Reviews* 1: 133–140.

Rohrscheib, C.E., Bondy, E., Josh, P. et al. (2015). *Wolbachia* influences the production of octopamine and affects *Drosophila* male aggression. *Applied Environmental Microbiology* 81: 4573–4580.

Rosenberg, E. and Zilber-Rosenberg, I. (2016). Microbes drive evolution of animal and plants: the hologenome concept. *mBio* 7: e01395-15.

Rostant, W.G., Bowyer, J., Coupland, J. et al. (2017). Pleiotropic effects of DDT resistance on male behaviour and size. *Behavioral Genetics* 47: 449–458.

Russell, J.A., Sanders, J.G., and Moreau, C.S. (2017). Hotspots for symbiosis: function, evolution, and specificity of ant-microbe associations from trunk to tip of the ant phylogeny (Hymenoptera: Formicidae). *Myrmecology News* 24: 43–69.

Schmid-Hempel, P. (2017). Parasites and their social hosts. *Trends in Parasitology* 3 (6).

Sender, R., Fuchs, S., and Milo, R. (2016). Revised estimates for the number of human and bacteria cells in the body. *PLoS Biology* 14: e1002533.

Sharon, G., Segal, D., Ringo, J.M. et al. (2010). Commensal bacteria play a role in mating preference of *Drosophila melanogaster*. *Proceedings of the National Academy of Sciences USA* 107: 20051–20056.

Shawkey, M.D., Pillai, S.R., and Hill, G.E. (2009). Do feather-degrading bacteria affect sexually selected plumge colour? *Naturwissenschaften* 96: 123–128.

Shi, W., Guo, Y., Xu, C. et al. (2014). Unveiling the mechanism by which microsporidian parasites prevent locust swarm behavior. *Proceedings of the National Academy of Sciences USA* 11: 1343–1348.

Smith, D.T., Hosken, D.J., Rostant, W.G. et al. (2011). DDT resistance, epistasis and male fitness in flies. *Journal of Evolutionary Biology* 24: 1351–1362.

Soghigian, J., Valsdottir, L.R., and Livdahl, T.P. (2017). A parasite's modification of host behavior reduces predation on its host. *Ecology and Evolution* 7: 1453–1461.

Starks, P.T., Blackie, C.A., and Seeley, T.D. (2000). Fever in honeybee colonies. *Naturwissenschaften* (5): 229–231.

Strunov, A., Schneider, D.I., Albertson, R., and Miller, W.J. (2017). Restricted distribution and lateralization of mutualistic *Wolbachia* in the *Drosophila* brain. *Cellular Microbiology* 19: e12639.

Sylvia, K.E., Jewell, C., Rendon, N. et al. (2017). Sex-specific modulation of the gut microbiome and behavior in Siberian hamsters. *Brain, Behavior, and Immunity* 60: 51–62.

Tan, J., McKenzie, C., Potamitis, M. et al. (2014). The role of short-chain fatty acids in health and disease. *Advances in Immunology* 121: 91–119.

Tan, S.-Q., Zhang, K.-Q., Chen, H.-X. et al. (2015). The mechanism for microsporidian parasite suppression of the hindgut bacteria of the migratory locus *Locusta migratoria manilensis*. *Scientific Reports* 5: 17365.

Temple, N. and Richard, F.-J. (2015). Intra-cellular bacterial infections affect learning and memory capacities of an invertebrate. *Frontiers in Zoology* 12: 36.

Vale, P.F. and Jardin, M.D. (2015). Sex-specific behavioural symptoms of viral gut infection and *Wolbachia* in *Drosophila melanogaster*. *Journal of Insect Physiology* 82: 28–32.

Van Zweden, J.S. and d'Ettorre, P. (2010). Nestmate recognition in social insects and the role of hydrocarbons. In: *Insect Hydrocarbons: Biology, Biochemistry and Chemical Ecology* (ed. G.J. Blomquist and A. Bagnères), 222–243. Cambridge: Cambridge University Press.

Venu, I., Durisko, Z., Xu, J., and Dukas, R. (2014). Social attraction mediated by fruit flies' microbiome. *Journal of Experimental Biology* 217: 1346–1352.

Vilcinskas, A., Stoecker, K., Schmidtberg, H., and Rohrich, C.R. (2013). Invasive harlequin ladybird carries biological weapons against native competitors. *Science* 340: 862–863.

Voung, H.E., Yano, J.M., Fung, T.C., and Hsiao, E.Y. (2017). The microbiome and host behavior. *Annual Review of Neuroscience* 40: 21–49.

Wan, G., Jiang, S., Wang, W. et al. (2015). Rice stripe virus counters reduced fecundity in its insect vector by modifying insect physiology, primary endosymbionts and feeding behavior. *Scientific Reports* 5: 12527.

Wedell, N. (2013). The dynamic relationship between polyandry and selfish genetic elements. *Philosophical Transactions of the Royal Society B: Life Sciences* 368: 20120049.

Weeks, A.R., Turelli, M., Harcombe, W.R. et al. (2007). From parasite to mutualist: rapid evolution of *Wolbachia* in natural populations of *Drosophila*. *PLoS Biology* 5: e177.

Wei, J.Q., Mo, J.C., Pan, C.Y. et al. (2007). The intestinal microbes inducing the agonistic behavior of inter-colonial individuals in *Coptotermes formosanus* (Isoptera: Rhinotermitidae). *Sociobiology* 50: 245–256.

Weinert, L., Werren, J.H., Aebi, A. et al. (2009). Evolution and diversity of *Rickettsia* bacteria. *BMC Biology* 7 (6).

Weinert, L.A., Araujo-Jnr, E.V., Ahmed, M.Z., and Welch, J.J. (2015). The incidence of bacterial endosymbionts in terrestrial arthropods. *Proceedings of the Royal Society B: Biological Sciences* 282: 20150249.

Weisfeld, G.E., Czilli, T., Phillips, K.A. et al. (2003). Possible olfaction-based mechanisms in human kin recognition and inbreeding avoidance. *Journal of Experimental Child Psychology* 85: 279–295.

Werren, J.H., Baldo, L., and Clark, M.E. (2008). *Wolbachia*: master manipulators of invertebrate biology. *Nature Reviews Microbiology* 6: 741–751.

Wilkinson, G.S., Presgraves, D.C., and Crymes, L. (1998). Male eye span in stalk-eyes flies indicates genetic quality by meiotic drive suppression. *Nature* 391: 276–279.

Wilson, A.C.C. and Duncan, R.P. (2015). Signatures of host/symbiont genome coevolution in insect nutritional endosymbiosis. *Proceedings of the National Academy of Sciences USA* 112: 10255–10261.

Wong, A.C.-N., Holmes, A., Ponton, F. et al. (2015). Behavioral microbiomics: a multi-dimensional approach to microbial influence on behavior. *Frontiers in Microbiology* 6: 1359.

Wong, A.C.-N., Wong, Q.-P., Morimoto, J. et al. (2017). Gut microbiota modifies olfactory-guided microbial preferences and foraging decisions in *Drosophila*. *Current Biology* 27: 2397–2404.

Yang, C.-C., Yu, Y.-C., Valles, S.M. et al. (2010). Loss of microbial (pathogen) infections associated with recent invasions of the red imported fire ant *Solenopsis invicta*. *Biological Invasions* 12: 3307–3318.

Zheng, H., Powell, J.E., Steele, M.I. et al. (2017). Honeybee gut microbiota promotes host weight gain via bacterial metabolism and hormonal signaling. *Proceedings of the National Academy of Sciences USA* 114: 4775–4780.

9

The Nature and Nurturing of Animal Minds

Alex Thornton and Neeltje J. Boogert

Centre for Ecology & Conservation, University of Exeter, Penryn Campus, Penryn, TR10 9EZ, UK

This chapter deals with ideas as old as western philosophy itself. What is the nature of the mind, and how is it shaped? What is humanity's place in nature? In Aristotelian philosophy, nature was conceived as a linear, ladder-like progression of forms, from the lowly to the divine. During the Middle Ages, beautiful tableaux depicted this *scala naturae* as a glorious ladder of life with God and heavenly beings followed in descending order down the rungs by noblemen (not women) and commoners and then in turn by wild animals, domesticated animals, plants, and minerals (Figure 9.1). Under this view, the human mind was uniquely endowed with the capacity for thought, a capacity that separates us from the rest of the animal kingdom and links us to the divine. Whereas animals were mere automata, Descartes (1637/1994) taught that humanity had a dual nature: a material body inhabited by a divine soul (residing in the pineal gland). Through this duality, we alone could reason and think.

Thus, our mental lives – our thoughts, emotions and virtues – were a God-given part of our nature. However, as the Enlightenment dawned, a different view began to prevail, in which humans were increasingly seen as agents of their own destiny. In an essay published in 1690, the father of empiricism John Locke argued that the human mind was a blank slate (or *tabula rasa*), whose capacity for reason and knowledge was shaped by experience. Thus, the seeds of the nature–nurture debate were sown, but the debate concerned the human mind alone; the mere notion that other animals might have mental lives would have been laughable to most scholars at the time. Darwin's insights (1859, 1872) fundamentally overturned our conceptions of nature, revealing life not as a ladder but as a branching tree, its tips inhabited by 'endless forms most beautiful', each adapted physically, behaviourally and mentally to the specific challenges of its environment. Thus, by arguing that the intellectual difference between humans and other animals was one of degree and not of kind, Darwin revolutionized our conception of humanity's place in nature and opened up the possibility of scientific investigation of animal minds.

Despite the triumphs of Darwinism, pre-Darwinian thinking regarding the minds of animals has continued to cloud the perceptions of the public and scientists alike. As recently as 1980, the eminent psychologist Arthur Jensen echoed the *scala naturae* view of life, claiming that intelligence increases progressively across 'different levels of the phyletic scale – that is, earthworms, crabs, fishes, turtles, pigeons, rats, and monkeys' (Jensen 1980, p. 177). While we now may scoff at Jensen's naïveté, researchers often

Genes and Behaviour: Beyond Nature-Nurture, First Edition.
Edited by David J. Hosken, John Hunt and Nina Wedell.

Figure 9.1 The Great Chain of Being, reproduced from the Retorica Christiana by Didacus Valdes in 1579. Similar conceptions of a ladder-like progression of intellectual abilities continue to dog the study and public perception of animal cognition.

continue to rank animals' cognitive performance based on the extent to which they match humans' achievements. There also remains a residual assumption that animal behaviour is predominantly instinctive, or at most controlled by primitive learning processes, in contrast to the rational decision making thought to underlie human behaviour. However, the history of the field of comparative cognition reads as a litany of the demise of supposedly uniquely human traits: tool manufacture, teaching, imitation, episodic memory, theory of mind, and so on (Goodall 1986; Thornton and McAuliffe 2006; Whiten et al. 1996; Clayton and Dickinson 1998; Krupenye et al. 2016).

In this chapter we will probe the nature of the animal mind. We will begin by considering the evidence that cognition evolves and is shaped by genetic inheritance. We then turn our attention to the ways in which experiences throughout development shape animal minds. Finally, we consider the interplay between nature and nurture. Research in comparative cognition lags far behind other areas of biology in our understanding of gene by environment interactions, but recent developments offer some promise that we may at last begin to move away from asking whether geniuses are born or made.

9.1 Cognition Evolves

While psychologists commonly use rather restrictive and anthropocentric definitions of cognition, animal behaviour researchers tend to favour Shettleworth's (2010) broader conception of cognition as all the neural processes involved in acquiring, processing, storing and using information to guide decisions. At its core, cognition involves neuronal processing to reduce informational uncertainty. This allows individuals to track statistical regularities in the environment and modify their behaviour in response to changing conditions. If the environment were either entirely stable and predictable or completely unpredictable, there would be no benefit in tracking environmental information to update one's behaviour, and reflexive stimulus–response mechanisms would suffice.

From this perspective, all animals may be considered cognitive, but natural selection shapes and constrains the manifestation of cognitive traits depending on the environmental problems individuals encounter in the same way as it shapes morphological, physiological and behavioural characters. However, in contrast to other biological traits, cognition is unobservable and can only be inferred through observation of behavioural or neural responses in carefully designed experiments. As a result of the difficulties in identifying and quantifying variation in cognitive traits, progress has lagged far behind other areas of biology, and methodological and conceptual debates continue to rage (Penn et al. 2008; Rowe and Healy 2014a; Thornton and Lukas 2012; Thornton et al. 2014).

9.1.1 Adaptive Cognitive Specializations

The view of cognitive traits as being subject to natural selection may strike us now as obvious, but not so long ago it would have been seen by many psychologists as heretical. The behaviourist school of thought, exemplified by Skinner and Watson, effectively extended Locke's *tabula rasa* view of human behaviour to all animals. According to the behaviourists, all behaviour could be explained through universal processes of associative learning, whereby any previously neutral stimulus could become associated with positive or negative outcomes. Thus, behaviour was entirely moulded by experience. As a morbid illustration of the point, a hapless nine-month-old infant, Albert, was trained through Pavlovian conditioning to fear fluffy, white objects, a fear which later seemed to generalize to other similar stimuli, including Father Christmas's beard (Harris 1979).

This view of universal, limitless learning was turned upside down by John Garcia's discovery that some associations are easier to learn than others. In a series of classic experiments, rats were exposed to ionizing radiation to induce nausea after drinking water that was paired with either a novel flavour or an audio-visual stimulus ('tasty' water or 'bright noisy water') (Garcia and Koelling 1966). While the rats readily learned to avoid the flavoured, 'tasty' water, they did not appear to learn to avoid the 'bright noisy' water.

This conditioned taste aversion (also known as the 'Garcia effect') provides a beautiful illustration of how selection shapes cognition, favouring learning of biologically relevant information: while flavour is often a reliable cue of the palatability of food, noise and light are not. We now know that such biological constraints of learning are common, serving to limit learning of functionally irrelevant contingencies. A monkey seeing a conspecific appearing to respond in alarm to a snake or a bunch of flowers, for instance, will subsequently learn to fear the former but not the latter (Mineka and Cook 1988).

While some evolutionary influences on learning may be universal (after all, flowers are unlikely to launch a vicious attack on any creature), in many cases cognitive demands will differ depending on the species' ecological niche. Arguably, the most compelling examples of cognitive adaptations occur in food-caching birds, which store food during the autumn for later consumption in the winter. Storing food is of little use if you cannot remember where you stored it, so food-caching species typically have a larger hippocampus (the region of the brain associated with spatial memory) and are able to remember spatial locations for longer than non-cachers (Biegler et al. 2001; Krebs 1990).

Similar species differences in cognitive specializations may be linked to variation in social ecology. In tests of transitive inference, for example, individuals must infer that if A beats B and B beats C, then A must beat C. Such an ability would be useful in a hierarchically structured society as a means for individuals to gauge their positions relative to other group members without having to fight every individual in the group. Accordingly, within the corvids (birds of the crow family), the highly social pinyon jay (*Gymnorhinus cyanocephalus*) outperforms the less social Western scrub jay (*Aphelocoma californica*) in transitive inference tasks (Bond et al. 2003). Similar cognitive differences related to socio-ecology have now been reported across a range of taxa, including fish, other birds and primates (Bshary et al. 2002; Maclean et al. 2008; Scheid and Bugnyar 2008).

Variation across ecological niches may also drive species differences in the *sources* of information on which animals rely. Animals living in social groups may gather information either through their own interactions with the environment ('personal information') or by observing and interacting with others ('social information'). While personal information is likely to be more accurate, using social information can allow individuals to bypass the costs of trial and error learning and effectively parasitize information from others, but at the risk that the information may be inaccurate, irrelevant or outdated.

Consequently, one would predict that the costs of information gathering will determine the extent to which individuals rely on personal versus social information (Boyd and Richerson 1985). In accordance with this prediction, three-spined sticklebacks (*Gasterosteus aculeatus*), which are armoured with robust defences, will ignore social information when assessing the relative quality of foraging patches. In contrast, the poorly defended and vulnerable nine-spined stickleback (*Pungitius pungitius*) prefers to remain under cover, observing the foraging success of conspecifics before choosing a patch (Coolen et al. 2003). Whether this species difference reflects underlying differences in learning mechanisms or simply differential attention to social and non-social stimuli remains unclear (Heyes and Pearce 2015; Webster and Laland 2015). Either way, this example neatly illustrates how varying ecological pressures may drive species differences in information gathering.

Varying ecological pressures also have the potential to generate differences not only between species but within species. In an elegant series of experiments, Pravosudov and Clayton showed that in black-capped chickadees (*Poecile atricapilla*), climatic variation across the species' range is associated with cognitive differences between allopatric populations. In one study, chickadees were caught from two different populations: one in the harsh Arctic climate of Alaska, where food availability is limited and unpredictable, and the other in the relatively benign environment of Colorado. When tested under identical lab conditions, the Alaskan birds cached more food, were more efficient in retrieving their caches and showed greater accuracy in one-trial spatial learning performance, but did not differ in non-spatial associative learning from their Colorado conspecifics (Pravosudov and Clayton 2002).

This result appears consistent with the suggestion that natural selection acting on heritable genetic variation has shaped spatial cognition in the two populations, allowing birds to survive in harsh Arctic climates by accurately retrieving the food they have cached for the barren winter months. However, as the birds were trapped as adults, it remains possible that population differences could be linked to developmental rather than genetic effects. To address this possibility, a subsequent study reared 10-day-old chickadee chicks from the northern and southern extremes of their range (Alaska and Kansas, respectively) in a common garden environment. When subsequently tested as adults, the Alaskan birds outperformed those from Kansas in spatial memory tasks, and had substantially more neurons in the hippocampus (Roth et al. 2012) (Figure 9.2a,b). Moreover, later work indicates that the two populations exhibit differential expression of genes, some of which are thought to be associated with hippocampal function (Pravosudov et al. 2013).

While these results are consistent with the argument that neural and cognitive differences have evolved in response to local climatic conditions, it is important to note that maternal or other environmental effects prior to 10 days (when the chicks were caught from the wild) cannot be ruled out. It thus remains unclear whether differential gene expression results from genetic or epigenetic differences. Indeed, this work has also highlighted the potential importance of environmental effects, as both captive populations had lower overall hippocampal volumes than their wild counterparts (Roth et al. 2012; Figure 9.2c). The extent to which genetic and environmental effects interact to influence cognition remains very much an open question, which we shall return to later.

9.1.2 Heritability of Cognitive Traits

If selection is to cause evolution of cognitive traits, they must of course be heritable. A small number of studies have revealed the importance of specific genetic loci (e.g. for associative learning in *Drosophila* (Mery et al. 2007) and in human psychiatric disorders (Skuse et al. 1997)), but it is highly likely that most cognitive traits will be influenced by the expression of multiple different genes (Deary et al. 2009) (see Chapter 5 for evidence of the heritability of behavioural traits in humans). Thus, research has tended to focus not on the identification of specific loci but rather on estimating the heritability of cognitive or neuroanatomical traits (see, for example, Table 1.1, Chapter 1). To date, the majority of work has focused on humans and other primates, with a small but growing body of evidence showing substantial heritability in general intelligence (positively covarying performance across batteries of diverse cognitive tests), memory and even educational attainment (Croston et al. 2015). In non-primates, a handful of studies provide some evidence for heritability in traits including avoidance conditioning and learning ability as well as avian song traits and their associated neural correlates (Croston et al. 2015).

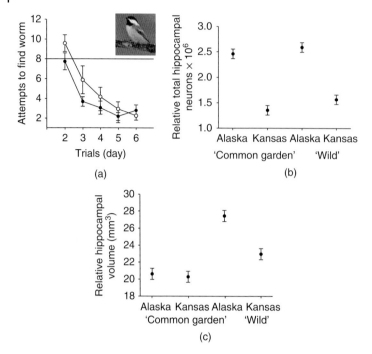

Figure 9.2 (a) Black-capped chickadees from Alaska (*black circles*) were faster and more accurate in an associative spatial learning task than conspecifics from Kansas (*white circles*) raised in the same common garden environment. The horizontal line indicates the number of attempts to locate the food expected by chance. (b) Alaskan birds had more hippocampal neurons (relative to total number of brain neurons) than those from Kansas, regardless of whether they were raised in a common garden environment or in their natural environments in the wild. This finding is suggestive of genetically controlled differences between the populations. However, environmental effects are also important (c) as wild birds from both populations showed greater hippocampal volumes than their counterparts raised in captivity. Source: Figures reproduced with permission from Roth et al. (2012). Photo of black-capped chickadee reproduced under GNU Free Documentation License.

Through these studies, our understanding of how cognitive traits are inherited across generations is beginning to grow. However, it is important to note that as selection does not act on traits in isolation, heritability estimates of single traits may be of limited value. Moreover, the manifestation of an individual's cognitive performance is likely to be influenced by a multitude of other variables including personality traits and life history strategies (Sih and Del Giudice 2012; Griffin et al. 2015; Thornton and Lukas 2012) (see also Chapter 2). Thus, to fully understand how selection acts on cognitive traits, future work must begin to consider the genetic covariance between a host of cognitive, behavioural and life history traits and their associated fitness outcomes (Thornton and Wilson 2015).

In recent years, studies have begun to reveal how artificial selection on heritable cognitive traits may generate associated responses in other traits. *Drosophila* lines selected for increased associative learning ability, for instance, show reduced longevity (Burger et al. 2008) and their larvae show reduced competitive ability (Mery and Kawecki 2003). In female guppies (*Poecilia reticulata*), selection for large brains appears to generate similar trade-offs, resulting in elevated performance in a numerical discrimination test (but see Healy and Rowe 2013), but also in reduced gut size (Kotrschal et al. 2013). A key priority for future research is to determine the heritability of cognitive traits in wild populations, and potential covariance between different traits.

9.2 Cognition Develops

As we age, we acquire knowledge, skills, and habits as a result of our interactions with the environment. But can environmental influences shape not just what we know, but the underlying processes through which we acquire and use our knowledge?

A growing body of evidence indicates that cues and stressors in the physical and social environment may have profound influences on the development of cognitive processes. Research has typically focused on developmental influences early in life, as these are thought to have particularly pronounced effects. This is because young, inexperienced individuals are likely to have high levels of uncertainty as to the state of the world (Fawcett and Frankenhuis 2015; English et al. 2016). Nevertheless, it is important to remember that development is a continuous process, and may continue to shape the expression of cognitive traits throughout life. One particularly striking example of this is found in London taxi drivers, whose extensive experience of daily navigation through the city's maze of streets appears to be reflected in a larger posterior hippocampus than control subjects who do not drive taxis (Maguire et al. 2000). Moreover, in contrast to bus drivers, who drive fixed routes, taxi drivers' need to navigate unpredictable routes throughout the city also appears to be reflected in improved spatial cognitive performance such as better landmark recognition (Maguire et al. 2006).

9.2.1 Cognitive Consequences of a Poor Start in Life

Environmental conditions may have a variety of developmental effects on cognitive traits, ranging from the pathological to the adaptive. Neural tissue is extremely metabolically expensive to form and maintain (Aiello and Wheeler 1995), so the availability of resources, particularly during early life, may constrain neuroanatomical and cognitive development. Research into the effects of early nutritional stress on cognitive development was largely driven by attempts to understand the adaptive value of song learning in birds (Nowicki et al. 1998). Male songbirds learn to sing by copying the songs of adult males in the vicinity, and females prefer males with elaborate songs (Catchpole and Slater 2003). Chicks that experience early nutritional stress cannot afford to invest as much in growing the brain nuclei underlying song learning (primarily the higher vocal centre) relative to their better-fed counterparts. Consequently, males that have had a poor start in life become poor singers and suffer reduced reproductive success (Spencer et al. 2005).

Later work has shown that the negative effects of poor developmental conditions are not restricted to song learning, but may influence a range of cognitive traits including spatial memory and associative learning across taxa (Buchanan et al. 2013). For instance, in Western scrub jays, a food-caching corvid species, experimental food deprivation in early life led to substantially impaired performance in spatial memory tasks relative to control birds. These impairments were reflected in reduced hippocampal volume and neuron numbers (Pravosudov et al. 2005).

Cognitive development may be influenced not only by how much you eat, but also by what you eat. Seabirds, for example, thrive on a diet of lipid-rich fish. In a study of red-legged kittiwakes (*Rissa brevirostris*), chicks reared on lipid-poor diets showed important impairments in an associative colour discrimination task, taking substantially longer than control birds to learn that food could be found only in either black or white dishes. This reduced ability to use visual cues to learn about the locations of food has been argued to play an important role in recent, sharp population declines in seabirds as result of human-induced changes in the availability of lipid-rich fish (Kitaysky et al. 2006).

Although nutritional deprivation can have major negative effects on the development of cognitive function, in social species early-life exposure to social interactions may be no less important for later cognitive function. Familiar, but happily rare, examples of 'feral' human children, who have managed to survive outside society, highlight the severe linguistic and cognitive impairments that can result from a life devoid of social interaction (Newton 2002). Better studied are tragic cases of large-scale institutional abuse. Children who suffered severe socio-emotional deprivation in Romanian orphanages, for instance, often showed impairments in brain function and sociocognitive ability (Nelson 2007; Chugani et al. 2001). Similar effects of social deprivation have been reported in other primates and in rodents (Würbel 2001; Winslow et al. 2003). Nevertheless, in many, if not all of these cases, it is difficult to disentangle the specific effects of social deprivation from nutritional effects and other stressors induced by impoverished rearing conditions.

9.2.2 Cognitive Silver Spoons

Thus far, we have painted a gloomy picture of reduced cognitive function resulting from a poor start in life. One flip side of this picture is that benign conditions may induce the so-called 'silver spoon' effect: those individuals lucky enough to be raised in times of plenty may reap cognitive rewards. The majority of research on cognitive silver spoons to date has focused on humans. For example, growing up in bilingual environments has been shown to be associated with a range of positive outcomes, including not only linguistic abilities but also cognitive traits such as enhanced working memory, attentional control and abstract thinking (Adesope et al. 2010). In recent years, whole 'brain training' industries have sprung up, claiming to harness the effects of a plethora of interventions including nutritional supplements, yoga, massage, and dance to promote infants' cognitive development. While most of these claims lack any clear scientific basis, there is some evidence that musical training in childhood can enhance cognitive function (Moreno et al. 2011; Schlaug et al. 2005).

In non-human animals, research has tended to focus on the negative cognitive effects of deprivation, but there is some evidence for silver spoon effects. For instance, in some passerine birds, parents provision their young with large numbers of spiders early in life. Spiders are particularly high in the amino acid taurine, which is thought to be vital for normal brain growth and development in mammals (Aerts and van Assche 2002). In blue tits (*Cyanistes caeruleus*), experimental supplementation of taurine has been shown to be associated with improved ability to learn to remove an obstacle to uncover hidden seed and to subsequently remember the location of the food (Arnold et al. 2007). Similar positive cognitive effects may also arise if individuals engage in numerous and varied social interactions. In captive rhesus macaques (*Macaca mulatta*), for example, individuals that had been living in larger social groups had increased grey matter and increased neural connectivity in some brain regions than those from small groups (Sallet et al. 2011). Whether these neuroanatomical differences are associated with particular benefits for cognitive function remains to be investigated. However, recent work indicates that early-life social conditions can have substantial impacts on cognitive development, with knock-on consequences for reproductive success. In Western Australia, Australian magpies (*Cracticus tibicen dorsalis*) live in stable groups. Using a battery of cognitive tasks, Ashton et al. (2018) showed that individuals that grow up in larger social groups show elevated cognitive performance and that females that perform well in tasks have elevated reproductive success. Thus, social factors may influence the development of cognitive abilities, which in turn affect fitness.

9.2.3 Adaptive Developmental Plasticity in Cognition

An important question that is receiving increasing attention from evolutionary biologists is whether developmental responses to environmental conditions early in life may be adaptive, effectively preparing animals for circumstances they will encounter later. Adaptive developmental plasticity occurs if there has been selection for the expression of a particular phenotype conditional on having experienced particular developmental inputs (Nettle and Bateson 2015). If environments show temporal autocorrelation, such that conditions in early life are likely to be predictive of those encountered later, then cues encountered during early development may help to shape phenotypes so as to maximize later gains (Buchanan et al. 2013; Fawcett et al. 2014; Monaghan 2008). Although such adaptive phenotypic plasticity has attracted a great deal of attention, a recent meta-analysis of examples in plants and animals suggests that the evidence is weak at best (Uller et al. 2013). Nevertheless, there are some tantalizing suggestions that the means by which animals gather information may be shaped by early developmental experiences. In particular, evidence for such developmental effects is beginning to accumulate in the field of animal social learning, although in virtually all cases it remains to be established whether this phenotypic plasticity is actually adaptive.

There is extensive evidence that animals across a wide range of taxa are able to learn socially by observing and interacting with others (Hoppitt and Laland 2013). Social learning can provide substantial benefits by allowing individuals to bypass the costs associated with learning through individual experience. However, learning from others is not always beneficial: individuals that copy others blindly are liable to acquire outdated or irrelevant information. Theoretical models show that the solution to this problem is to use 'social learning strategies'; that is, rules that determine when, how and from whom to learn (Boyd and Richerson 1985; Hoppitt and Laland 2013). Such strategies have been documented in taxa from insects to birds and humans (Laland 2004; Heyes 2016), and are generally assumed to have evolved through natural selection. The possibility that social learning strategies are moulded by experience has received relatively little attention until recently.

In birds, mothers are able to modify the yolk composition of their eggs. For example, mothers exposed to unpredictable food availability may deposit increased levels of the avian stress hormone corticosterone (CORT) into their eggs (Henriksen et al. 2011). These changes in CORT levels can be mimicked experimentally by injecting the hormone directly into the eggs. Adult Japanese quail (*Coturnix japonica*) that had been exposed to experimentally elevated CORT in the egg were more likely to copy the choices of conspecific demonstrators trained to feed from one of two novel food sources. In contrast, chicks exposed to unpredictable food availability in early life were more inclined to choose the container that demonstrators did not feed from (Boogert et al. 2013). These results suggest that the nature and timing of developmental stressors may influence later information-gathering strategies. They are also consistent with theoretical and empirical evidence that individuals are more likely to use social information when uncertain (Laland 2004; Rafacz and Templeton 2003), as elevated egg CORT levels may be associated with maternal uncertainty about environmental conditions. The chicks exposed to unpredictable food, on the other hand, may have chosen the unpopular food container so as to avoid food competition. Further work is needed to determine whether these effects necessarily provide fitness benefits.

Similar experiments suggest that early-life conditions may modify information-gathering strategies in highly gregarious animals such as colonial songbirds. While unmanipulated juvenile zebra finches (*Taeniopygia guttata*) preferentially copied their parents to solve a novel foraging task, their CORT-fed siblings copied only unrelated adults (Farine et al. 2015). Similar patterns were observed when juvenile males learned their songs, with controls learning from their fathers while CORT-fed birds were less inclined to do so (Boogert et al., unpublished data). Together, these findings raise the possibility that stressed juveniles use the fact that they are stressed as a cue that their parents have made poor choices in life and are therefore not to be copied. However, the alternative explanation, that the parents of stressed chicks are less tolerant (and so more difficult to copy), cannot yet be excluded.

While there is mounting evidence that social learning strategies may be influenced by stress, they may also be shaped through learning from previous experiences. In other words, social learning strategies may themselves be learned (Heyes 2016; Mesoudi et al. 2016). Human populations, for instance, show cultural differences in patterns of social learning. In computer-based experiments, people from mainland China showed a higher tendency to copy others than did people from Hong Kong, the UK and Chinese immigrants in the UK. This difference was attributed to cultural differences in social norms between collectivist and individualist societies (Mesoudi et al. 2015). There is also evidence from other species that associative learning processes shaped by past experience may help to determine social learning strategies. In a study of house sparrows (*Passer domesticus*), chicks were assigned to two experimental groups: in one, a parent model (a stuffed female adult conspecific) visited locations containing food while in the other treatment group the model visited unprofitable locations. After five days of training in which they followed the parent model, chicks in the former group were significantly more likely to join others when searching for food compared to chicks in the unhelpful mother model group, suggesting that past experience shapes reliance on social information (Katsnelson et al. 2008). Similarly, in fringe-lipped bats (*Trachops cirrhosis*), a tendency to use the social learning rule 'copy others when dissatisfied' can be explained through learned associations from previous interactions with food sources in the presence or absence of conspecifics (see Heyes' (2016) interpretation of experiments by Jones et al. 2013).

The potential for adaptive phenotypic plasticity in response to information acquired during development may also help to explain some of the more puzzling aspects of human and non-human psychology, including the prevalence of patterns of behaviour that appear to violate economically rational expectations (Fawcett et al. 2014). For instance, economically rational decisions ought to be based on accurate estimates of the current value of alternative options, but both human and non-human animals commonly exhibit distinct 'optimistic' or 'pessimistic' biases, judging outcomes as better or worse than they really are. Recent theoretical work suggests that such cognitive or emotional biases may generate important benefits and have evolved under natural selection (Fawcett et al. 2014; Nettle and Bateson 2012). Animals in poor environments should be risk averse, seeking to avoid the negative consequences of poor decisions that can edge them closer to death, so they may benefit from 'playing it safe' and from interpreting ambiguous stimuli unfavourably. If an animal experiences poor environmental conditions, and environmental quality remains stable over time, this could induce long-term pessimistic biases. For instance, in an elegant experiment, honeybees (*Apis mellifera*) were trained that one odour was associated with sucrose rewards, inducing them to extend their mouthparts, while another was associated with bitter-tasting quinine, inducing them to withhold their mouthparts. Some bees were then shaken vigorously to simulate a vicious attack by a nest predator. When subsequently presented with novel odours that were intermediate in composition between the previously trained positive and negative stimuli, shaken bees were substantially more likely to withhold their mouthparts compared to the controls (Bateson et al. 2011). These results suggest that negative experiences in bees can induce pessimistic emotion-induced biasing of information processing analogous to that seen in humans. Similar results have been reported in a number of bird and mammal species (Bateson 2016). Conversely, several studies have shown that environmental enrichment may be linked to optimistic overestimates of the outcomes of ambiguous stimuli (Bateson 2016). To date, no study has yet documented the occurrence of such cognitive biases in wild animals, so their potential adaptive value remains to be confirmed.

While it is becoming increasingly clear that cognitive traits are shaped by developmental processes, we still know little about whether this developmental plasticity is necessarily adaptive. Even less is known about the means by which an individual's genetic endowments interact with its developmental experiences to influence the expression of its cognitive traits. We turn our attention to this mysterious issue in the next section.

9.3 Cognitive Reaction Norms: Mind-Moulding Gene-by-Environment Interactions

Thus far, we have presented evidence that cognitive traits are shaped by both genetic inheritance and developmental factors, but how do nature and nurture intertwine? Evolutionary biologists working on other traits, including physiology, morphology and behaviour, are increasingly exploring this issue (Hunt and Hosken 2014) (see Chapters 1, 4, and 7). For example, phenotypic 'reaction norms' are used to depict how the phenotypic consequences of a given genotype are altered by particular environmental conditions (Figure 9.3c). However, in a recent review, Buchanan et al. highlighted the fact that not a single study has yet addressed how such reaction norms may underlie the expression of cognitive traits in non-human animals (Buchanan et al. 2013). To our knowledge, this remains the case today.

The core of the problem for researchers is that individual variation in cognitive traits is not directly observable, and thus extremely difficult to quantify. Research in comparative cognition has therefore focused primarily on detecting the existence of particular cognitive traits in the first place, rather than probing whether and why these traits may vary within species (Thornton and Lukas 2012). Nevertheless, there is a growing consensus that if we are to understand how cognitive traits evolve, we must move to the individual level of analysis to determine how gene–environment interactions give rise to the phenotypes that are exposed to natural selection (Rowe and Healy 2014b; Thornton et al. 2014; Morand-Ferron et al. 2015).

9.3.1 The Mystery of (the Lack of) Cognitive Resilience

It is commonly assumed (but seldom tested) that elevated cognitive performance provides fitness benefits. For example, a food-caching bird that can accurately remember the location of its caches ought to be more likely to survive the winter than a more forgetful peer. A major outstanding question, therefore, is why, if cognitive traits are so important, they appear to be so sensitive to developmental stressors. For example, as we considered in the previous section, early-life food deprivation in Western scrub jays resulted in reduced hippocampal volume and spatial memory (Pravosudov et al. 2005). Given this species' reliance on food caching, why has selection not acted to constrain plasticity, thus safeguarding the neurocognitive traits that individuals rely upon for survival? A similar argument holds for bird song which, although a critical component of reproductive success, seems particularly vulnerable to impairments resulting from developmental stressors (Buchanan et al. 2013). One possibility is that these apparent negative effects are artefacts of experimental conditions. First, it is possible that the levels of environmental perturbation imposed in experiments would reduce the probability of survival in the wild to such an extent that the effects on cognition are irrelevant, so selection could not drive cognitive resilience. Alternatively, as there are no studies of the long-term effects of these developmental stressors, it is possible that animals are able to compensate or bounce back under benign conditions later in life.

If it is confirmed that early-life stresses produce long-lasting cognitive impairments in wild animals, one potential explanation might be that cognitive processes must, by their nature, be plastic to cope with varying informational demands. However, this may generate a double-edged sword, as this need for plasticity renders cognition particularly vulnerable to perturbation. In seasonally breeding songbirds, for example, the size of the song control nuclei (HVC and robust nucleus of the arcopallium (RA)) increases during the breeding season and declines afterwards when courtship and territorial song are no longer needed (Catchpole and Slater 2003). Here neuronal plasticity allows individuals to invest in song production when required, but this very plasticity may render the song system vulnerable to stresses. To examine the possibility that variation in the benefits of plasticity may generate differences in developmental resilience between species or populations, one valuable approach may be to compare reaction norms in animals that are more or less reliant on a particular cognitive function (e.g. bird species or populations varying in their reliance on cached food in winter). Similarly, it would be interesting to compare reaction norms within populations for traits that differ in their assumed importance for fitness.

9.3.2 Practice Makes Perfect: Genetic Quality and Cognitive Silver Spoons

As we have seen, just as poor environmental conditions may impair cognitive development, good conditions may promote cognitive performance. One as yet unexplored possibility is that, under the right conditions, individuals may be able to drive their own cognitive development. Young animals often spend time and energy in behaviour that appears to have no current benefit. Meerkat pups (*Suricata suricatta*), for example, spend a great deal of time digging ineffectually in the sand, but very rarely find prey on their own, and are reliant on adults to feed them (Figure 9.3a). Pups that are in good body condition tend to spend more time digging and less time begging for food than pups in poor condition (Figure 9.3b), and show elevated foraging efficiency later in life (Thornton 2008). Given extensive evidence that prior experience in cognitive tasks boosts later cognitive performance (Thornton and Lukas 2012), it is possible that such practice-make-perfect effects could allow individuals to promote their own cognitive development. Indeed, this may help to explain the adaptive function of play, which is observed at higher frequencies in individuals in good condition (Sharpe et al. 2002) and

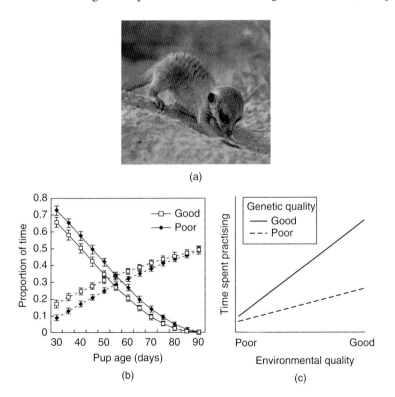

(a)

(b)

(c)

Figure 9.3 (a) A meerkat pup digging in the sand, and failing to find any food. Photo credit: Alex Thornton. (b) Meerkat pups in good body condition spend less time begging to adults for food (*solid lines*) and more time practising foraging (*dashed lines*) than those in poor condition. Source: Reproduced with permission from Thornton (2008). (c) A hypothesized reaction norm illustrating how individuals may drive their own cognitive development. If high-quality individuals in good environmental conditions can afford to invest time and effort in practising for the future, this may boost their future cognitive performance.

may help to facilitate brain and cognitive development (Ferchmin and Eterovic 1982). Here, the interplay between genetic quality and current condition may be critical. For any given level of environmental quality, individuals of higher genetic quality may be expected to invest more in practising to improve their cognitive performance. For individuals of high genetic quality in benign conditions, this could generate strong positive feedback loops whereby early-life silver spoon conditions are amplified, resulting in particularly 'clever' individuals (Figure 9.3c).

9.3.3   Cultural and Epigenetic Inheritance of Cognitive Traits

When biologists think of inheritance, they typically think of genetic inheritance. However, there is now abundant evidence that behavioural traits may be inherited culturally (see Chapters 3 and 10), as a result of social learning between generations. Examples of this in nature include song learning in passerine birds (Fehér et al. 2009; Catchpole and Slater 2003), as well as daily activity budgets in meerkats (Thornton et al. 2010) and food preferences and foraging techniques in a range of vertebrates (Galef and Giraldeau 2001; Thornton and Clutton-Brock 2011; Slagsvold and Wiebe 2007; Aplin et al. 2015; Allen et al. 2013). These behavioural traits are the *products* of cognition (specifically social learning), but it remains to be seen whether cognitive *processes* themselves can be similarly culturally inherited; in other words, might individuals socially learn how to learn and think?

There is some evidence to suggest that this may be the case in humans, where, for example, people in regions with a cultural history of collectivism tend to show higher levels of holistic thinking than people from regions with a history of individualism (Talhelm et al. 2014; Mesoudi et al. 2016). There is also the potential for cognitive traits to be passed on through the generations through epigenetic inheritance. A recent example is found in Japanese quail, where mothers exposed in the egg to experimentally elevated CORT levels showed changes in stress physiology, neurotransmitter expression and reduced neophobia, facilitating the discovery of food in a novel environment. Remarkably, the same stress-coping physiological, neurological, and behavioural phenotypes were observed in the offspring of these mothers, who themselves had not been exposed experimentally to CORT (Zimmer et al. 2017). Some researchers have recently claimed that such examples of non-genetic trait inheritance call for a dramatic extension of the Modern Synthesis in evolutionary biology (Laland et al. 2015). Although many feel that such claims are overblown (see debates in Laland et al. 2014), understanding the interplay between genetic and non-genetic inheritance is a clear priority for future research.

9.3.4   Gene by Environment and Methodological Issues in Comparative Cognition

Acknowledging the potential interplay between genetic and environmental factors in shaping cognition has fundamental implications for the way in which cognitive research is conducted and interpreted. In the field of comparative cognition, the results of comparisons between species are typically interpreted as genetically controlled traits shaped by natural selection. These comparisons are then used, either implicitly or explicitly, to rank species according to their supposed similarity to humans. However, such comparisons typically confound genetic differences with developmental effects related to factors such as age, rearing conditions and prior experimental experience. For instance, a large body of research focuses on comparisons between human children and apes, but almost invariably uses adult apes. This age-confound, combined with radical differences in the physical and social rearing conditions of the test subjects, renders any meaningful interpretation of the results very difficult. Moreover, such comparisons tend to ignore the variation that exists within species. For example, enculturated apes that have been raised from a young age in human environments typically outperform conspecifics across a range of cognitive tasks (Lyn et al. 2010; Thornton and Lukas 2012). These tasks often require apes to interact with, or learn from, human demonstrators. This suggests that performance on these tasks is more reflective of developmental conditions than of genetically endowed adaptive cognitive abilities.

Comparing animals to humans may also carry the assumption that the trait in question is universal and adaptive in humans. For example, mirror self-recognition is used as a test of self-awareness in animals. A small and select number of species are thought to have the 'capacity' to recognize themselves in mirrors, but often on the basis of a very small proportion of test subjects actually passing the test (Thornton and Lukas 2012). Exacerbating this issue, the development of mirror self-recognition appears to be far from universal in our own species: while children from western societies display self-oriented behaviours in front of a mirror from 1.5–2 years old, children in a host of non-western societies do not display these behaviours until much later (Broesch et al. 2011). This raises the question of what the underlying trait is that mirror self-recognition tests actually capture.

Rather than focusing on crude differences between species irrespective of developmental effects, on the assumption that these species differences have adaptive meaning, a more valuable approach would be to address what drives variation in the developmental trajectory of attentional, learning or reasoning processes that determine performance on the test. Theoretical models are beginning to consider how seemingly complex cognitive traits such as Theory of Mind, which are often assumed to have arisen *de novo* as distinct cognitive modules, may instead be shaped gradually through the coevolution of information acquisition (e.g. the particular environmental stimuli an animal attends to) and processing (e.g. learning) (Goldstein et al. 2010; Lotem and Halpern 2012; van der Vaart et al. 2012). Comparative empirical studies considering both animals' evolutionary history and developmental influences are now critical to test theoretical predictions.

9.4   Conclusion

Perhaps the clearest theme that has emerged throughout this chapter is the current lack of understanding of how nature and nurture combine to shape cognition. Although this may sound rather gloomy, we see it instead as an exciting challenge. The evolution of cognition remains one of the most mysterious aspects of biology. The fact that neurocognitive traits are strikingly plastic makes them extremely difficult to quantify, but also extremely interesting. Understanding cognitive reaction norms is not only of fundamental scientific importance, but also of immense potential practical value, with applications ranging from the treatment of human psychiatric disorders to improving the welfare of captive animals. As animals' cognitive responses to stimuli in their environment may have dramatic consequences for individual fitness and population dynamics, understanding these cognitive processes is also critical in a world dominated by human-induced environmental change (Greggor et al. 2014). The tools we need, ranging from advances in molecular genetics (see Chapter 11) and automated behavioural testing to the more humble but no less important workhorses of field biology such as cross-fostering, exist to tackle this challenge head on.

References

Adesope, O.O., Lavin, T., Thompson, T., and Ungerleider, C. (2010). A systematic review and meta-analysis of the cognitive correlates of bilingualism. *Review of Educational Research* 80: 207–245.

Aerts, L. and van Assche, F.A. (2002). Taurine and taurine-deficiency in the perinatal period. *Journal of Perinatal Medicine* 30: 281–286.

Aiello, L.C. and Wheeler, P. (1995). The expensive tissue hypothesis: the brain and the digestive system in human and primate evolution. *Current Anthropology* 36: 199–221.

Allen, J., Weinrich, M., Hoppitt, W., and Rendell, L. (2013). Network-based diffusion analysis reveals cultural transmission of lobtail feeding in humpback whales. *Science* 340: 485–488.

Aplin, L.M., Farine, D.R., Morand-Ferron, J. et al. (2015). Experimentally induced innovations lead to persistent culture via conformity in wild birds. *Nature* 518: 538–541.

Arnold, K.E., Ramsay, S.L., Donaldson, C., and Adam, A. (2007). Parental prey selection affects risk-taking behaviour and spatial learning in avian offspring. *Proceedings of the Royal Society B: Biological Sciences* 274: 2563–2569.

Ashton, B.J., Ridley, A.R., Edwards, E.K., and Thornton, A. (2018). Cognitive performance is linked to group size and affects fitness in Australian magpies. *Nature* 554: 364–367.

Bateson, M. (2016). Optimistic and pessimistic biases: a primer for behavioural ecologists. *Current Opinion in Behavioral Sciences* 12: 115–121.

Bateson, M., Desire, S., Gartside, S.E., and Wright, G.A. (2011). Agitated honeybees exhibit pessimistic cognitive biases. *Current Biology* 21: 1070–1073.

Biegler, R., McGregor, A., Krebs, J.R., and Healy, S.D. (2001). A larger hippocampus is associated with longer-lasting spatial memory. *Proceedings of the National Academy of Sciences USA* 98: 6941–6944.

Bond, A.B., Kamil, A.C., and Balda, R.P. (2003). Social complexity and transitive inference in corvids. *Animal Behaviour* 65: 479–487.

Boogert, N.J., Zimmer, C., and Spencer, K.A. (2013). Pre- and post-natal stress have opposing effects on social information use. *Biology Letters* 9: 20121088.

Boyd, R. and Richerson, P.J. (1985). *Culture and the Evolutionary Process*. Chicago: University of Chicago Press.

Broesch, T., Callaghan, T., Henrich, J. et al. (2011). Cultural variations in children's self-recognition. *Journal of Cross-Cultural Psychology* 42: 1018–1029.

Bshary, R., Wickler, W., and Fricke, H. (2002). Fish cognition: a primate's eye view. *Animal Cognition* 5: 1–13.

Buchanan, K.L., Grindstaff, J.L., and Pravosudov, V.V. (2013). Condition dependence, developmental plasticity, and cognition: implications for ecology and evolution. *Trends in Ecology & Evolution* 28: 290–296.

Burger, J.M.S., Kolss, M., Pont, J., and Kawecki, T.J. (2008). Learning ability and longevity: a symmetrical evolutionary trade-off in *Drosophila*. *Evolution* 62: 1294–1304.

Catchpole, C.K. and Slater, P.J.B. (2003). *Bird Song: Biological Themes and Variations*. Cambridge: Cambridge University Press.

Chugani, H.T., Behen, M.E., Muzik, O. et al. (2001). Local brain functional activity following early deprivation: a study of post-institutionalized Romanian orphans. *NeuroImage* 14: 1290–1301.

Clayton, N.S. and Dickinson, A. (1998). Episodic-like memory during cache recovery by scrub jays. *Nature* 395: 272–274.

Coolen, I., van Bergen, Y., Day, R.L., and Laland, K.N. (2003). Species difference in adaptive use of public information in sticklebacks. *Proceedings of the Royal Society B: Biological Sciences* 270: 2413–2419.

Croston, R., Branch, C.L., Kozlovsky, D.Y. et al. (2015). Heritability and the evolution of cognitive traits. *Behavioral Ecology* 26: 1447–1459.

Darwin, C. (1859). *On the Origin of Species by Means of Natural Selection*. London: John Murray.

Darwin, C. (1872). *The Expression of the Emotions in Man and Animals*. London: John Murray.

Deary, I.J., Johnson, W., and Houlihan, L.M. (2009). Genetic foundations of human intelligence. *Human Genetics* 126: 215–232.

Descartes, R. (1994). *Discourse on the Method* (ed. G. Heffeman). Notre Dame: University of Notre Dame Press (Original work published 1637).

English, S., Fawcett, T.W., Higginson, A.D. et al. (2016). Adaptive use of information during growth can explain long-term effects of early life experiences. *American Naturalist* 187: 620–632.

Farine, D.R., Spencer, K.A., and Boogert, N.J. (2015). Early-life stress triggers juvenile zebra finches to switch social learning strategies. *Current Biology* 25: 2184–2188.

Fawcett, T.W. and Frankenhuis, W.E. (2015). Adaptive explanations for sensitive windows in development. *Frontiers in Zoology* 12: S3.

Fawcett, T.W., Fallenstein, B., Higginson, A.D. et al. (2014). The evolution of decision rules in complex environments. *Trends in Cognitive Sciences* 18: 153–161.

Fehér, O., Wang, H., Saar, S. et al. (2009). De novo establishment of wild-type song culture in the zebra finch. *Nature* 459: 564–568.

Ferchmin, P.A. and Eterovic, V.A. (1982). Play stimulated by environmental complexity alters brain and improves learning-abilities in rodents, primates, and possibly humans. *Behavioral and Brain Sciences* 5: 164.

Galef, B.G. and Giraldeau, L.A. (2001). Social influences on foraging in vertebrates: causal mechanisms and adaptive functions. *Animal Behaviour* 61: 3–15.

Garcia, J. and Koelling, R.A. (1966). Relation of cue to consequence in avoidance learning. *Psychonomic Science* 4: 123–124.

Goldstein, M.H., Waterfall, H.R., Lotem, A. et al. (2010). General cognitive principles for learning structure in time and space. *Trends in Cognitive Sciences* 14: 249–258.

Goodall, J. (1986). *The Chimpanzees of Gombe: Patterns of Behaviour*. Cambridge: Belknap Press.

Greggor, A.L., Clayton, N.S., Phalan, B., and Thornton, A. (2014). Comparative cognition for conservationists. *Trends in Ecology & Evolution* 29: 489–495.

Griffin, A., Healy, S.D., and Guillette, L.M. (2015). Cognition and personality: an analysis of an emerging field. *Trends in Ecology & Evolution* 30: 207–214.

Harris, B. (1979). Whatever happened to little Albert? *American Psychologist* 34: 151–160.

Healy, S.D. and Rowe, C. (2013). Costs and benefits of evolving a larger brain: doubts over the evidence that large brains lead to better cognition. *Animal Behaviour* 86: e1–e3.

Henriksen, R., Rettenbacher, S., and Groothuis, T.G.G. (2011). Prenatal stress in birds: pathways, effects, function and perspectives. *Neuroscience and Biobehavioral Reviews* 35: 1484–1501.

Heyes, C. (2016). Who knows? Metacognitive social learning strategies. *Trends in Cognitive Sciences* 20: 204–213.

Heyes, C. and Pearce, J.M. (2015). Not-so-social learning strategies. *Proceedings of the Royal Society B: Biological Sciences* 282: 20141709.

Hoppitt, W. and Laland, K.N. (2013). *Social Learning: An Introduction to Mechanisms, Methods and Models*. Princeton: Princeton University Press.

Hunt, J. and Hosken, D.J. (2014). *Genotype-By-Environment Interactions and Sexual Selection*. Chichester: Wiley Blackwell.

Jensen, A.R. (1980). *Bias in Mental Testing*. New York: Free Press.

Jones, P.L., Ryan, M.J., Flores, V., and Page, R.A. (2013). When to approach novel prey cues? Social learning strategies in frog-eating bats. *Proceedings of the Royal Society B: Biological Sciences* 280: 20132330.

Katsnelson, E., Motro, U., Feldman, M.W., and Lotem, A. (2008). Early experience affects producer-scrounger foraging tendencies in the house sparrow. *Animal Behaviour* 75: 1465–1472.

Kitaysky, A.S., Kitaiskaia, E.V., Piatt, J.F., and Wingfield, J.C. (2006). A mechanistic link between chick diet and decline in seabirds? *Proceedings of the Royal Society B: Biological Sciences* 273: 445–450.

Kotrschal, A., Rogell, B., Bundsen, A. et al. (2013). Artificial selection on relative brain size in the guppy reveals costs and benefits of evolving a larger brain. *Current Biology* 23: 168–171.

Krebs, J.R. (1990). Food-storing birds: adaptive specialization in brain and behaviour? *Philosophical Transactions of the Royal Society B: Biological Sciences* 329: 153–160.

Krupenye, C., Kano, F., Hirata, S. et al. (2016). Great apes anticipate that other individuals will act according to false beliefs. *Science* 354: 110–114.

Laland, K.N. (2004). Social learning strategies. *Learning & Behavior* 32: 4–14.

Laland, K., Uller, T., Feldman, M. et al. (2014). Does evolutionary theory need a rethink? *Nature* 514: 161–164.

Laland, K.N., Uller, T., Feldman, M.W. et al. (2015). The extended evolutionary synthesis: its structure, assumptions and predictions. *Proceedings of the Royal Society B: Biological Sciences* 282: 20151019.

Locke, J. (1690). *An Essay Concerning Human Understanding*. London: Bassett.

Lotem, A. and Halpern, J.Y. (2012). Coevolution of learning and data-acquisition mechanisms: a model for cognitive evolution. *Philosophical Transactions of the Royal Society B: Life Sciences* 367: 2686–2694.

Lyn, H., Russell, J.L., and Hopkins, W.D. (2010). The impact of environment on the comprehension of declarative communication in apes. *Psychological Science* 21: 360–365.

Maclean, E.L., Merritt, D.J., and Brannon, E.M. (2008). Social complexity predicts transitive reasoning in prosimian primates. *Animal Behaviour* 76: 479–486.

Maguire, E.A., Gadian, D.G., Johnsrude, I.S. et al. (2000). Navigation-related structural change in the hippocampi of taxi drivers. *Proceedings of the National Academy of Sciences USA* 97: 4398–4403.

Maguire, E.A., Woollett, K., and Spiers, H.J. (2006). London taxi drivers and bus drivers: a structural MRI and neuropsychological analysis. *Hippocampus* 16: 1091–1101.

Mery, F. and Kawecki, T.J. (2003). A fitness cost of learning ability in *Drosophila melanogaster*. *Proceedings of the Royal Society B: Biological Sciences* 270: 2465–2469.

Mery, F., Belay, A.T., So, A.K. et al. (2007). Natural polymorphism affecting learning and memory in Drosophila. *Proceedings of the National Academy of Sciences USA* 104: 13051–13055.

Mesoudi, A., Chang, L., Murray, K., and Lu, H.J. (2015). Higher frequency of social learning in China than in the west shows cultural variation in the dynamics of cultural evolution. *Proceedings of the Royal Society B: Biological Sciences* 282: 20142209.

Mesoudi, A., Chang, L., Dall, S.R.X., and Thornton, A. (2016). The evolution of individual and cultural variation in social learning. *Trends in Ecology & Evolution* 31: 215–225.

Mineka, S. and Cook, M. (1988). Social learning and the acquisition of snake fear in monkeys. In: *Social Learning: Psychological and Biological Perspectives* (ed. T.R. Zentall and B.G. Galef), 51–73. Hillsdale: Erlbaum.

Monaghan, P. (2008). Early growth conditions, phenotypic development and environmental change. *Philosophical Transactions of the Royal Society B: Life Sciences* 363: 1635–1645.

Morand-Ferron, J., Cole, E.F., and Quinn, J.L. (2015). Studying the evolutionary ecology of cognition in the wild: a review of practical and conceptual challenges. *Biological Reviews* 91: 367–389.

Moreno, S., Bialystok, E., Barac, R. et al. (2011). Short-term music training enhances verbal intelligence and executive function. *Psychological Science* 22: 1425–1433.

Nelson, C.A. (2007). A neurobiological perspective on early human deprivation. *Child Development Perspectives* 1: 13–18.

Nettle, D. and Bateson, M. (2012). The evolutionary origins of mood and its disorders. *Current Biology* 22: R712–R721.

Nettle, D. and Bateson, M. (2015). Adaptive developmental plasticity: what is it, how can we recognize it and when can it evolve? *Proceedings of the Royal Society B: Biological Sciences* 282: 20151005.

Newton, M. (2002). *Savage Girls and Wild Boys: A History of Feral Children*. London: Faber & Faber.

Nowicki, S., Peters, S., and Podos, J. (1998). Song learning, early nutrition and sexual selection in songbirds. *American Zoologist* 38: 179–190.

Penn, D.C., Holyoak, K.J., and Povinelli, D.J. (2008). Darwin's mistake: explaining the discontinuity between human and nonhuman minds. *Behavioral and Brain Sciences* 31: 109–130.

Pravosudov, V.V. and Clayton, N.S. (2002). A test of the adaptive specialization hypothesis: population differences in caching, memory, and the hippocampus in black-capped chickadees (*Poecile atricapilla*). *Behavioral Neuroscience* 116: 515–522.

Pravosudov, V.V., Lavenex, P., and Omanska, A. (2005). Nutritional deficits during early development affect hippocampal structure and spatial memory later in life. *Behavioral Neuroscience* 119: 1368–1374.

Pravosudov, V.V., Roth, T.C., Forister, M.L. et al. (2013). Differential hippocampal gene expression is associated with climate-related natural variation in memory and the hippocampus in food-caching chickadees. *Molecular Ecology* 22: 397–408.

Rafacz, M. and Templeton, J.J. (2003). Environmental unpredictability and the value of social information for foraging starlings. *Ethology* 109: 951–960.

Roth, T.C., LD, L.D., Freas, C.A., and Pravosudov, V.V. (2012). Variation in memory and the hippocampus across populations from different climates: a common garden approach. *Proceedings of the Royal Society B: Biological Sciences* 279: 402–410.

Rowe, C. and Healy, S.D. (2014a). Measuring cognition will be difficult but worth it: a response to comments on Rowe and Healy. *Behavioral Ecology* 25: 1298–1298.

Rowe, C. and Healy, S.D. (2014b). Measuring variation in cognition. *Behavioral Ecology* 25: 1287–1292.

Sallet, J., Mars, R.B., Noonan, M.P. et al. (2011). Social network size affects neural circuits in macaques. *Science* 334: 697–700.

Scheid, C. and Bugnyar, T. (2008). Short-term observational spatial memory in jackdaws (*Corvus monedula*) and ravens (*Corvus corax*). *Animal Cognition* 11: 691–698.

Schlaug, G., Norton, A., Overy, K., and Winner, E. (2005). Effects of music training on the child's brain and cognitive development. *Annals of the New York Academy of Sciences* 1060: 219–230.

Sharpe, L.L., Clutton-Brock, T., Brotherton, P.N.M. et al. (2002). Experimental provisioning increases play in free-ranging meerkats. *Animal Behaviour* 64: 113–121.

Shettleworth, S.J. (2010). *Cognition, Evolution and Behaviour*, 2e. Oxford: Oxford University Press.

Sih, A. and del Giudice, M. (2012). Linking behavioural syndromes and cognition: a behavioural ecology perspective. *Philosophical Transactions of the Royal Society B: Life Sciences* 367: 2762–2772.

Skuse, D.H., James, R.S., Bishop, D.V. et al. (1997). Evidence from Turner's syndrome of an imprinted X-linked locus affecting cognitive function. *Nature* 387: 705–708.

Slagsvold, T. and Wiebe, K.L. (2007). Learning the ecological niche. *Proceedings of the Royal Society B: Biological Sciences* 274: 19–23.

Spencer, K.A., Wimpenny, J.H., Buchanan, K.L. et al. (2005). Developmental stress affects the attractiveness of male song and female choice in the zebra finch (*Taeniopygia guttata*). *Behavioral Ecology and Sociobiology* 58: 423–428.

Talhelm, T., Zhang, X., Oishi, S. et al. (2014). Large-scale psychological differences within China explained by rice versus wheat agriculture. *Science* 344: 603–638.

Thornton, A. (2008). Early body condition, time budgets and the acquisition of foraging skills in meerkats. *Animal Behaviour* 75: 951–962.

Thornton, A. and Clutton-Brock, T. (2011). Social learning and the development of individual and group behaviour in mammal societies. *Philosophical Transactions of the Royal Society B: Life Sciences* 366: 978–987.

Thornton, A. and Lukas, D. (2012). Individual variation in cognitive performance: developmental and evolutionary perspectives. *Philosophical Transactions of the Royal Society B: Life Sciences* 367: 2773–2783.

Thornton, A. and McAuliffe, K. (2006). Teaching in wild meerkats. *Science* 313: 227–229.

Thornton, A. and Wilson, A.J. (2015). In search of the Darwinian Holy Trinity in cognitive evolution: a comment on Croston et al. *Behavioral Ecology* 26: 1460–1461.

Thornton, A., Samson, J., and Clutton-Brock, T. (2010). Multi-generational persistence of traditions in neighbouring meerkat groups. *Proceedings of the Royal Society B: Biological Sciences* 277: 3623–3629.

Thornton, A., Isden, J., and Madden, J.R. (2014). Toward wild psychometrics: linking individual cognitive differences to fitness. *Behavioral Ecology* 25: 1299–1301.

Uller, T., Nakagawa, S., and English, S. (2013). Weak evidence for anticipatory parental effects in plants and animals. *Journal of Evolutionary Biology* 26: 2161–2170.

Van der Vaart, E., Verbrugge, R., and Hemelrijk, C.K. (2012). Corvid re-caching without "theory of mind": a model. *PLoS One* 7: e32904.

Webster, M.M. and Laland, K.N. (2015). Public information use by foraging ninespine sticklebacks: social learning or an unlearned social influence on travel direction? *Behaviour* 152: 1569–1584.

Whiten, A., Custance, D.M., Gomez, J.C. et al. (1996). Imitative learning of artificial fruit processing in children (*Homo sapiens*) and chimpanzees (*Pan troglodytes*). *Journal of Comparative Psychology* 110: 3–14.

Winslow, J.T., Noble, P.L., Lyons, C.K. et al. (2003). Rearing effects on cerebrospinal fluid oxytocin concentration and social buffering in rhesus monkeys. *Neuropsychopharmacology* 28: 910–918.

Würbel, H. (2001). Ideal homes? Housing effects on rodent brain and behaviour. *Trends in Neurosciences* 24: 207–211.

Zimmer, C., Larriva, M., Boogert, N.J., and Spencer, K.A. (2017). Transgenerational transmission of a stress-coping phenotype programmed by early-life stress in the Japanese quail. *Scientific Reports* 7: 46125.

10

Evolution and Human Behaviour: Helping to Make Sense of Modern Life

Louise Barrett[1] and Gert Stulp[2]

[1]*Department of Psychology, University of Lethbridge, T1K3M4, Canada*
[2]*Department of Sociology, University of Groningen / Inter-university Center for Social Science Theory and Methodology (ICS), Groningen, The Netherlands*

Biological or genetic explanations of human behaviour tend to make people nervous, and not without reason. The promotion of eugenics and racist science by leading biologists and anthropologists of the nineteenth and early twentieth century produced some deeply shameful outcomes, culminating in the Nazis' appropriation of eugenic ideals to justify the Holocaust (Beckwith 1993; Marks 2017). Following World War II, UNESCO issued a statement disavowing the idea of biological/genetic differences in behaviour across different racial groups, emphasizing instead that all variation was environmental in origin (UNESCO 1952); we are all nurture, with no apparent nature. While understandable and, in some ways, laudable, such statements cannot erase the fact that we are biological organisms – for what else could we be? Like all other life forms, we are evolved creatures, the product of both genes and our environments (see Chapter 1). If you doubt this, try building an organism of any kind without using DNA. Clearly, then, the objections raised cannot be about biology or genetics *per se* – even the most dedicated environmentalist accepts that they will die without oxygen, and that babies are not delivered by the stork. Instead, resistance lies in the way that genetic and biological processes are theorized, studied and interpreted in the human context.

Similar objections are made to the application of evolutionary theory to modern humans (although there is no apparent problem applying it to the extinct versions of our genus). Again, there is a general acceptance that humans are evolved creatures while, at the same time, there is entrenched resistance to the idea that evolutionary processes influence contemporary human behaviour. At least some of this resistance reflects the way that evolutionary thinking has been applied to humans, rather than resistance to evolutionary thinking itself. In both cases, there remains a worry that labelling a trait or behaviour as genetic/biological or evolved is to suggest that it is immutable and predetermined. There is also the tendency for lay people to equate the words 'genetic' and 'biological' with the word 'natural', which can, and often does, lead to the conflation of what just happens to be the case with normative claims of what should be the case, otherwise known as the naturalistic fallacy (also see Chapter 1).

Evolutionary biology and genetics are, of course, inextricably linked – one of the most prominent definitions of evolution is that of changes in allele frequencies over time – hence concerns about genetic and evolutionary explanations go hand in hand. There is a distinction to be made between the two, however: talk of a genetic influence on behaviour does not, in itself, imply any kind of adaptive evolutionary explanation (or indeed any evolutionary explanation at all). Discussions of the value of evolutionary thinking as applied to humans thus address a different set of issues to those concerning the value of genetic studies of humans, including behavioural genetics. That is, although there is often the assumption that the heritable traits under study have been 'bred into' people, behavioural geneticists typically do not engage with the question of whether the traits are or were evolutionarily adaptive. Instead, behavioural genetics is concerned with the extent to which variance in a trait across people can be attributed to genetic versus environmental differences, and what such differences might mean.

In our view, the persistence of the nature-nurture debate into the twenty-first century often has more to do with a misunderstanding of the aims of behavioural genetic studies than with the application of evolutionary theory to humans in a broader sense. Consequently, in what follows, we first discuss briefly why studies in behavioural genetics seem to fuel the nature-nurture debate. We then go on to consider how evolutionary thinking can help improve our understanding of human behaviour, as well as showing why non-evolutionary thinking can sometimes go awry. Note that we make no attempt to be comprehensive in our assessment, rather we use a few key examples to illustrate the value of evolutionary thinking to real-world issues.

10.1 Understanding Interaction

One reason why nature and nurture just won't go away is because behavioural genetics seemingly promotes such a distinction (Tabery 2014; Fox Keller 2010). The reasons for this can be found in its origins in animal breeding and crop domestication. Artificial selection requires an accurate assessment of the likely response to controlled breeding; if most of the differences between individuals reflect variation in the environmental conditions encountered during growth and development, then artificial selection may prove ineffectual at producing a (suite of) desired trait(s). Hence, experimental and analytical techniques were developed that could partition the variance across individuals into its genetic and environmental components, enabling an assessment of the likely response to selection.

Of course, the analytical partitioning of variance in this way (explained more in Chapter 1) tells us nothing about the actual developmental process – the specific interactions of genes and environment – that give rise to a particular trait. Indeed, when used analytically, the notion of an 'interaction' is a purely statistical construct that complicates the interpretation of main effects (i.e. an interaction is a non-linear effect interfering with the attempt to cleanly partition the variance) and, as R.A. Fisher himself made clear, an inconvenience that can either be ignored (if the effect is very small) or else transformed to meet linear expectations (Fisher and Mackenzie 1923). This stands in contrast to the views of evolutionary developmentalists and geneticists, who are interested in the causal mechanisms that produce traits, and thus consider interactions between genes, gene products, the cellular environment and the broader ecological (and, in the case of humans, socio-economic) environment as both fundamental and central to any understanding of how a given gene might exert its effects (Tabery 2014).

This difference in how the term *interaction* is understood and used in behavioural genetics has helped generate the impression that the discipline is concerned entirely with separating nature from nurture, despite the fact that behavioural geneticists, just like developmentalists, recognize that traits are produced in the individual by complex interactions that make it impossible to cleanly assign the outcome to genes or environment.

As Tabery (2014) suggests, one way to reconcile these two views is to understand that what gets identified in population-based behavioural genetic studies are 'actual difference makers'. That is, we can view all genes as potential difference makers, but only some will make an actual difference to the actual differences seen across individuals. Partitioning the variance then gives us some idea of whether *the* difference maker has been identified, whether we have identified *a* difference among *many* difference makers, or whether the difference made depends on environmental context (i.e. whether there is a gene-by-environment interaction). Once identified, actual difference makers can be studied by evolutionary developmentalists who seek to understand the causal mechanisms by which differences across individuals are produced. In other words, the two approaches can and do dovetail neatly. An emphasis on genes as 'difference makers', rather than 'genes for' particular traits, also lessens the temptation to see genes as somehow opposed to environmental influences, rather than operating in concert with them (Tabery 2014).

10.2 Understanding the Scope and Limits of an Evolutionary Approach

Evolutionary thinking applied to human behaviour is sometimes seen as addressing similar issues to those of behavioural genetics; that is, the extent to which a behaviour is influenced by our genes as opposed to our environment, combined with a more explicit assumption that those genes persist in the population because they enhance (or once enhanced) the fitness of their bearers. This represents a misunderstanding, if not an outright distortion, of an evolutionary approach, not least because the interactive process of development requires that organisms inherit certain stable features of their environment in addition to genes.

Given this, it makes no sense to insist that behaviour can be driven by genes alone, particularly not in large multicellular animals like ourselves. Rather, our nature is nurtured by our developmental environments. It is also important to note that demonstrating that a current behaviour is fitness enhancing (i.e. increases survival and lifetime reproductive success) does not entail that the behaviour also increased fitness in the past, nor does the failure to identify fitness-enhancing behaviour in contemporary environments mean that a trait does not have an evolved component. It is also the case that, even if one can demonstrate natural selection phenotypically (i.e. that individuals engaging in certain behaviours or possessed of certain traits experience higher survival and reproductive success relative to those without the trait or behaviour), it does not necessarily mean that adaptive evolution has or will occur; to do so, it is necessary to demonstrate a genetic response to such selection (i.e. one must demonstrate that the trait is inherited by offspring and increases in frequency in the population).

Stulp et al. (2015), for example, were able to show that height is currently under phenotypic selection in The Netherlands, with tall males and average height women experiencing higher levels of reproductive success (i.e. greater numbers of surviving offspring). The size of the effect was, however, very small, making it highly unlikely that natural selection alone explains much of the startling 20 cm increase in height shown by the Dutch over the past century (Stulp et al. 2015; Tarka et al. 2015). In addition, there was no evidence presented to support a genetic response to selection. What was apparent, however, was that the small effect of natural selection is acting in concert with the environmental conditions that promote increased height (e.g. widespread access to health services, low levels of social inequality, and a diet high in dairy products). This stands in contrast to the USA, where natural selection is acting to reduce height (Stulp et al. 2012; Byars et al. 2010), possibly mediated via a relationship between height and age at marriage and first birth (Stulp and Barrett 2016). In the US, natural selection therefore operates in opposition to environmental influences. Thus, even when effects are small, it is possible that they can help explain differences in trait values across different populations (Stulp and Barrett 2016).

Finally, the fact that social and cultural practices are themselves inherited and undergo transformation (see Chapter 3), and interact with genetic evolution, can make human evolutionary processes subject to more complex evolutionary dynamics (Richerson and Boyd 2005; Henrich and McElreath 2003). Evolutionary analyses must therefore be conducted with care, and interpreted cautiously, especially as ethical concerns mean that confirmatory experiments cannot be conducted.

10.3 Evolutionary Thinking as Puzzle Solving

So, how does evolutionary thinking help sharpen our understanding of human behaviour? One way is that it can help make sense of findings that would otherwise seem counterintuitive or puzzling. For example, a long-standing puzzle concerns the 'Neolithic transition' – the shift to agriculture from a hunting and gathering lifestyle. Agricultural lifestyles are associated with sedenterization, food storage, wealth accumulation and increased population growth (Piperno and Pearsall 1998; Price and Gebauer 1995). At the same time, the shift from hunting and gathering to sedenterization and cultivation demonstrably resulted in poorer health and increased mortality, as revealed by reductions in stature, poorer oral health, and evidence of diseases, such as tuberculosis, plague and syphilis (Cohen and Crane-Kramer 2007). Why then did agriculture succeed in replacing hunting and gathering lifestyles given that it poses a much greater threat to survival? The short, and flippant, answer is that natural selection does not care about health, only fitness. Of course, natural selection doesn't care about anything at all, it is simply the outcome of a process of differential survival and reproduction. This, however, is the key point.

As Page et al. (2016) demonstrated, in a study of Agta people of the Philippines, a shift from hunting and gathering to settled agriculture is indeed associated with increased morbidity and mortality, including higher rates of child mortality. This is offset, however, by an increased reproductive rate, due to the energy savings that result from a more settled lifestyle. This results in a larger number of surviving offspring for settled compared to hunting and gathering Agta. Settled Agta thus display a shift in life history strategy, from producing a small number of high-quality offspring to producing a larger number of poorer quality offspring (Stearns 1992). This quantity–quality trade-off is precisely in line with life history theory, where an increase in child mortality (and hence a reduced chance of surviving to reproductive age) selects for an increased rate of reproduction, as this increases the probability that at least some offspring will survive and ultimately reproduce. Based on their findings, Page et al. (2016) therefore suggest that an analogous situation may have characterized the Neolithic transition: the cultural innovation of agriculture proved to be fitness enhancing and hence biologically adaptive, despite its costs in terms of human health. Thus, the invention and adoption of agriculture, as a cultural trait, is argued to have selected for a faster life history strategy, resulting in agriculturalists outcompeting hunter-gatherers, which in turn helped increase the spread of agriculture via a process of intertwined cultural and biological evolution. Applying an explicitly evolutionary perspective can thus help explain the otherwise counterintuitive spread of a behaviour that actively reduced the health and well-being of those who practised it.

Another example of this phenomenon is provided by Howard and Gibson's (2017) recent analysis of female genital mutilation or cutting (FGC) (defined by the WHO as the partial or total removal of the external genitalia or other injury to the female genital organs for non-medical reasons). FGC has clear and obvious health costs, both physically and psychologically (Iavazzo et al. 2013). It is equally obvious that it has negative consequences for women's reproductive success, due to the obstetric problems it creates (Banks et al. 2006; Adam et al. 2010; Berg and Underland 2013). Despite this, the practice remains widespread in many areas of Africa and the Middle East (where its prevalence ranges from 1% to 99%), and continues in the face of concerted and long-standing political efforts to eradicate it (Howard and Gibson 2017). One suggestion why eradication is so difficult is that cultural evolutionary processes take precedence here, promoting a behaviour that would otherwise fail to persist given its apparent negative biological fitness consequences for individuals. Specifically, the idea is that FGC persists either due to a conformity bias (i.e. where people copy the majority without regard to outcome) (Hayford 2005) or because it represents a co-ordination problem (i.e. men and women believe that FGC is necessary for marriage because the other sex demands it, and hence they reinforce the behaviour, despite the fact that both parties would be better off without it) (Mackie 1996). This has led to the suggestion that it should be possible, via a co-ordinated effort to change behaviour, to move a given population to a 'tipping point' whereby everyone switches rapidly from a cutting to a non-cutting norm. Howard and Gibson (2017) proposed that, in addition to considering these purely cultural evolutionary explanations, it was worth considering whether FGC confers fitness benefits on those adopting the practice, rather than simply assuming that FGC jeopardizes fitness; if so, FGC might prove to be adaptive within specific local ecologies and hence persist, despite its heavy costs in terms of female health and psychological well-being.

Using data from the demographic and health surveys (DHSs) of five West African countries, Howard and Gibson (2017) found that FGC did accrue fitness benefits to women, but in a highly frequency-dependent fashion. In areas with high FGC, cut women produced a larger number of surviving offspring at age 40 than women who had not undergone FGC. Conversely, in areas where the prevalence of FGC was low, women who had not been cut had a reproductive advantage over those who had been. Put into absolute terms, in areas with 0% FGC prevalence, cut women were predicted to have 0.34 fewer surviving offspring, whereas in areas with 100% FGC prevalence, cut women were predicted to have 0.44 more surviving offspring.

The persistence of FGC thus requires a more nuanced understanding of how cultural and biological evolutionary processes intersect. It is not simply the case that a strange and harmful idea has somehow become entrenched, persisting despite its fitness costs. Instead, performing FGC in an FGC-prevalent environment seems to promote the production and survival of offspring. As importantly, rejecting the practice under conditions of high FGC prevalence doesn't automatically lead to a boost in women's reproductive success (although, of course, they will be more healthy in other ways). Howard and Gibson (2017) argue that copying the majority in a given FGC environment might enhance a woman's 'mate value' in the marriage market, which may then translate into higher fertility and/or better child survival by ensuring better access to resources. FGC status relative to the majority may also be associated with entry into social networks that can provide enhanced access to resources and support, which could similarly be reflected in higher fertility and child survivorship.

These fitness-related differences are thus relevant to policy makers. Howard and Gibson (2017) suggest that the pattern shown in FGC prevalence across DHS countries is indicative of a move towards either 0% or 100% prevalence for a given group, that is, the tendency to copy the most frequent behaviour eventually leads one or other to dominate. If eradication policies were thus able to drive FGC below 50%, then a tendency to copy the most frequent behaviour, combined with the relative fitness-related benefits of doing so, should mean that FGC will gradually decrease over time. In other words, the notion of a sudden 'tipping point' that shifts the entire system rapidly is not the only mechanism by which FGC can be pushed into decline.

One final point to note here is that this study generated some negative responses on social media when published, as people interpreted the notion of fitness benefits to somehow suggest a tacit approval of FGC as a practice, or at least an attempt to deny its negative effects. What should be apparent is that Howard and Gibson's (2017) analysis of fitness-related benefits was aimed at providing an explanation of why eradication proves so difficult, and potentially identifying alternative means by which to drive the practice into decline.

10.4 Recognizing the Consequences of Our Actions

Another way in which evolutionary thinking can help sharpen our understanding of human behaviour is by predicting and explaining the consequences of certain courses of action that non-evolutionary approaches would not anticipate. Gibson and Mace (2006) provide one such example, with respect to the provision of labour-saving wells. Drawing on evolutionary life history theory, Gibson and Mace (2006) predicted that labour-saving initiatives designed to improve maternal and child health would also have the inevitable, but unintended, effect of increasing women's fertility, raising questions of whether such interventions represent an unalloyed good. Much like the Agta example described above, reduced workloads would automatically translate into more energy available for reproduction and, in the absence of reliable contraception, larger families.

Their study focused on a rural agro-pastoralist community in Arsi, southern Ethiopia, which suffered from both regular water shortages and food insecurity. Between 1996 and 2000, tap-stands were introduced into a number of villages as part of a water development scheme. During the driest months of the year, this reduced the amount of time women spent carrying water from just over three hours to only 15 minutes per day. The water development scheme thus represented a natural experiment into how shifts in women's workloads and energy budgets influenced fertility. As predicted, women with access to taps were three times more likely to give birth in any given month relative to women without access, and water access was also associated with a 50% lower risk of children dying. However, there was no evidence to suggest that the intervention improved maternal and child nutritional status and health outcomes – the ostensible aim of the initiative. In fact, children in villages with access to water were significantly more likely to be malnourished, and access to water taps was unrelated to women's body fat levels. Although the intervention was successful at reducing child mortality, it came at the cost of increasing childhood malnutrition. Gibson and Mace (2006) suggest this latter effect could be due to increased sibling competition for limited resources (as seen in other studies, such as Lawson and Mace 2009) or, perhaps more likely, as a consequence of reduced mortality among low-birthweight babies (which, potentially, could represent a relaxation of selection on low birthweight, if such a trait were heritable). The effects of malnutrition were seen only in the children born following tap installation, rather than across all age groups, suggesting that improved water access increased the likelihood of low-birthweight babies coming to full term and surviving critical early periods of childhood.

Another example of a potentially misapplied intervention is the fortification with iron of infant milk formula. Infant iron stores are largely accumulated over the course of gestation, and decline in the first few months post partum in breast-fed infants, as human milk is low in iron ($0.2–0.5 \, \text{mg} \, \text{l}^{-1}$). This decline generally has been viewed as pathological, thus resulting in a debate over the appropriate amount of iron fortification in infant formula: this can range from 4 to $12 \, \text{mg} \, \text{l}^{-1}$, so even the lowest levels are an order of magnitude higher than those found in breast milk. Quinn (2014) has suggested that we have things the wrong way around, and high levels of fortification may, in fact, be pathological. Her hypothesis is that iron depletion over the course of early life is an adaptive response to the onset of weaning and the introduction of non-milk foods. Many bacteria require iron for growth and replication, so low iron levels may limit the duration and severity of any infection to which an infant is exposed during the introduction of solid foods. Selection may thus have favoured mothers who produced low-iron milk, along with offspring with decreased iron stores at weaning, as this would increase the chance of surviving the weaning period. Although we currently lack the data needed to fully test this hypothesis, Quinn (2014) cites data from a number of studies demonstrating that low levels of iron intake during infancy were not associated with higher levels of anaemia and, in one study, low levels of iron were associated with both lower rates of infection and greater head growth. In this same study, iron supplementation did nothing to further reduce the risk of anaemia, but was associated with higher rates of gastrointestinal disease and reduced growth (Domellöf et al. 2001). Contemporary fortification practices may therefore undermine these adaptive mechanisms, and increase the chances of children falling ill, rather than providing them with a health-giving boost.

The application of an evolutionary perspective thus helps makes sense of why technological developments aimed at improving maternal and infant quality of life can backfire potentially. These examples also have policy implications. For example, introducing technological advances that fail to consider impacts on women's fertility may exacerbate problems of population growth, so development initiatives need to include provisions for family planning and access to effective contraception. Similarly, if low iron levels are adaptive in early infancy, this has implications for government and WHO guidelines concerning the nutritional content of infant formula.

10.5 Thinking Differently about Fertility Control

Contraceptive uptake itself is also something that has been studied via an evolutionary perspective. This often strikes people as rather odd, given that the availability of reliable contraception has been given as a reason why people no longer maximize their fitness in modern, industrial societies. For example, Lutz et al. (2006, p. 172) state that 'through the introduction of modern contraception, the evolutionary link between sex drive and procreation has been broken and now reproduction is merely a function of individual preferences and culturally determined norms' (see also Vining 1986; Pérusse 1993). Indeed, in our experience, people often assume that the term 'natural fertility population' refers to those in which contraceptive uptake is low or absent, when it simply refers to a lack of parity-specific control of fertility (crudely speaking, there is no target family size or 'stopping rule'). Natural fertility populations can and do practise fertility control (but often with less reliable methods than the contraceptive pill) and there is widespread evidence for the use of contraception and abortifacients reaching all the way back to antiquity (Colleran and Mace 2015; McLaren 1990; Riddle 1994).

Attempting to control fertility should not, therefore, automatically be seen as a break between sex and procreation, rendering evolutionary explanations irrelevant. The existence of fundamental trade-offs between investment in continued reproduction and investment in other domains, such as growth, bodily maintenance and care of existing offspring, prevents any organism from producing the maximum number of offspring of which they are physiologically capable. Accordingly, attempts to control certain aspects of these trade-offs, namely the number and timing of offspring, should be seen as integral to human life history and parental investment strategies, regardless of whether these are achieved by physiological or technological means. Indeed, in some cases, a life history perspective provides a more satisfactory explanation than one that assumes that contraceptive uptake reflects only the transmission and adoption of new arbitrary cultural norms.

For example, Alvergne et al. (2011) found no evidence that membership of social or spatial networks influenced the initial uptake of contraception by Ethiopian women during the early stages of a demographic transition. Rather, the initial adoption of contraceptives was a means by which women could regulate their fertility and increase parental investment per child (Alvergne et al. 2013). Indeed, 96% of the women adopting contraceptives for the first time had already reproduced: women had nearly four children on average at the point of contraceptive uptake (Alvergne et al. 2011). Moreover, those children born before their mothers adopted contraception were less likely to die before the age of five than the children of non-contracepting mothers. It is therefore plausible to argue that the use of contraception can have fitness-enhancing effects, via increased investment in offspring and increased survivorship. Looking at patterns within women, before and after they adopted contraception, it was found that mothers were less likely to reproduce after short interbirth intervals once they had adopted contraception, but this had no subsequent effect on their children's mortality risk. This latter result possibly reflects the fact that increases in contraceptive uptake among women also coincided with other developmental initiatives in the region that served to reduce child mortality.

Overall, then, women in this sample apparently began using contraception to lengthen the spacing between births, enabling them to invest more in their current children. Moreover, this was a product of individual behavioural innovation by women, rather than by social diffusion of new (arbitrary) cultural norms (ultimately, this must, of course, be the case, given that new norms require a source from which they can diffuse). Alvergne et al.'s (2013) results thus speak to theories concerning fertility decline more generally. While shifts in cultural norms of desired family size undoubtedly contribute to fertility decline, it is also possible that, as Alvergne et al. (2013) point out, when mortality declines before fertility, and family size increases more rapidly than expected, the use of contraceptives may, in fact, represent adherence to a stable, locally adaptive cultural norm for intermediate family sizes, albeit via novel, technological means (see also Winterhalder and Leslie 2002; Carey and Lopreato 1995).

Of course, there are certain situations in which the tight control of fertility clearly does not serve to maximize fitness: voluntarily choosing to remain childless and investing resources in luxury holidays or other consumer goods is one such example (but it is also important to remember that individuals of all species fail to maximize their fitness at times; such variability is, after all, the engine of natural selection). Equally, the demographic transitions that have taken place over the course of the nineteenth and twentieth centuries in Europe and North America, where populations have shifted from high-fertility/high-mortality regimes to low-fertility/low-mortality regimes suggest that people no longer maximize fitness in the industrialized west (Sear et al. 2016). These shifts in behaviour have been argued to reflect the adoption of cultural norms, and/or evolved psychological mechanisms aimed at optimizing parental investment in children, which lead to extreme quantity-quality trade-offs in modern industrial settings, neither of which serves to enhance fitness (see the papers in Lawson et al. 2016 for a thorough exploration of these issues).

Even in such apparently clear-cut cases of maladaptive behaviour, some caution is warranted, however. Burnside et al. (2012), for example, argue that the dramatic increases in non-metabolic energy use of the last 10 000 years, which made possible the agricultural, industrial and technological revolutions, can and should be factored into assessments of human life history as well. This is because the fecundity and reproductive rates of human females are not constrained by their metabolism as in other mammals, but instead vary with total energy use (Moses and Brown 2003). Burnside et al. (2012) thus demonstrated that if our 'extrasomatic energy' in the form of fossil fuel consumption were to be considered as a contribution to human metabolism, then the total energy use of a modern-day American woman would be equivalent to that predicted for a hypothetical 30 000 kg primate. Seen in this context, the lifetime reproductive output of modern industrial humans is precisely that predicted by life history theory for an animal of such vast size, thereby complicating the notion that modern industrial humans are capable of rearing many more children than they actually produce (and according to DeLong et al. (2010) such considerations also mean it is unlikely that the rest of the planet will undergo a demographic transition in just the same way: the energy cost of doing so far exceeds the fossil fuel energy that will be available, and consequently cannot slow population growth).

The relationship between energy use and fertility is not only apparent when cross-species comparisons are made, but also across cultures and across time within human cultures. Indeed, the amount of extrasomatic energy used by a population is a good predictor of fitness components like fertility, child mortality and age at first reproduction (Burger et al. 2011). That is, these traits seem to be highly plastic, and respond readily to increases in extrasomatic energy. As Burger et al. (2011) also show, however, the same is not true of lifespan and age at menopause; these are only weakly related to energy use, and appear to be more constrained physiologically. What this means is that some industrial humans have a much shorter reproductive lifespan than predicted for an ape of our (King Kong) size. That is, the increase in age at first reproduction (which follows from an increase in extrasomatic energy, and hence an effective increase in body size) is not compensated for by a longer period of productivity, as life history theory would predict. This may be a reason why industrial humans' fertility falls to below-replacement levels (i.e. fewer than two children per woman) in some instances.[1] Having said this, the greater plasticity of age at first birth and fertility, both of which show moderate heritability, suggests selection could continue to act – by favouring lower age at first birth, for example – and fertility levels could increase in the future.

There is no reason, then, to believe that current 'lowest low' levels of below-replacement fertility are irreversible. Indeed, as Burger and DeLong (2016) point out, there is no genuine theoretically grounded demographic principle for accepting this to be the case; it merely follows from the assumption that sustained economic growth will continue, and returns to human capital will remain high. The application of evolutionary theory to demographic issues provides one very powerful means to understand contemporary shifts in fertility behaviour, and hence to generate more accurate population forecasts.

10.6 Modern Contraception and Mate Choice

Contraceptive use has also been studied in the context of mate preferences and mate choice, and the work we discuss below also helps reinforce our point concerning the need for a careful and cautious interpretation of particular findings. In this case, evolutionary thinking was used to provide a new perspective regarding the unforeseen side effects of oral contraceptives (OCs) on women's sexual functioning.

Ever since the introduction of the oral contraceptive pill, concerns have been raised about its negative effects on women's libido. It has proved difficult, however, to identify the mechanism responsible for such effects, and it is equally difficult to predict which women will suffer from them (Burrows et al. 2012; Sanders et al. 2001). One possible evolutionary explanation, known as the 'congruency hypothesis', is that OC use influences aspects of the mate choice process, with the result that a woman's attraction to her partner may change over time if she either discontinues or begins taking the pill, because her

---

1 Some caution is needed here, however, as the life history predictions are made at the individual level, whereas the relationship between energy use and fertility occurs at the population level. We thus need to be aware of the ecological fallacy: relationships at the aggregate level do not necessarily translate to the same relationship at the individual level (see Pollet et al. (2014, 2015) for more details).

current contraceptive status no longer matches her status during partnership formation (Roberts et al. 2012, 2013, 2014; Alvergne and Lummaa 2010). There is evidence to suggest, for example, that the major histocompatibility complex (MHC) exerts an influence on mate preferences (specifically those related to odour), with preferences shown for MHC-dissimilar partners (Wedekind et al. 1995; Havlicek and Roberts 2009). The use of OCs has been shown to shift preferences towards MHC similarity, however (Roberts et al. 2008). Thus, couples that form when a woman is taking OCs should be more likely to show MHC similarity. If the woman then stops taking OCs as the relationship develops, this may lead to a shift in women's preferences for MHC dissimilarity, a reduction in attraction to her partner, and a loss of libido, one that is mediated by her use of OCs but is not directly caused by the pill in and of itself (Roberts et al. 2012).

In support of this, Roberts et al. (2012) found that women who were using OCs during partnership formation, and then went on to have their first child with this same partner, scored lower on a number of measures of sexual satisfaction with their partners compared to non-users, and they also rated their partner's body as less attractive. In a second study, Roberts et al. (2014) presented a more powerful test of the congruency hypothesis, and more convincing evidence of its effects. Here, they asked both members of a couple to assess their relationship satisfaction, and obtained detailed information on women's current and previous OC use. In this way, they could obtain a measure of congruency between a woman's OC use during partner choice and her current OC use, and test whether this better predicted her relationship satisfaction than either OC use during partner formation or current OC alone. This design also allowed Roberts et al. (2014) to explore different routes to congruency in OC use (i.e. users who remained users versus non-users who remained non-users). Finally, asking male partners for their views enabled Roberts et al. (2014) to investigate how congruency in OC use affected men's versus women's assessments of satisfaction.

In line with the congruency hypothesis, women's sexual satisfaction was predicted by congruent OC status, but congruency did not affect women's non-sexual satisfaction, nor did it predict men's satisfaction with their partners. The latter finding thus supports the idea that changes in women's OC status directly influences women's satisfaction, rather than being a consequence of changes in men's attraction to their partners due to a shift in her OC status. In addition, there was moderate evidence for an interaction between previous and current OC use, where non-congruency via either of the two possible routes was associated with lower satisfaction than women whose status remained congruent. Importantly, neither previous nor current OC status alone could explain any aspect of women's sexual satisfaction. These results were thus argued to explain why the negative effects of OCs on women's libido and behaviour often prove elusive; understanding the effects of hormonal contraceptives requires an assessment of how women's OC use changes in relation to partnership formation, and potential shifts in women's hormonally mediated partner-specific levels of attraction.

A recent replication of Roberts' 2014 study, however, failed to find any support for the congruency hypothesis (Jern et al. 2018). The study was conducted on a large sample of Finnish women (n = 948), which gave it sufficient statistical power to detect the effect sizes previously reported. The only effects found were differences in sexual satisfaction between women who consistently used OCs and those who did not, with consistent OC users showing higher sexual satisfaction scores than non-users (and no effect at all for relationship satisfaction scores). Jern et al. (2018) concluded that the congruency effects reported in Roberts' previous study most likely arose from the fact that there was an unequal distribution of current OC users among the congruent and incongruent user groups. That is, in the congruent user group, participants were equally distributed between current users and non-users, but in the incongruent user group, 115 of the participants were current non-users (i.e. had stopped taking OCs following partnership formation), and only 35 participants were current users (i.e. had begun taking OCs following partnership formation). As a result, the difference between the groups that was taken to be an effect of congruency in OC use may simply have reflected consistency in OC use: the congruent group had far more consistent users in it than the incongruent group.

As Jern et al. (2018) point out, studies of the congruency hypothesis received widespread attention in both the academic community and the mainstream media (including *Time*, *Women's Health* magazine and the *Guardian* newspaper). It is therefore possible that these studies may have influenced women's decisions regarding contraceptive use, or influenced their perceptions of their likely long-term relationship success. These new findings, which suggest that the congruency effect may not exist, illustrate both the vital importance of replication and the dangers of accepting effects as real in advance of replication taking place. At the same time, a single non-replication does not necessarily rule out the existence of the congruency effect (e.g. it is possible, if unlikely, that Finnish women alone are not susceptible to these effects). Further replications are needed before we can conclusively state that the congruency hypothesis fails to explain the pill's negative effects on libido, and we should consider the jury still out, rather than accept any firm conclusions one way or another.

10.7 Evolution and Assisted Reproductive Technologies

Humans have not only developed technologies for limiting or controlling fertility, but also for overcoming the problem of infertility in both men and women – something that is estimated to affect ~10% of couples in industrial nations (Gnoth et al. 2003), and ~80 million people world-wide (Vayena et al. 2002), with the highest prevalence in the so-called 'infertility belt' of central and southern Africa (Inhorn and van Balen 2002). The first 'test tube baby', Louise Brown, was conceived via *in vitro* fertilization (IVF) and born in 1978. Since then, over five million children have been born via assisted reproductive technologies (ARTs) (Inhorn and Birenbaum-Carmeli 2008), and in some countries, 4% of all newborns are conceived by IVF (Kupka et al. 2014). ARTs now include not just IVF for women with blocked fallopian tubes, but also intrauterine insemination (IUI), insemination by a donor (DI), and intracytoplasmic sperm injection (ICSI) to combat male infertility, the latter of which involves the injection of 'weak' sperm directly into oocytes. In addition, egg donation and gestational surrogacy are increasing in frequency.

As well as challenging traditional views of kinship (children can potentially have three parents), ARTs expand the option of parenthood to gay couples (via donor insemination or surrogacy), and shift what people consider to be the 'natural' response to infertility. In Europe and America, where adoption was once seen as the most acceptable, and natural, way for infertile individuals to become parents, there is an increasing desire for people to 'chase the blood tie' (Ragoné 1996), and use ARTs to conceive their own biological children. This is now regarded as the more 'natural' option, even though it requires a massively choreographed technological intervention by a wide variety of medical professionals (Thompson 1996, 2005). There are also increasing numbers of postmenopausal women over the age of 50 achieving IVF pregnancies using hormonal assistance and donor eggs (whether their own previously frozen eggs or those of a younger woman) – a technological, extrasomatic means to circumvent the fixed limit to the human reproductive lifespan identified above. Perhaps more than any other human invention, ARTs blur our long-standing notions of nature and nurture, biology and culture. In principle, embryos can be 'designed, stored, exchanged and implanted in any womb' (Hanevik et al. 2016) and reproduction is therefore likely to become increasingly independent of age, gender, and sexual orientation (for a comprehensive and fascinating review of how ARTs affect not only our understandings of kinship and parenthood, but also gender equity, social inequality and bioethics, see Inhorn and Birenbaum-Carmeli 2008).

Thus, although the link between sex and reproduction has been broken by ARTs, they nevertheless have the capacity potentially to influence evolutionary processes. IVF involves a number of artificial environments, procedures and selection criteria that are distinctively different from those that occur in the womb. Under natural conditions, large follicles with high sensitivity to both follicle-stimulating hormone (FSH) and luteinizing hormone (LH) are selected, and there is a single oocyte that undergoes ovulation. These pressures are relaxed under conditions of IVF, where all but the very smallest follicles are suctioned out, and a number of oocytes are subsequently selected for further manipulation. In turn, this manipulation exerts an array of new selection pressures, including the ability to withstand needle puncture and removal of cytoplasm in the case of ICSI (Hanevik et al. 2016).

The same is, of course, true for spermatozoa. In the absence of human intervention, successful spermatozoa are excellent long-distance swimmers, that can make their way through the vaginal mucus and locate the oocyte via chemotactic means, whereas the IVF selection processes favour sprinters, that can swim faster over shorter distances and rapidly penetrate the egg. In the case of ICSI, the normal selection pressures are almost entirely relaxed, as low-motility spermatozoa that clearly could not fertilize an egg under typical conditions are given the opportunity to do so, and do not even need to possess the capacity to penetrate the egg (Hanevik et al. 2016).

Conversely, IVF imposes new selection barriers on laboratory-created embryos. These bypass the barrier imposed by the fallopian tubes, but such embryos need to make it through the computerized process by which embryo morphology and development are visually analysed and assessed. As Hanevik et al. (2016) point out, this means that computer algorithms now comprise a selection barrier that has no analogue to the normal process of embryo transport to the womb. Similarly, oocytes and embryos are increasingly subject to freezing – another laboratory process that must be withstood if they are to be selected for future fertilization and implantation. The increased robustness required of IVF spermatozoa, oocytes, and embryos may come at the cost of other traits that we have yet to identify (Hanevik et al. 2016); this is not to say the outcome will be negative or detrimental, merely that we need to recognize that how these new selective regimes represent a particularly potent form of gene-culture coevolution; these new technological interventions occur at the point where natural selection has operated at its strongest.

Finally, Hanevik et al. (2016) point out that ARTs select for certain people to reproduce by these technological means. They give the example of Norway, where a woman with a BMI of 25 will qualify for state-funded IVF treatment, but one with a BMI of 45 will not. The high cost of IVF therefore represents another form of selection barrier to those who cannot access public funds, and selects against those with low income, both within economically developed countries, as well as setting up a differential between low- and high-income nations.

One possible – if extreme – outcome of ARTs, then, is that we will generate a distinct population of humans that is dependent entirely on IVF to reproduce, with certain traits linked to infertility no longer coming under selection, or experiencing a form of positive selection. Both endometriosis (which leads to blocked fallopian tubes) and polycystic ovary syndrome (which causes anovulation) are known to be heritable (Treloar et al. 1999; Vink et al. 2006), such that an increased prevalence of IVF may lead to an increased prevalence of these conditions in particular populations. Similarly, boys conceived by ICSI are more likely to show a malformation of the penis known as hypospadia (Ericson and Källén 2001) (in a neat historical twist of fate, the very first record of artificial insemination was performed to assist a man who suffered from this particular condition) (Home, 1799).

As Hanevik et al. (2016) point out, we are still very far from understanding the potential long-term evolutionary consequences of ARTs, given that the oldest IVF individual is still only in her early forties. ARTs do, however, force a recognition that humans are becoming biologically, as well as culturally, dependent on our own technology. Hanevik et al. (2016) suggest that, to date, an evolutionary perspective may have received limited attention because of the implications that ARTs allow 'inferior' individuals to reproduce, leading to odious eugenics-style arguments against it. Obviously, no such normative conclusions are warranted, not least because natural selection does not lead to the 'best' or 'fittest' solutions in the colloquial sense, but simply to those outcomes that are most successful in a given environment. The main message is that, as we mentioned at the start, an evolutionary approach requires an certain degree of caution, and an appropriate level of humility.

10.8   No Free Lunch

We can illustrate this latter point with a final cautionary tale concerning the treatment of stomach ulcers. We now know that *Helicobacter pylori* is the cause of stomach ulcers, thanks to the efforts of Barry Marshall and Robin Wallace, who won the Nobel Prize for Medicine in 2005 for this discovery. *H. pylori* is one of the most prominent gut bacteria we acquire during childhood and, among other things, is involved in the regulation of gastric hormones, affecting gastric pH (Blaser and Falkow 2009). Over 50% of all cells found in the gastrointestinal niche are *H. pylori* cells.

When Marshall and Wallace first suggested that the cause of stomach ulcers was bacterial, back in the early 1980s, the idea was met with scepticism: no one believed that any bacteria could survive the acidic environment of the stomach. When experiments that attempted to infect piglets with *H. pylori* failed, Barry Marshall, in a marvellous display of self-experimentation, (in)famously drank some cultured *H. pylori*, determined to prove that it was the cause of peptic ulcers (he expected this to occur years later).

Within a few days, he began to feel nauseated and about a week later developed major gastritis, and *H. pylori* was cultured from his stomach. After two weeks, he began taking antibiotics, the inflammation cleared up and he thus demonstrated that the gastritis was due to the bacterium (you can read the whole story in Barry Marshall's Nobel Prize speech) (Marshall and Warren 2005; Marshall and Adams 2008). Following this discovery, *H. pylori* has also been shown to substantially increase the risk of a particular kind of stomach cancer (Blaser et al. 1995; Helicobacter and Cancer Collaborative Group 2001).

Given its status as a human pathogen linked to devastating and fatal diseases, pharmaceutical companies have spent a lot of time and effort developing a means to eradicate *H. pylori* from the human gut. Why wouldn't we want to remove a potential source of mortality, particularly one as nasty as gastric cancer? As Blaser (2006) points out, however, the fact that all mammals have one or other kind of *Helicobacter* species residing in their guts, combined with our own long history of coevolution with *H. pylori*, should perhaps have given us pause. Genetic studies of *H. pylori* have shown that diversity among strains decreases with distance from East Africa, much like human genetic diversity does, suggesting that *H. pylori* has coevolved with humans over the last 60 000 years (at least) (Linz et al. 2007). *H. pylori* is thus found across all human populations, and (until recently) its prevalence was around 80%, but only a relatively small number of people suffered any ill effects due to pathogenic strains.

The potential consequences of ridding our microbiota of its most dominant player didn't feature in these efforts to 'cure' us of our *H. pylori* infections but, as Blaser (2010) notes, we are now receiving something of a wake-up call, as we begin to recognize some of the costs associated with our 'disappearing microbiota' (Blaser and Falkow 2009). For example, Sonnenberg et al. (2010), in a study of over 78 000 US patients, found that the absence of *H. pylori* was associated with the presence of a condition called Barrett's oesophagus, which involves abnormal changes in the cells of the lower oesophagus, a condition that can turn malignant and lead to oesophageal adenocarcinoma (EAC) (a particularly nasty form of throat cancer). Barrett's oesophagus, in the first instance, develops from a condition called gastro-oesophageal reflux disease (GERD), which also seems to be linked to the absence of *H. pylori*. As *H. pylori* is the most common microbe, it is also possible that it isn't the loss of *H. pylori* itself, but of another species that occurs at lower density, and so isn't picked up by such studies. *H. pylori* could therefore just be a marker for the loss of some other micro-organism.

Over the past century, then, the prevalence of *H. pylori* has been dropping steadily among economically developed countries to the extent that less than 6% of US children now carry it, and the incidence of both peptic ulcers and gastric cancer has dropped accordingly in the adult population (Blaser 2006). At the same time, GERD, Barrett's oesophagus and EAC have all increased. Essentially, we seem to have swapped one set of unpleasant, but curable, diseases for another set of equally nasty, but currently less curable, diseases, and generated a new selection pressure on humans.

10.9 Conclusion

The examples we have covered here illustrate that modern evolutionary explanations go well beyond the kinds of caricatured biological/genetic explanations that critics fear. Evolutionary theories enable us to test well-formulated hypotheses and offer convincing explanations for a variety of human traits and behaviours in contemporary populations, particularly those that seem counterintuitive. Most of our examples here have concerned fertility in one way or another, and this was no accident. Fertility lies at the core of evolutionary biology, and using an evolutionary lens offers an effective aid to understanding the reproductive strategies of contemporary humans, including how and why contraceptive uptake occurs, patterns of fertility decline through history and across the globe, as well as forcing the recognition that current levels of fertility need not persist indefinitely, and that technological interventions in the reproductive process are likely to have evolutionary consequences.

Moreover, we hope to have shown how human agency and conscious decision making do not, in fact, allow us to transcend our biology in the way we often imagine: active human decision making designed to improve health and increase well-being has itself generated a new set of biologically based health and population problems we now have to solve. An even broader lesson is that an evolutionary perspective on human behaviour, especially in contemporary populations, means more than just focusing on individual circumstances. We must also account for the cultural historical and sociological factors that influence how governments and industry act in the face of technological advances of various kinds.

The final take-home message, then, is that an evolutionary approach to human behaviour requires an understanding of the entanglements of nature and nurture, culture and biology, and has much to offer the study of societies that differ widely from those of ancestral populations. Via a variety of 'extrasomatic' capacities, including extensive social learning, our use of tools and technology, and the harnessing of fossil fuels and other forms of energy, we can and do adapt to the environments we create for ourselves much faster than is usually believed.

References

Adam, T., Bathija, H., Bishai, D. et al. (2010). Estimating the obstetric costs of female genital mutilation in six African countries. *Bulletin of the World Health Organization* 88: 281–288.

Alvergne, A. and Lummaa, V. (2010). Does the contraceptive pill alter mate choice in humans? *Trends in Ecology & Evolution* 25: 171–179.

Alvergne, A., Gibson, M.A., Gurmu, E., and Mace, R. (2011). Social transmission and the spread of modern contraception in rural Ethiopia. *PLoS One* 6: e22515.

Alvergne, A., Lawson, D.W., Clarke, P.M. et al. (2013). Fertility, parental investment, and the early adoption of modern contraception in rural Ethiopia. *American Journal of Human Biology* 25: 107–115.

Banks, E., Meirik, O., Farley, T., and Akande, O. (2006). Female genital mutilation and obstetric outcome: WHO collaborative prospective study in six African countries. *Lancet* 367: 1835.

Beckwith, J. (1993). A historical view of social responsibility in genetics. *BioScience* 43 (5): 327–333.

Berg, R.C. and Underland, V. (2013). The obstetric consequences of female genital mutilation/cutting: a systematic review and meta-analysis. *Obstetrics and Gynecology International* 2013: 1–15.

Blaser, M.J. (2006). Who are we? Indigenous microbes and the ecology of human diseases. *EMBO Reports* 7: 956.

Blaser, M.J. (2010). *Helicobacter pylori* and esophageal disease: wake-up call? *Gastroenterology* 139 (6): 1819.

Blaser, M.J. and Falkow, S. (2009). What are the consequences of the disappearing human microbiota? *Nature Reviews Microbiology* 7: 887–894.

Blaser, M.J., Perez-Perez, G.I., Kleanthous, H. et al. (1995). Infection with *Helicobacter pylori* strains possessing cagA is associated with an increased risk of developing adenocarcinoma of the stomach. *Cancer Research* 55: 2111–2115.

Burger, O. and DeLong, J.P. (2016). What if fertility decline is not permanent? The need for an evolutionarily informed approach to understanding low fertility. *Philosophical Transactions of the Royal Society B: Biological Sciences* 371: 20150157.

Burger, O., DeLong, J.P., and Hamilton, M.J. (2011). Industrial energy use and the human life history. *Scientific Reports* 1: 56.

Burnside, W.R., Brown, J.H., Burger, O. et al. (2012). Human macroecology: linking pattern and process in big-picture human ecology. *Biological Reviews* 87: 194–208.

Burrows, L.J., Basha, M., and Goldstein, A.T. (2012). The effects of hormonal contraceptives on female sexuality: a review. *Journal of Sexual Medicine* 9: 2213–2223.

Byars, S.G., Ewbank, D., Govindaraju, D.R., and Stearns, S.C. (2010). Natural selection in a contemporary human population. *Proceedings of the National Academy of Sciences USA* 107 (Suppl 1): 1787–1792.

Carey, A.D. and Lopreato, J. (1995). The evolutionary demography of the fertility-mortality quasi-equilibrium. *Population and Development Review* 21: 613–630.

Cohen, M.N. and Crane-Kramer, G.M.M. (2007). *Ancient Health: Skeletal Indicators of Agricultural and Economic Intensification*. Gainesville: University Press of Florida.

Colleran, H. and Mace, R. (2015). Social network- and community-level influences on contraceptive use: evidence from rural Poland. *Proceedings of the Royal Society B: Biological Sciences* 282: 20150398.

DeLong, J.P., Burger, O., and Hamilton, M.J. (2010). Current demographics suggest future energy supplies will be inadequate to slow human population growth. *PLoS One* 5: e13206.

Domellöf, M., Cohen, R.J., Dewey, K.G. et al. (2001). Iron supplementation of breast-fed Honduran and Swedish infants from 4 to 9 months of age. *Journal of Pediatrics* 138: 679–687.

Ericson, A. and Källén, B. (2001). Congenital malformations in infants born after IVF: a population-based study. *Human Reproduction* 16: 504–509.

Fisher, R.A. and Mackenzie, W.A. (1923). Studies in crop variation. II. The manurial response of different potato varieties. *Journal of Agricultural Science* 13: 311–320.

Gibson, M.A. and Mace, R. (2006). An energy-saving development initiative increases birth rate and childhood malnutrition in rural Ethiopia. *PLoS Medicine* 3 (4): e87.

Gnoth, C., Godehardt, D., Godehardt, E. et al. (2003). Time to pregnancy: results of the German prospective study and impact on the management of infertility. *Human Reproduction* 9: 1959–1966.

Hanevik, H.I., Hessen, D.O., Sunde, A., and Breivik, J. (2016). Can IVF influence human evolution? *Human Reproduction* 31: 1397–1402.

Havlicek, J. and Roberts, S.C. (2009). MHC-correlated mate choice in humans: a review. *Psychoneuroendocrinology* 34: 497–512.

Hayford, S.R. (2005). Conformity and change: community effects on female genital cutting in Kenya. *Journal of Health and Social Behavior* 46: 121–140.

Helicobacter and Cancer Collaborative Group (2001). Gastric cancer and *Helicobacter pylori*: a combined analysis of 12 case control studies nested within prospective cohorts. *Gut* 49: 347–353.

Henrich, J. and McElreath, R. (2003). The evolution of cultural evolution. *Evolutionary Anthropology: Issues, News, and Reviews* 12: 123–135.

Home, E. (1799). An account of the dissection of an hermaphrodite dog. To which are prefixed, some observations on hermaphrodites in general. By Everard Home, Esq. FRS. *Philosophical Transactions of the Royal Society of London* 89: 157–178.

Howard, J.A. and Gibson, M.A. (2017). Frequency-dependent female genital cutting behaviour confers evolutionary fitness benefits. *Nature Ecology & Evolution* 1: 0049.

Iavazzo, C., Sardi, T.A., and Gkegkes, I.D. (2013). Female genital mutilation and infections: a systematic review of the clinical evidence. *Archives of Gynecology and Obstetrics* 287: 1137–1149.

Inhorn, M.C. and van Balen, F. (eds.) (2002). *Infertility around the Globe: New Thinking on Childlessness, Gender, and Reproductive Technologies*. Berkeley: University of California Press.

Inhorn, M.C. and Birenbaum-Carmeli, D. (2008). Assisted reproductive technologies and culture change. *Annual Review of Anthropology* 37: 177–196.

Keller, E.F. (2010). *The Mirage of a Space Between Nature and Nurture*. Durham: Duke University Press.

Kupka, M.S., Ferraretti, A.P., de Mouzon, J. et al. (2014). Assisted reproductive technology in Europe, 2010: results generated from European registers by ESHRE. *Human Reproduction* 29: 2099–2113.

Lawson, D.W. and Mace, R. (2009). Trade-offs in modern parenting: a longitudinal study of sibling competition for parental care. *Evolution and Human Behavior* 30: 170–183.

Linz, B., Balloux, F., Moodley, Y. et al. (2007). An African origin for the intimate association between humans and *Helicobacter pylori*. *Nature* 445: 915–918.

Mackie, G. (1996). Ending footbinding and infibulation: a convention account. *American Sociological Review* 61: 999–1017.

Marks, J. (2017). *Is Science Racist?* New York: Wiley.

Marshall, B. and Adams, P.C. (2008). *Helicobacter pylori*: a Nobel pursuit? *Canadian Journal of Gastroenterology and Hepatology* 22: 895–896.

Marshall, B.J. and Warren, J.R. (2005). *The Bacterium Helicobacter pylori and Its Role in Gastritis and Peptic Ulcer Disease*. The Nobel Prize in Physiology or Medicine, Press Release.

McLaren, A. (1990). *A History of Contraception: From Antiquity to the Present Day*. Oxford: Basil Blackwell.

Moses, M.E. and Brown, J.H. (2003). Allometry of human fertility and energy use. *Ecology Letters* 6: 295–300.

Page, A.E., Viguier, S., Dyble, M. et al. (2016). Reproductive trade-offs in extant hunter-gatherers suggest adaptive mechanism for the Neolithic expansion. *Proceedings of the National Academy of Sciences USA* 113: 4694–4699.

Pérusse, D. (1993). Cultural and reproductive success in industrial societies: testing the relationship at the proximate and ultimate levels. *Behavioral and Brain Sciences* 16: 267–283.

Piperno, D. and Pearsall, D.M. (1998). *The Origins of Agriculture in the Lowland Neotropics*. London: Academic Press.

Pollet, T.V., Tybur, J.M., Frankenhuis, W.E., and Rickard, I.J. (2014). What can cross-cultural correlations teach us about human nature? *Human Nature* 25: 410–429.

Pollet, T.V., Stulp, G., Henzi, S.P., and Barrett, L. (2015). Taking the aggravation out of data aggregation: a conceptual guide to dealing with statistical issues related to the pooling of individual-level observational data. *American Journal of Primatology* 77: 727–740.

Price, D.T. and Gebauer, A.B. (1995). New perspectives on the transition to agriculture. In: *Last Hunters-First Farmers* (ed. D.T. Price and A.B. Gebauer), 3–20. Santa Fe: School of American Research Press.

Quinn, E.A. (2014). Too much of a good thing: evolutionary perspectives on infant formula fortification in the United States and its effects on infant health. *American Journal of Human Biology* 26: 10–17.

Ragoné, H. (1996). Chasing the blood tie: surrogate mothers, adoptive mothers and fathers. *American Ethnologist* 23: 352–365.

Richerson, P.J. and Boyd, R. (2005). *Not by Genes Alone: How Culture Transformed Human Evolution*. Chicago: University of Chicago Press.

Riddle, J.M. (1994). *Contraception and Abortion from the Ancient World to the Renaissance*. Cambridge: Harvard University Press.

Roberts, S.C., Gosling, L.M., Carter, V., and Petrie, M. (2008). MHC-correlated odour preferences in humans and the use of oral contraceptives. *Proceedings of the Royal Society B: Biological Sciences* 275: 2715–2722.

Roberts, S.C., Klapilová, K., Little, A.C. et al. (2012). Relationship satisfaction and outcome in women who meet their partner while using oral contraception. *Proceedings of the Royal Society B: Biological Sciences* 279: 1430–1436.

Roberts, S.C., Cobey, K.D., Klapilová, K., and Havlíček, J. (2013). An evolutionary approach offers a fresh perspective on the relationship between oral contraception and sexual desire. *Archives of Sexual Behavior* 42: 1369–1375.

Roberts, S.C., Little, A.C., Burriss, R.P. et al. (2014). Partner choice, relationship satisfaction, and oral contraception: the congruency hypothesis. *Psychological Science* 25: 1497–1503.

Sanders, S.A., Graham, C.A., Bass, J.L., and Bancroft, J. (2001). A prospective study of the effects of oral contraceptives on sexuality and well-being and their relationship to discontinuation. *Contraception* 64: 51–58.

Sear, R., Lawson, D.W., Kaplan, H., and Shenk, M.K. (2016). Understanding variation in human fertility: what can we learn from evolutionary demography? *Philosophical Transactions of the Royal Society B: Biological Sciences* 371: 20150144.

Sonnenberg, A., Lash, R.H., and Genta, R.M. (2010). A national study of *Helicobactor pylori* infection in gastric biopsy specimens. *Gastroenterology* 139: 1894–1901.

Stulp, G. and Barrett, L. (2016). Wealth, fertility and adaptive behaviour in industrial populations. *Philosophical Transactions of the Royal Society B: Biological Sciences* 371 (1692): 20150153.

Stulp, G., Verhulst, S., Pollet, T.V., and Buunk, A.P. (2012). The effect of female height on reproductive success is negative in western populations, but more variable in non-western populations. *American Journal of Human Biology* 24: 486–494.

Stulp, G., Barrett, L., Tropf, F.C., and Mills, M. (2015). Does natural selection favour taller stature among the tallest people on earth? *Proceedings of the Royal Society B: Biological Sciences* 282: 20150211.

Tabery, J. (2014). *Beyond Versus: The Struggle to Understand the Interaction of Nature and Nurture.* Cambridge: MIT Press.

Tarka, M., Bolstad, G.H., Wacker, S. et al. (2015). Did natural selection make the Dutch taller? A cautionary note on the importance of quantification in understanding evolution. *Evolution* 69: 3204–3206.

Thompson, C. (1996). Ontological choreography: agency through objectification in infertility clinics. *Social Studies of Science* 26: 575–610.

Thompson, C. (2005). *Making Parents: The Ontological Choreography of Reproductive Technologies.* Cambridge: MIT Press.

Treloar, S.A., O'Connor, D.T., O'Connor, V.M., and Martin, N.G. (1999). Genetic influences on endometriosis in an Australian twin sample. *Fertility and Sterility* (4): 701–710.

UNESCO (1952). *The Race Concept: Results of an Inquiry.* Paris: UNESCO.

Vayena, E., Rowe, P.J., and Griffin, P.D. (eds.) (2002). *Current Practices and Controversies in Assisted Reproduction.* Geneva: World Health Organization.

Vining, D.R. (1986). Social versus reproductive success: the central theoretical problem of human sociobiology. *Behavioral and Brain Sciences* 9: 167–187.

Vink, J.M., Sadrzadeh, S., Lambalk, C.B., and Boomsma, D.I. (2006). Heritability of polycystic ovary syndrome in a Dutch twin-family study. *Journal of Clinical Endocrinology & Metabolism* (6): 2100–2104.

Wedekind, C., Seebeck, T., Bettens, F., and Paepke, A.J. (1995). MHC-dependent mate preferences in humans. *Proceedings of the Royal Society B: Biological Sciences* 260: 245–249.

Winterhalder, B. and Leslie, P. (2002). Risk-sensitive fertility: the variance compensation hypothesis. *Evolution and Human Behavior* 23: 59–82.

11

Next-Gen and the Study of Behaviour

Simone Immler

School of Biological Sciences, University of East Anglia, Norwich Research Park, Norwich, NR4 7TJ, UK

Linking behaviour to genetics is notoriously difficult for reasons including the challenge to define and quantify behaviour (Leshner and Pfaff 2011), the plasticity and condition dependence of any behavioural trait (Snell-Rood 2013) and the involvement of many genes being expressed differentially in different tissues at different times (Liedvogel et al. 2011; Wahlsten 2012). Nevertheless, the field of behavioural genetics is incredibly active and has seen significant advancements (Sokolowski 2001). Part of this advancement can be explained by the onset of the next-generation sequencing (NGS) era where the possibilities of answering questions related to the genetic underpinnings of virtually any trait, and of complex traits such as behavioural phenotypes in particular, are greatly expanded (see Chapter 1). Sequencing technologies and the related field of bioinformatics are developing at the speed of light and keep opening new avenues to studying different aspects of the genome at an incredible rate. While less than 10 years ago, most of the available NGS approaches seemed to be reserved for typical model organisms, this has radically changed in the mean time (Ellegren 2014) and in theory, any organism can be used for NGS studies.

One may feel as happy as a flea in a doghouse at the choice of NGS avenues to take but this is also where the potential trouble starts. What NGS avenue (if any) one should take to link a behavioural phenotype to the underlying genotype may be a crucial question at the start of such an endeavour and it may not always have a straight answer. In this chapter, I would like to give an overview of the current sequencing technologies and how they can be applied to studying the genetics of behaviour. I will discuss where the potential limitations of the currently available NGS technologies and bioinformatics tools lie and present examples where NGS has been successfully used for linking genes and behaviour. I will round up with a glimpse into where we might be going next and how recently developed tools may be incorporated into the study of behavioural phenotypes.

11.1 Current Sequencing Technologies

Sequencing has become an increasingly solid part of the study of behaviour and with ever diminishing prices, the data output from sequencing projects has been generated at an exponential rate over the past 10–15 years. The exponential increase is directly

Genes and Behaviour: Beyond Nature-Nurture, First Edition.
Edited by David J. Hosken, John Hunt and Nina Wedell.
© 2019 John Wiley & Sons Ltd. Published 2019 by John Wiley & Sons Ltd.

associated with the development of what is generally referred to as next-generation or massively parallel sequencing technology (see Figure 11.1 for a description of the technology). NGS technologies allow the retrieval of more data from less material at a faster rate and a lower price. While 10 years ago, obtaining DNA or RNA from a single cell was unthinkable and the objective of a project on its own, aims like single cell genomics and transcriptomics are increasingly feasible. In fact, so much so that in some places it is already performed at an automated level, where hundreds of single cells are being fed into a robot at one end and the data on the genome or transcriptome of each cell is coming out at the other end. The rapid development of single-cell sequencing technology is just an example to illustrate the leaps sequencing technology has taken within a relatively short period of time, and it may continue at this speed, opening up more options.

Nevertheless, while sequencing technologies are faster and cheaper than ever before, the financial aspect is likely to affect the decision about which approach to take for a given question as the different approaches do vary in price. Whole-genome sequencing is an attractive but naturally expensive option and may not always be the most efficient way to answer a specific question. A range of alternatives to whole-genome sequencing are available, which may yield the appropriate amount of information to provide satisfying results and information. The approaches may differ by how they cover the entire genome (see Figure 11.2) and may target specific loci or regions of interest distributed across the genome (e.g. microarrays, restriction site-associated DNA (RAD) sequencing or exome sequencing). They also differ by the nature of the results one obtains, as some approaches will provide information about the actual genotype but little about gene activity (e.g. RAD sequencing, whole-genome sequencing), whereas some approaches provide information about which genes are expressed at what levels and how they are regulated (e.g. RNA sequencing, chromatin immunoprecipitation (CHiP) sequencing or bisulfite sequencing). While the availability of a sequenced reference genome is by no means a must for all of these approaches, it will help substantially with the interpretation and identification of relevant genes and regions associated with a trait of interest, particularly when dealing with a complex trait such as a behavioural phenotype. The reason is that knowing the genome structure and location of genes relative to each other may provide information about regulatory and epistatic interactions.

Here, I provide an introduction to the most important sequencing technologies currently in use with a glimpse into what we may expect in the near future. Many excellent reviews of NGS technologies can be found in the published literature (e.g. Buermans and Dunnen 2014; van Dijk et al. 2014) and hence I will not delve into the methodological details of each approach, but mention the aspects I think are relevant to consider when aiming to link genes and behaviour. The different technologies are arranged by type of analysis: (i) genome-wide association studies (GWAS)/linkage mapping, (ii) expression analyses, and (iii) epigenetic analyses. Key articles for further reading are cited in each section accordingly.

11.1.1 Genome-Wide Association Study (GWAS) and Linkage Mapping

One widely used approach to link phenotypic traits to underlying genes is the GWAS. GWAS aims to screen many individuals varying for a specific phenotype at a genome-wide scale without any pre-existing bias towards any regions or genes in

From extraction to the mapped genome

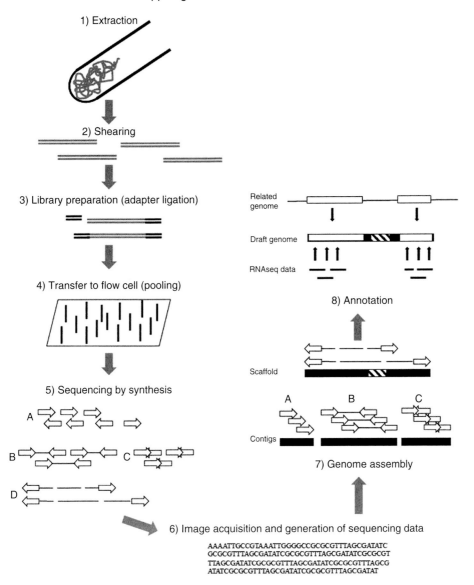

Figure 11.1 Simplified illustration of the steps involved from extraction to genome annotation during the process of Next generation Sequencing. (1) Extraction of RNA or DNA from the sample. (2) Shearing leads to the fragmentation of the DNA to be sequenced. (3) Library preparation refers to the ligation of adapters for sample identification during data analyses. (4) Libraries are pooled according to the coverage/read depth aimed for where more samples per lane/flow cell lead to lower coverage/read depth per sample. (5) Sequencing of the DNA fragments by synthesis using 3′ blocked labelled nucleotides. (6) Light signals sent from the flow cell are digitized and translated into bases. (7) Genome assembly by aligning and mapping contigs from A – single end, B – paired end, C – paired end overlapping sequencing or scaffolds from D – mate-pair sequencing and (8) gene annotation and description of possible functional connections based on the reference genome.

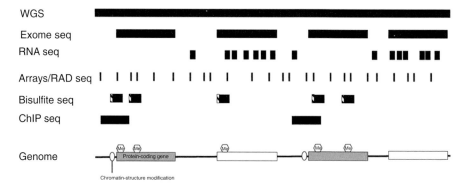

Figure 11.2 A schematic demonstration of the respective sections covered by all the methods described in this chapter. Whole-genome sequencing (WGS) provides the most complete information about the genome sequence, followed by exome sequencing, which targets only protein-coding regions. RADseq and arrays are agnostic methods sampling random sites along the genome. Bisulfite sequencing targets areas enriched with methylated cytosine (CpG islands) and CHiPseq targets methyl-modified proteins.

order to link the phenotypic variation with underlying genetic variation (Bush and Moore 2012). Initially, GWAS was used in humans in connection with the International HapMap Project (International HapMap Consortium 2003), which aimed at creating a haplotype map of the human genome to be used to assess genetic variation associated with disease and health issues. The map contained more than 1 million genotyped single nucleotide polymorphisms (SNPs) and many GWAS studies in humans based their SNP arrays on the HapMap. More generally, in the past, GWAS studies were largely based on targeted sequencing approaches such as RAD sequencing or microarrays, but with the decreasing costs of high-throughput whole-genome sequencing, it is increasingly common that genomes sequenced at relatively low coverage are used instead.

Linkage mapping (or genetic mapping) is a statistical approach used to assess recombination rates between different genes (Sturtevant 1913). Linkage maps provide information about the likelihood of genes being inherited across generations and, in combination with phenotypic traits, reveal information about the association between specific phenotypes and underlying genes. Technologies such as microarrays, restriction site-associated DNA sequencing, exome sequencing and whole-genome sequencing are all used for GWAS and linkage mapping.

11.1.1.1 Microarrays

Microarrays are a collection of single nucleotides (e.g. SNP arrays) or very short sequences of RNA or DNA (probes) associated with specific genes located across the genome (see Heller 2002 for review of methods). The number of probes may vary substantially between arrays and can be anywhere from ten probes to over 2 million probes. Needless to say, the more probes an array contains, the more of the genome will be covered. While it is true that microarrays can be manufactured for pretty much any study organism, it is also true that pre-existing microarrays offered by commercial companies often are a substantial short-cut in terms of time and energy as the compilation of a microarray from scratch in itself will take quite a bit of both. Another possible caveat to the use of microarrays is that only a limited number of regions in the genome are covered and hence one may risk losing out on vital information by not including them into an array. In other words, the compilation of a new microarray requires a certain level of knowledge of the genome and the genes to be included.

While microarrays were widely used in the past, they have become somewhat unfashionable and are now increasingly replaced by other technologies and by resequencing and RNA sequencing in particular. Part of the reason for this decline is that issues may arise from background levels of hybridization, making the analysis of the analogue data more problematic than the analysis of digital data obtained from NGS.

11.1.1.2 RAD Sequencing

Restriction site-associated DNA sequencing combines the activity of restriction enzymes to cut the genome into fragments of varying lengths and subsequent sequencing of the regions flanking the restriction sites (see Miller et al. 2007 for review of the method). Any polymorphisms (usually SNPs) in the flanking regions are used as markers and are referred to as RAD tags. The density of RAD tags depends strongly on the restriction enzyme used and the sites are randomly distributed across the genome. This method allows for the interrogation of anywhere between 0.1% and 10% of a genome and is generally used to identify regions in the genome that are particularly divergent between groups (or individuals) exhibiting differences in traits of interest.

11.1.1.3 Exome Sequencing

The protein-coding regions of the genome are referred to as the exome and this method's focus is on sequencing this particular part of the genome (see Teer and Mullikin 2010 for review of the method). Sequencing the exome substantially reduces the sequencing costs and the amount of data to analyse, as it generally occupies only a small part of the genome – in humans, for example, the exome constitutes only about 1% of the whole genome (Ng et al. 2009). The process involves a target enrichment step in which regions of interest are identified and specified prior to sequencing. Microarrays can be one way to capture the exomic regions, which are then sheared and amplified for subsequent sequencing. Alternatively, a custom-made selection of oligonucleotides is used to capture the specific regions in solution, which are then purified and sequenced without an amplification step. Exome sequencing is an efficient way of identifying genetic variants in an individual's protein-coding genes.

11.1.1.4 Whole-Genome Sequencing

Next-generation sequencing technology can be employed to sequence the whole genome (see Figure 11.2) (see Ekblom and Wolf 2014 for review of the method). Whole-genome sequencing (WGS) can be used for a *de novo* assembly of a target organism's genome that has not been sequenced previously or for genotyping-by-sequencing where the genomes of individuals of an organism with a reference genome are being (re-) sequenced.

One of the main challenges that one may encounter especially during *de novo* genome assemblies is the trouble with non-coding regions containing large amounts of repetitive elements and hence the assessment of copy number variation (CNV) and transposable elements (TEs), which are notoriously difficult to map. The current read length of short read technologies typically ranges between 50 (SOLiD) and 150 bp

(Illumina HiSeq), but methods allowing for read lengths of up to 2×300 bp (MiSeq), ~500 bp (IonTorrent), 5 kb (Pacific Biosciences), 10 kb (Illumina Moleculo) and 150 kb (10X Genomics Chromium Technology) at high throughput are available and are being further developed and improved.

11.1.2 Gene Expression Analyses

The analysis of gene expression targets not the genome directly but the actual transcripts in the form of RNAs and hence can be used to obtain insight into gene activity and regulation. In the past, methods such as real-time or quantitative polymerase chain reactions (qPCR) were predominantly used for the quantification of the transcription activity of specific genes through the use of designed primers. NGS allows the quantification of RNA at a genome-wide scale, allowing the targeting of many genes in one go in relatively little time compared to any qPCR experiment.

11.1.2.1 RNA Sequencing (RNAseq)

RNA sequencing uses NGS technology to capture the RNA content in terms of presence and quality of a given sample at a given time (see Wang et al. 2015 for review of the method). It is important to bear in mind that RNA profiles may change relatively rapidly and drastically even over short time intervals and hence the timing and tissue specificity during sample collection are crucial. Since some RNA families are particularly redundant but may not convey the required information (e.g. ribosomal RNA), most extraction and library prep protocols include steps to get rid of such by-products.

RNAseq as such is generally unspecific and targets any RNA sequences encountered and additional steps, such as running libraries on a gel prior to sequencing and cutting out bands within a specific length, will further allow the focusing on the RNA families of interest (e.g. messenger ribonucleic acids (mRNAs), microribonucleic acids (miRNAs), Piwi-interacting ribonucleic acids (piRNAs), etc.). The transcriptome profiles obtained provide information on alternative splicing products and posttranscriptional modification as well as alterations of the underlying genome such as mutations (SNPs) and gene fusion. In addition, they allow the quantification of specific RNAs and hence the assessment of differential expression and activity between target groups of interest.

11.1.3 Epigenetic Analyses

Analyses that go beyond the genome sequence and its transcriptome are increasing rapidly with growing interest in epigenetic mechanisms that not only contribute to cell differentiation but also have the potential to transfer non-genetic information across generations. Such analyses assess changes in the DNA and chromatin structure, which in turn are linked with gene activity.

11.1.3.1 CHiP Sequencing (CHiPseq)

The combination of CHiP with NGS technology is used to enrich DNA associated with a specific marker protein *in vivo* for subsequent sequencing (see Park 2009 for review of the method). CHiPseq allows identifying epigenetic marks in the form of binding sites of DNA-associated marker proteins such as transcription factors, polymerases and other DNA modifications along the genome. The information on histone methylation obtained from CHiPseq data reveals details of the possible activity of the genes at the respective loci. However, to reveal the true activity and expression level, the combination with RNAseq is a good way forward. Because CHiPseq can only detect one or a few specific proteins at a time, a pre-existing knowledge of the proteins in question is necessary.

11.1.3.2 Bisulfite Sequencing

This method combines a bisulfite treatment of extracted DNA with subsequent NGS and provides information about the cytosine methylation patterns across the genome (see Smith et al. 2009 for review of the method). The bisulfite treatment converts non-methylated cytosine molecules into uracil whereas all methylated cytosines are protected and hence can be easily identified during subsequent sequencing. A cheaper version of bisulfite sequencing is reduced representation bisulfite sequencing (RRBS), in which restriction enzymes fragment the genome to enrich for areas with high CpG content (Cytosine followed by Guanine: regions that generally show high rates of methylation) (Meissner et al. 2005). Cytosine methylation is the first epigenetic mark that has been discovered and it plays a pivotal part in regulating gene expression. However, it is worth noting that cytosine methylation may be absent in some taxa including *Caenorhabditis elegans* and several yeast species or is present at such a low level (e.g. *Drosophila melanogaster*) that bisulfite sequencing will not allow its detection (Capuano et al. 2014).

The above technologies are great tools to link phenotypes with possible underlying molecular mechanisms. Because these technologies cover different mechanisms, combining two or more of them will probably yield more satisfying results, particularly when assessing complex phenotypes such as those encountered in the study of behaviour. However, the technologies do not only differ in the mechanisms they cover but also in the overall coverage along the genome (see Figure 11.2), and while whole-genome approaches may be a desirable standard, reduced representation methods such as RAD-seq, exome sequencing and RRBS will often go a long way at a significantly reduced financial cost. This reduced cost means such technologies also allow the sequencing of many more individuals, resulting in higher sample sizes than whole-genome approaches.

11.2   Caveats and Challenges and Some Solutions

The benefits of NGS for the study of behavioural genomics are undeniable and I am the first one to encourage anybody to use these fantastic tools to address important and unanswered questions such as the role of genetic and epigenetic mechanisms in determining behavioural phenotypes. Nevertheless, having ventured into the area of NGS myself only recently and learning all about it as a non-molecular biologist, it may be worth pointing out some of the most important caveats and challenges that one may encounter when starting an NGS project. Below, I list some aspects I consider worth thinking about when planning a project that involves NGS.

11.2.1   Solid Phenotype

Linking a behavioural phenotype to the underlying genotype may be done using one of two directions: (i) by starting with a well-defined behavioural phenotype and finding its underlying molecular mechanisms (forward genetics) or (ii) by starting with a specific gene and testing its effect on a range of behavioural phenotypes (reverse genetics). Forward genetics is likely to be more relevant for the study of behaviour – at least in non-model organisms – as our knowledge of the genome as a functional entity is still limited and assessing the effect of candidate genes on phenotypes is still largely limited to model organisms where knock-down and knock-out mutants can be generated (but see section 11.1.1). Forward genetics is also where NGS will be most useful.

Studying genetic or epigenetic underpinnings of a phenotype will be easier the clearer and more extreme the phenotype of interest. In other words, having a well-defined and highly repeatable phenotype to start with is more likely to result in the identification of related genes and/or epigenetic patterns. It is for good reason that many studies in humans focus on psychoses and mental illnesses as these can readily be distinguished from a control group (see 'Case 11.3: Human psychoses' for examples), whereas more subtle phenotypes such as personality and individual variation in behaviour will be more difficult to determine. Nevertheless, a carefully designed experiment may be able to deal with more complex phenotypes and hence a good *a priori* understanding of the phenotypic variation in the trait of interest will help significantly with planning.

11.2.2   Sample Quality

Best results are always obtained from samples containing a lot of well-preserved material to start with. Careful sampling is therefore crucial if one wants to avoid major headaches further down the line. This means that appropriate equipment needs to be available during sampling and for long-term storage after sampling. RNA molecules, for example, are highly unstable and tend to decay within minutes after sampling. Special measures to maintain a high quality in samples collected for RNAseq include treating them with special buffers such as RNAlater (Thermo Fisher Scientific) that protect RNA molecules through the inactivation of RNAses where possible. Alternatively, snap freezing in liquid nitrogen and careful storage will maintain high quality of the RNA content and increase the chance of obtaining optimal results. DNA is much more stable than RNA but fresh tissue samples being frozen and stored at $-80\,°C$ will probably yield much better results than ethanol-stored samples kept at room temperature for many years.

11.2.3   Sampling

While DNA sequences will be highly conserved across different tissues, some tissues may be easier to process for extractions than others. In contrast, the situation for RNAs and epigenetic marks is more complex, as these will vary strongly across different tissues and will also probably change with time. It is therefore important to consider carefully, which tissue to target and to standardize the tissue sections used as well as the timing of the sampling across samples and pools.

11.2.4   Libraries and Sample Pools

The quantity of extracted material will determine how easy library preparation will be. Library preparation is the barcoding of the samples by ligation of sample-specific adapters to the fragments in one sample. This process allows the recognition of each sample later during the bioinformatics analyses by its specific barcoding sequence. Commercial sequencing platforms generally offer to do library prep as a part of the package but they usually require a substantial amount of material to work with – partly to have enough material available for backup should any of the subsequent steps fail. This may not always be possible or desirable and you may choose to perform the library preps in your own lab, which opens up the opportunity to work with much lower yields (e.g. from fewer cells and smaller tissues or samples containing small amounts and highly decayed DNA such as museum samples or ancient tissue samples). Developing a library prep protocol for specific samples may require a bit of time but may pay off as the quality of the data may be improved. It is worth consulting a molecular lab with experience in preparing samples for sequencing when planning on performing library preps in your own lab.

Whether to pool samples or not and if so, how many samples to include are important decisions and depend on the scientific question. Pooling may be an efficient way to reduce biological variance and if one is not specifically interested in this variance it may be a good solution, also because it may reduce sample sizes required for sequencing and hence costs. Biological replication is important and at least three biological replicate pools are recommended but the number of pools also depends on factors such as statistical power. The question of the number of samples per pool is mostly influenced by the DNA/RNA yield obtained from individual samples and by the expected biological variance. Pools can contain anywhere from less than ten up to several hundreds of individual samples.

11.2.5 Reference Genome

A reference genome is not mandatory for all methods but it can help with the interpretation of results. The more resolved the reference genome is, the more information can be retrieved from the dataset. When designing a project, one should check whether the study organism of interest already has a sequenced genome (of any suitable quality) as the existence of a reference genome makes life much easier for any subsequent studies. Reference genomes should come from the organism in question itself (the same or a different strain) but can sometimes come from a closely related organism, if they share large regions of their genome. If the reference genome comes from a closely related species, it is worth spending some time to understand how similar the two genomes are. While typically, closely related species will share much of their genome, it can happen that the genome of one species may be rather different. An example is the finding that the genome of *C. elegans* is about 30% smaller than that of its close relative *C. remanei*, which has a genome containing many more protein-coding genes (Fierst et al. 2015).

For organisms where a reference genome is not available, one can proceed without and use technologies that do not require one (e.g. RADseq, RNAse). Alternatively, one can switch to an organism with a sequenced genome or decide to do a *de novo* assembly of a new reference genome. The former may be significantly easier and is worth serious consideration. The latter can prove to be a major hurdle and some major classic study organisms that have been widely used for the study of genetics and experimental studies are notoriously difficult to sequence (e.g. Atlantic salmon *Salmo salar* due to its recent polyploidization or the white campion *Silene latifolia* due to its unusually large amount of non-coding regions and repetitive elements). Most genomes contain regions with non-coding DNA as structural challenges and regions around the centromeres are particularly prone to be repetitive rich. Long-read technologies such as PACBio and more advanced technologies in the future should help to alleviate this problem.

11.2.6 Sample Size

The question about sample size in the context of NGS studies is important as it usually directly affects the financial commitments required to realize the project. What can be said first and foremost is that whatever sample sizes are needed for accurate statistical results and the estimation of biological effect sizes in other studies are needed just as much in NGS studies. It may have been legitimate to draw fundamental conclusions from NGS data based on very small sample sizes just a few years ago when the use of NGS *per se* was still a novelty, but these examples can no longer serve as the gold standard when designing a study. Factors such as the strength of the phenotype of interest and the genetic variation in the sequenced samples (and hence the noise in the resulting data) will strongly affect the sample size needed for accurate statistics. An experiment where two groups belonging to highly inbred strains are exposed to two different treatments to compare their transcriptomes is likely to require fewer samples than a study where samples have been collected randomly from a wild population to identify the genes underlying a continuous and highly variable phenotype. Nevertheless, this may not always be true (see, for example, Dean et al. 2017) and the better the information about genetic and phenotypic variation in any population, the easier it will be to identify an adequate sample size.

11.2.7 Replication

A design aspect related to sample size is the question about biological and technical replicates to be included in the study. Biological replication generally refers to samples that help to determine the biological variation that can be found in the samples of interest. This may be of particular importance whenever samples need to be pooled for sequencing, which means that the extracted material from more than one individual is included in one library and hence it will no longer be possible to distinguish between different individuals. As discussed above, the use of pools may reduce variance across replicates and may allow for a reduction in sample size.

Besides biological replication, one may also need to consider technical replication, as sequencing biases or variation due to technical error across flow cells/lanes may add noise to the data. A common strategy to alleviate variance due to technical differences is to barcode all samples and run all of them across all lanes in a batch. Dividing samples across batches is problematic, as is combining data from multiple sequencing centres or technology platforms.

11.2.8 Coverage

Finally, the third key aspect to consider is the number of reads ('coverage' for DNA sequencing or 'read depth' for RNA sequencing) to be obtained per sample. Coverage will have a direct impact on the project costs as it will determine the number of lanes/flow cells needed to reach the required data quality. The answer about coverage/read depth is highly project specific and may be rather low in large-scale GWAS studies where thousands of individuals are being sequenced or rather high in studies where allele frequencies or number of small RNA copies are to be determined. For

WGS, the recommended coverage ranges from 1× to 8× for the identification of copy number variants (Xie and Tammi 2009; Medvedev et al. 2010) up to 60× for the identification of INDELs (Fang et al. 2014) and around 35× for genotype calling (Ajay et al. 2011). For exome sequencing, 100× coverage with local depth between 3× and 13× (Clark et al. 2011; Meynert et al. 2013) is necessary. RNAseq recommendations are 10–15 M reads for differential expression profiling, 50–100 M reads for alternative splicing and allele-specific expression analyses and more than 100 M reads for *de novo* assembly (Liu et al. 2013, 2014; ENCODE 2011 RNAseq guidelines). For CHiPseq, recommendations are 10–14 M reads up to 20–40 M reads, depending on the sharpness of the peaks (Rozowsky et al. 2009; ENCODE 2011 Genome guidelines; Landt et al. 2012) and for bisulfite sequencing from 5× to 30× (Ziller et al. 2015; Roadmap Epigenomics: http://www.roadmapepigenomics.org) and 10× for RRBS (ENCODE 2011 Genome guidelines). ENCODE guidelines can be found at www.encodeproject .org/about/experiment-guidelines.

11.2.9 Pilot Studies

Before sending off a large batch of samples for sequencing, it may be wise to work with a subset of the samples and look at the data obtained from these. It can be worth sequencing the pilot samples at a slightly higher coverage in order to fully understand the data. Pilot data usually help to understand the possible technical and analytical challenges one may expect from the samples of interest and provide information about biological variance. Carefully performed pilot studies may save a lot of time, money and stress.

11.2.10 Time and Planning

A key factor that needs to be taken into account when embarking on an NGS project is that every step is time-consuming. Any experience with experimental studies and *ad hoc* sampling in the field or experimental set-up may prove useless when it comes to estimating the time needed to complete an NGS project. The main difference between projects involving NGS and other experimental projects is the time needed from sample collection to the final results. Most sequencing platforms have a queuing time due to the demand being higher than can be handled. It is worth assessing the situation at the platforms located in close proximity compared to commercial platforms. Most researchers I know have tried out several pathways and all seem to have their pros and cons.

Once the platform has sent the data back, another time-consuming step commences, namely NGS data analysis (see also section on 'Bioinformatics' below). It is time well spent to carefully consider and evaluate bioinformatic pipelines even if this will take several months. In any case, careful planning all the way through will pay off. This includes careful sample collection to the work in the wetlab, choosing the sequencing platform, NGS data delivery and storage and bioinformatics analyses.

11.2.11 Bioinformatics

The processing of data of the scale of NGS data is never fast, regardless of what is being analysed. The fastest methods are those where working pipelines are at hand and the person running the analyses has handled many datasets of the same type many times before. However, often several different pipelines may be used to analyse a specific dataset and analysing datasets using different pipelines may at times lead to quite different results. It is therefore important to spend time on understanding the concepts behind the different pipelines and how these affect the specific dataset, and to run datasets with more than just one pipeline to test the robustness of the results. Now, even if pipelines have been used many times in previous studies, it is always worth looking into the latest developments and newer methods, as bioinformatics tools are being developed as fast as NGS technologies. This is for a good reason as handling such huge datasets requires completely different processing capacity. Spending a substantial amount of time with the analyses is therefore unavoidable and this should be factored in when planning a project.

11.2.12 Collaboration

Many of the challenges described above are most efficiently addressed by making an NGS project a collaborative venture. With NGS, single-author papers are generally not feasible or desirable. The fundamentally multifaceted expertise needed to successfully complete an NGS project requires the collaboration of experts, each in their own field. Molecular methods are continuously being replaced by more efficient and accurate versions and the same is true for bioinformatics analyses. Only experts can truly keep up with their respective fields and will generally have a much better overview of currently available tools and technologies. What may work well for one species and one tissue may not work for another species or a different tissue from the same species, and instead of wasting weeks and months trying things out, it is worth contacting a lab that specializes in this particular aspect to benefit from their experience and knowledge. In turn, experts in molecular methodology or bioinformaticians may find it stimulating to collaborate with scientists studying specific biological phenotypes. A joint effort will most likely yield a more satisfying result for both sides.

11.3 Linking Behavioural Phenotypes to Genotypes using NGS

Behavioural phenotypes vary enormously in their complexity and while some behavioural traits are largely managed by one or a few genes, a majority of behavioural traits follow a pattern of complex traits and fall into the category of quantitative traits. While NGS can certainly be applied to any behavioural trait of interest, its application will benefit the study of complex traits based on alleles located at multiple quantitative trait loci (QTL) and even more so to the study if the role of non-genetic effects. In this section, I present a series of examples for the successful use of NGS for the study of increasingly complex traits.

Case 1 *Circadian Clocks and Migratory Behaviour*   With the aim of understanding daily and annual cycles in animals, research into understanding the genetic basis of rhythms in animals has produced a large body of information about the genetic architecture of such traits. The *Clock* genes and allelic variation in some of their regions are known to affect circadian rhythms and sleeping and feeding behaviours as well as the onset of the breeding season and migratory behaviour. The first circadian clock gene *per* (period) was discovered in *Drosophila* in the early 1970s by producing mutant strains that differed significantly in their 24-hour circadian behaviour (Konopka and Benzer 1971). However, the description of the actual gene had to wait until the mid-1980s when molecular methods were advanced enough (Bargiello et al. 1984; Zehring et al. 1984). Similarly, the vertebrate *Clock* gene was initially identified by screening mutations in mice strains (Vitaterna et al. 1994) and was cloned ten years later (Antoch et al. 1997).

The identification and description of the *Clock* gene opened the possibility to use a candidate gene approach in wild populations to study variation in behaviours associated with circadian rhythms such as the sleeping time preference behaviour in humans (Katzenberg et al. 1998; Mishima et al. 2005), variation in spawning times in different trout strains (Leder et al. 2006) and seasonal timing and reproduction in birds (Johnsen et al. 2007; Liedvogel et al. 2009). These studies used a combination of arrays and targeted gene approaches to test for an association of *Clock* genes with the behavioural patterns in question. More recently, WGS of ten males from each of two subspecies of Swainson's thrushes (*Catharus ustulatus ustulatus* and *C. u. swainsoni*) and subsequent comparison of the genomes resulted in the identification of candidate genes involved in the determination of migratory pathways of individuals belonging to these two groups (Delmore et al. 2015). A follow-up study where 190 birds from the hybrid zone between the two subspecies were genotyped using 59 485 SNPs confirmed the role of *Clock* genes and genes in the nervous system and cell signalling pathways in determining migratory patterns (Delmore et al. 2016).

The migration of the monarch butterfly *Danaus plexippus* across North America is another great example where novel sequencing techniques have provided invaluable insights into the physiological mechanisms and evolutionary history of the behavioural patterns. A high-density microarray of 9417 unique cDNA sequences was used to compare gene expression in the brains of summer butterflies and autumn migrants and revealed expression differences in 40 genes, including some *Clock* genes and structural genes but also several unannotated genes (Zhu et al. 2009). A later study sequenced the whole genomes of a total of 101 butterflies from around the globe and revealed unexpected aspects of the evolutionary history of the species such as a migratory ancestor of the species (Zhan et al. 2014). In the study of *Clock* genes, the combination of WGS , RNA sequencing and the targeted sequencing of candidate genes proved a very powerful tool to gain a better understanding of the molecular and genetic mechanisms underlying striking behavioural patterns.

Case 2 *Parental Behaviour*   Parental behaviour in mammals is one of the more complex behavioural traits and studying its genetic underpinning is likely to be complex and difficult to untangle. A recent study combined careful species selection with several genetic and genomic tools, which yielded fascinating insights into parental behaviour in deer mice (Bendesky et al. 2017).

Two species of deer mice located at the opposite ends of the scale of mating systems (*Peromyscus polionotus* is monogamous and *P. maniculatus* is promiscuous) also differ significantly in their parental care, with *P. polionotus* generally being more parental (Dewsbury 1981). The two species were crossed to create F1 and F2 offspring, and 769 F2 offspring were analysed behaviourally as well as genotypically at 406 611 loci using

RADseq. The combination of behavioural and genetic data allowed the identification of several QTLs located across the genome. One QTL on chromosome 4 accounted for most of the variation observed in behaviour and was therefore examined more carefully. The protein-coding genes in this QTL were identified and expression levels assessed by describing the transcriptome in the hypothalamus through RNA sequencing of males and females of both species and by assessing allele-specific expression in F1 hybrids. Allele-specific expression reveals information about the imprinting patterns (usually through methylation) inherited from the heterospecific parents and hence gene regulation and expression. The end-result is that one particular gene for arginine vasopressin is strongly associated with nest-building behaviour in deer mice. The same gene has been found previously to be an important modulator of social behaviour, including maternal care (Lim et al. 2004; Insel 2010). This study is a great example if an approach to identify candidate genes, and future studies may widen the scope and identify other key genes located potentially somewhere further up the enzymatic cascade with a regulatory effect on the gene identified in this study. Even though this study shows that candidate genes can be identified even for rather complex behavioural traits, non-genetic factors are likely to play a crucial role in determining parental behaviour and its inheritance and account for a large part of the variation observed across individuals. In the prairie vole *Microtus ochrogaster*, a lack of handling experience in very young offspring influenced social behaviour as adults and led to reduced participation in alloparenting in the offspring of the following two generations (Stone and Bales 2010). It will be highly interesting to improve our understanding of the relative importance of genetics and epigenetics by combining DNA sequencing technologies with technologies describing the expression patterns and epigenetic marks associated with them.

Case 3 *Human Psychoses* The use of sequencing technology may be particularly promising in the detection of genes underpinning extreme behavioural phenotypes. Studies investigating the genetics of human psychoses provide great examples for the successful use of NGS to pinpoint genes related to psychotic behavioural phenotypes. GWASs have been particularly widely used in the study of human psychoses (Hirschhorn and Daly 2005; Lee et al. 2012). Most of the human studies so far have used a SNP chip of several hundred thousand or even millions of known SNP variants across the human genome to look for associations with specific diseases by comparing a control group with a group showing the phenotype of interest. While many of the previous GWAS studies were based on SNP chips, decreasing sequencing prices result in WGS increasingly replacing the use of SNP chips.

An important aspect of GWAS is the planning and sampling prior to any NGS. A great example of how careful sample selection may lead to a sought-after breakthrough comes from research into the genetic marks for a mental illness called recurrent Major Depressive Disorder (MDD). Identification of the genetic markers underlying this syndrome has been the aim of many research labs for many years without the success scientists were hoping for. Even large-scale studies, or mega-analyses, which combined existing datasets from previous studies based on SNP arrays into one enormous dataset, were unable to clearly link the syndrome to a region and even less to a specific genetic maker (Major Depressive Disorder Working Group of the Psychiatric GWAS Consortium 2013). This was until a recent study employed low-coverage genome sequencing of 5303 Chinese women with MDD and compared these to 5337 women without MDD and were able to identify two loci associated with the mental illness (CONVERGE Consortium 2015). The success of the study is partly to be found in the careful sampling from a relatively homogenous pool of individuals keeping genetic variation within a certain limit and in the use of WGS as opposed to the arrays used in previous studies.

Case 4 *Inheritance of Stress* While the role of the genetic background in the inheritance of behavioural phenotypes is undeniable, the rising interest in non-genetic inheritance has provided striking evidence for the critical role of environmental factors in affecting not only the physiology and condition of individuals experiencing them, but also behaviour in subsequent generations. The discovery of small non-coding RNAs being inherited not only through the female gametes but also the male gametes has triggered strong interest in this field and with RNA sequencing methods becoming ever more sophisticated and fine tuned to specific RNA families, the study of such specific groups of RNAs is particularly interesting.

The study of the inheritance of stress is a field of research which has greatly benefited from the development of ever more sophisticated sequencing technologies. Initial evidence comes from a study in house mice *Mus musculus*, in which exposure to aggressive encounters resulting in chronic defeat stress in males led to increased depression and anxiety-like phenotypes in their offspring (Dietz et al. 2011). Similarly, early life trauma due to maternal separation during early postnatal development in house mice has been shown to cause behavioural disorders not only in the pups experiencing the trauma but also in their offspring (Franklin et al. 2010). To understand the underlying mechanisms, a subsequent study repeated the experiment, exposing male pups to the same maternal separation and assessing the small non-coding RNA profiles in sperm of treated and control males at sexual maturity (Gapp et al. 2014). Males were also allowed to reproduce and offspring were exposed to behavioural tests to assess the behavioural and physiological phenotypes in the offspring. Several micro-RNAs were upregulated and some Piwi-regulating RNAs were downregulated in the sperm of treated males. When microinjecting purified RNAs obtained from treated and control males into wild-type fertilized oocytes, the resulting offspring showed similar behavioural and physiological phenotypes as the offspring sired by treated and control males, confirming the role of RNAs in the transmission of the observed effects. The combination of experimental manipulations and NGS is an exciting and novel road to studying the mechanisms underlying the inheritance of non-genetic factors affecting behavioural traits across generations.

11.4 What's Next

Many of the technologies described in this chapter are still in their infancy and their development will continue into the future and make studies such as those described in Cases 1–4 ever more feasible. In order to get to grips with complex behavioural phenotypes such as personality, it will be necessary to combine the identification of candidate genes with information on the gene activity and non-genetic factors involved. With the ability to analyse the composition of a genome by its DNA sequence, its physical structure and the related expression patterns and resulting RNAs (and potentially the proteome as a next step in the cascade), a profound understanding of the molecular composition and inheritance of a behavioural trait is possible. However, because the mechanisms underlying behavioural traits are likely to be multifaceted and highly complex, combining efforts and bringing together the expertise of researchers from different fields is pivotal for a successful outcome. The case studies described above generally focus on one or maybe two steps of the entire process from genotype to phenotype and bringing together all the pieces of the puzzle will require collaboration between labs and people with varying strengths.

In addition to combining different technologies, there are a few recent developments that will probably play an important role in the study of behaviour in the not too distant future. One is the study of the non-coding regions of the genome and the elements contained in them and what role they play in the regulation of gene function. By generally improving our understanding of the genome as a whole, we will improve our understanding of its role in determining the phenotype. The other aspect is the ability to actively manipulate gene expression and some recent technologies allow for a highly targeted approach, which makes gene editing and silencing more feasible. I discuss these two aspects in a bit more detail below.

11.4.1 Understanding the Non-Coding Regions of the Genome

Recent advances in genetics and genomics have focused on understanding the role of non-coding regions of the genome and in particular of transposable elements and repetitive elements. With the development of NGS technologies that allow a more accurate description of such regions, generally packed with repetitive elements that are notoriously difficult to sequence, we will be able to assess variation in the structure of these regions. Our understanding of the role of transposable elements and repetitive elements in combination with an increasing knowledge of the epigenetic mechanisms and the significance of different RNA families and their interactions will improve greatly over the coming years.

Copy number variation (CNV) is a type of structural variation in the genome and is based on processes of duplication and deletion of short sequences consisting of several base pairs (short repeats) (Lee et al. 2012). The interest in CNVs and their association with phenotypes originated in the study of disease in humans, where disease phenotypes were reliably identified by an increased number of copies of a specific short repeat. However, the study of CNVs is on the rise as CNVs generally cover a substantial area of the genome (more than two-thirds in humans) (Koning et al. 2011) and are known to affect gene function in many different taxa. In addition, the recent understanding of the potential role of transposable elements for gene expression and regulation has triggered a wave of interest in the study of the nature and distribution as well as the functional role of repetitive elements in the genome (Lippman et al. 2004). Long-read high-throughput technologies are proving particularly useful in the study of CNVs.

11.4.2 Gene Knock-down and Knock-out in Non-Model Organisms

As mentioned above, the forward genetics approach starting with a phenotype and identifying the underlying genes is currently one of the more promising approaches in

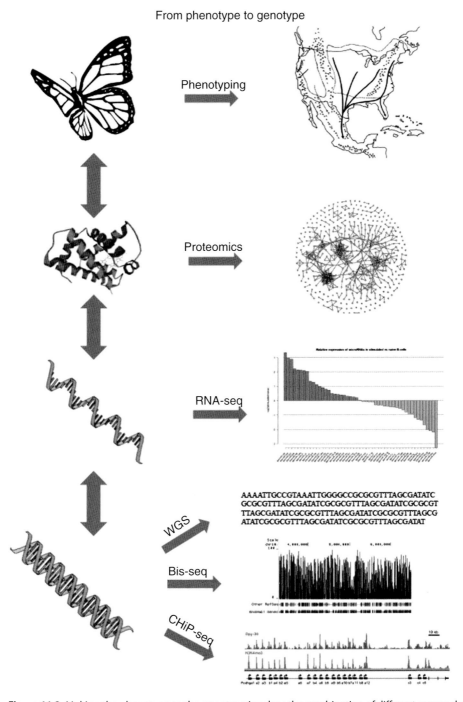

Figure 11.3 Linking the phenotype to the genotype involves the combination of different approaches from careful phenotype description, describing the proteome and enzymatic cascades and networks and comparing these to transcriptome profiles, assessing the level of gene expression by describing transcriptome profiles (RNAseq), assessing the functional structure of the underlying genome based on chromatin modifications (bisulfite seq) and histone modifications (CHiPseq) and ultimately identifying the underlying candidate genes (WGS) governing the entire cascade.

non-model organisms. Nevertheless, it is also true that technologies for the targeted knock-down and knock-out of candidate genes are improving continuously. It is therefore increasingly feasible that tools for gene knock-down and knock-out as well as experimental overexpression can be applied in projects using non-model organisms.

One of the technologies that is successfully applied in organisms as different as *C. elegans* and the house mouse is RNA interference (or short RNAi). RNAi is a key technology developed to silence target genes on a non-permanent basis, that is, without inducing a mutation. The silencing is obtained through the injection of double-stranded RNA, which interferes with the transcripts (mRNAs) of target genes (Fire et al. 1998). It is based on a biological process found in all cells where small RNAs bind to mRNAs, which results in interference with specific mRNAs and their incapacitation.

A recently developed promising tool for gene knock-out, the permanent silencing of a gene through its mutation, is based on the Clustered Regularly Interspaced Short Palindromic Repeats (CRISPR)/Cas-9 system (Cong et al. 2013). CRISPR is a mechanism applied in combination with activity of the enzyme Cas-9 to insert short repetitive sequences into the genome at predetermined sites in order to knock out genes in a highly targeted way. The CRISPR/Cas-9 technology is generally regarded as a significant methodological advancement for the generation of knock-out mutants as it has a much higher and more accurate success rate than previous technologies, where the production of mutants was a relatively random process.

11.5   Concluding Remarks

Behavioural phenotypes are among the most complex of traits to be understood and it is undeniable that only by employing a multitude of different approaches and combining the results from these approaches will we be able to fully understand them (Figure 11.3). With the onset of the NGS era, we are getting quite a bit closer to the goal of dissecting the genetic underpinnings of even complex behavioural traits more accurately. The combination of NGS approaches such as the search for candidate genes and the careful analysis of epigenetic marks and patterns of the genomes in question will provide exciting new answers to hitherto unanswered questions. The careful planning of experimental manipulations and/or highly specified sample collection in combination with the use of NGS technologies are currently some of the most powerful approaches available to understand the nature and evolution of complex traits.

References

Ajay, S.S., Parker, S.C., Abaan, H.O. et al. (2011). Accurate and comprehensive sequencing of personal genomes. *Genome Research* 21: 1498–1505.

Antoch, M.P., Song, E.-J., Chang, A.-M. et al. (1997). Functional identification of the mouse circadian *Clock* gene by transgenic BAC rescue. *Cell* 89: 655–667.

Bargiello, T.A., Jackson, F.R., and Young, M.W. (1984). Restoration of circadian behavioural rhythms by gene transfer in *Drosophila*. *Nature* 312: 752–754.

Bendesky, A., Kwon, Y.-M., Lassance, J.-M. et al. (2017). The genetic basis of parental care evolution in monogamous mice. *Nature* 544: 434–439.

Buermans, H.P.J. and Dunnen, J.T. (2014). Next generation sequencing technology: advances and applications. *Biochimica et Biophysic Acta* 1842: 1932–1941.

Bush, W.S. and Moore, J.H. (2012). Chapter 11: genome-wide association studies. *PLoS Computational Biology* 8: e1002822.

Capuano, A., Mülleder, M., Kok, R. et al. (2014). Cytosine DNA methylation is found in *Drosophila melanogaster* but absent in *Saccharomyces cerevisiae, Schizosaccharomyces pombe*, and other yeast species. *Analytical Chemistry* 86: 3697–3702.

Clark, M.J., Chen, R., Lam, H.Y. et al. (2011). Performance comparison of exome DNA sequencing technologies. *Nature Biotechnology* 29: 908–914.

Cong, L., Ran, F.A., Cox, D. et al. (2013). Multiplex genome engineering using CRISPR/Cas systems. *Science* 339: 819–823.

CONVERGE Consortium (2015). Sparse whole genome sequencing identifies two loci for major depressive disorder. *Nature* 523: 588–591.

Dean, R., Wright, A.E., Marsh-Rollo, S.E. et al. (2017). Sperm competition shapes gene expression and sequence evolution in the ocellated wrasse. *Molecular Ecology* 26: 505–518.

Delmore, K., Hübner, S., Kane, N.C. et al. (2015). Genomic analysis of a migratory divide reveals candidate genes for migration and implicates selective sweeps in generating islands of differentiation. *Molecular Ecology* 24: 1873–1888.

Delmore, K., Toews, D.P.L., Germain, R.R. et al. (2016). The genetics of seasonal migration and plumage color. *Current Biology* 26: 2167–2173.

Dewsbury, D.A. (1981). An exercise in the prediction of monogamy in the field from laboratory data on 42 species of muroid rodents. *Biologist* 63: 138–162.

Dietz, D., LaPlant, Q., Watts, E. et al. (2011). Paternal transmission of stress-induced pathologies. *Biological Psychiatry* 70: 408–414.

Ekblom, R. and Wolf, J.B.W. (2014). A field guide to whole-genome sequencing, assembly and annotation. *Evolutionary Applications* 7: 1026–1042.

Ellegren, H. (2014). Genome sequencing and population genomics in non-model organisms. *Trends in Ecology & Evolution* 29: 51–63.

ENCODE Project Consortium (2011). A user's guide to the encyclopedia of DNA elements (Encode). *PLoS Biology* https://doi.org/10.1371/journal.pbio.1001046.

Fang, H., Wu, Y., Narzisi, G. et al. (2014). Reducing INDEL calling errors in whole genome and exome sequencing data. *Genome Medicine* 6: 89.

Fierst, J.L., Willis, J.H., Thomas, C.G. et al. (2015). Reproductive mode and the evolution of genome size and structure in *Caenorhabditis* nematodes. *PLoS Genetics* 11: e1005323.

Fire, A., Xu, S., Montgomery, M. et al. (1998). Potent and specific genetic interference by double-stranded RNA in *Caenorhabditis elegans*. *Nature* 391: 806–811.

Franklin, T.B., Russig, H., Weiss, I.C. et al. (2010). Epigenetic transmission of the impact of early stress across generations. *Biological Psychiatry* 68: 408–415.

Gapp, K., Jawaid, A., Sarkies, P. et al. (2014). Implication of sperm RNAs in transgenerational inheritance of the effects of early trauma in mice. *Nature Neuroscience* 17: 667–669.

Heller, M.J. (2002). DNA microarray technology: devices, systems, and applications. *Annual Review of Biomedical Engineering* 4: 129–153.

Hirschhorn, J.N. and Daly, M.J. (2005). Genome-wide association studies for common diseases and complex traits. *Nature Reviews Genetics* 6: 95–108.

Insel, T.R. (2010). The challenge of translation in social neuroscience: a review of oxytocin, vasopressin, and affiliative behavior. *Neuron* 65: 768–779.

International HapMap Consortium (2003). The international HapMap project. *Nature* 426: 789–796.

Johnsen, A., Fidler, A., Kuhn, S. et al. (2007). Avian Clock gene polymorphism: evidence for a latitudinal cline in allele frequencies. *Molecular Ecology* 16: 4867–4880.

Katzenberg, D., Young, T., Finn, L. et al. (1998). A CLOCK polymorphism associated with human diurnal preference. *Sleep* 21: 569–576.

Koning, A.P.J.D., Gu, W., Castoe, T.A. et al. (2011). Repetitive elements may comprise over two-thirds of the human genome. *PLoS Genetics* 7: e1002384.

Konopka, R.J. and Benzer, S. (1971). Clock mutants of *Drosophila melanogaster*. *Proceedings of the National Academy of Sciences of the United States of America* 68: 2112–2116.

Landt, S.G., Marinov, G.K., Kundaje, A. et al. (2012). ChIP-seq guidelines and practices of the ENCODE and modENCODE consortia. *Genome Research* 22: 1813–1831.

Leder, E.H., Danzmann, R.G., and Ferguson, M.M. (2006). The candidate gene, Clock, localizes to a strong spawning time quantitative trait locus region in rainbow trout. *Journal of Heredity* 97: 74–80.

Lee, K.W., Woon, P.S., Teo, Y.Y., and Sim, K. (2012). Genome wide association studies (GWAS) and copy number variation (CNV) studies of the major psychoses: what have we learnt? *Neuroscience and Biobehavioral Reviews* 36: 556–571.

Leshner, A. and Pfaff, D.W. (2011). Quantification of behavior. *Proceedings of the National Academy of Sciences USA* 108: 15537–15541.

Liedvogel, M., Szulkin, M., Knowles, S.C.L. et al. (2009). Phenotypic correlates of Clock gene variation in a wild blue tit population: evidence for a role in seasonal timing of reproduction. *Molecular Ecology* 18: 2444–2456.

Liedvogel, M., Åkesson, S., and Bensch, S. (2011). The genetics of migration on the move. *Trends in Ecology & Evolution* 26: 561–569.

Lim, M.M., Wang, Z., Olazabal, D. et al. (2004). Enhanced partner preference in a promiscuous species by manipulating the expression of a single gene. *Nature* 429: 754–757.

Lippman, Z., Gendrel, A.-V., Black, M. et al. (2004). Role of transposable elements in heterochromatin and epigenetic control. *Nature* 430: 471–476.

Liu, Y., Ferguson, J.F., Xue, C. et al. (2013). Evaluating the impact of sequencing depth on transcriptome profiling in human adipose. *PLoS One* 8: e66883.

Liu, Y., Zhou, J., and White, K.P. (2014). RNA-seq differential expression studies: more sequence or more replication? *Bioinformatics* 30: 301–304.

Major Depressive Disorder Working Group of the Psychiatric GWAS Consortium (2013). A mega-analysis of genome-wide association studies for major depressive disorder. *Molecular Psychiatry* 18: 497–511.

Medvedev, P., Fiume, M., Dzamba, M. et al. (2010). Detecting copy number variation with mated short reads. *Genome Research* 20: 1613–1622.

Meissner, A., Gnirke, A., Bell, G.W. et al. (2005). Reduced representation bisulfite sequencing for comparative high-resolution DNA methylation analysis. *Nucleic Acids Research* 33: 5868–5877.

Meynert, A.M., Bicknell, L.S., Hurles, M.E. et al. (2013). Quantifying single nucleotide variant detection sensitivity in exome sequencing. *BMC Bioinformatics* 14: 195.

Miller, M.R., Dunham, J.P., Amores, A. et al. (2007). Rapid and cost-effective polymorphism identification and genotyping using restriction site associated DNA (RAD) markers. *Genome Research* 17: 240–248.

Mishima, K., Tozawa, T., Satoh, K. et al. (2005). The 3111T/C polymorphism of hClock is associated with evening preference and delayed sleep timing in a Japanese population sample. *American Journal of Medical Genetics B: Neuropsychiatric Genetics* 133: 101–104.

Ng, S.B., Turner, E.H., Robertson, P.D. et al. (2009). Targeted capture and massively parallel sequencing of 12 human exomes. *Nature* 461: 272–276.

Park, P.J. (2009). ChIP-Seq: advantages and challenges of a maturing technology. *Nature Reviews Genetics* 10: 669–680.

Rozowsky, J., Euskirchen, G., Auerbach, R.K. et al. (2009). PeakSeq enables systematic scoring of ChIP-seq experiments relative to controls. *Nature Biotechnology* 27: 65–75.

Smith, Z.D., Gu, H., Bock, C. et al. (2009). High-throughput bisulfite sequencing in mammalian genomes. *Methods* 48: 226–232.

Snell-Rood, E.C. (2013). An overview of the evolutionary causes and consequences of behavioural plasticity. *Animal Behaviour* 85: 1004–1011.

Sokolowski, M. (2001). Drosophila: genetics meets behaviour. *Nature Reviews Genetics* 2: 879–890.

Stone, A.I. and Bales, K.L. (2010). Intergenerational transmission of the behavioral consequences of early experience in prairie voles. *Behavioral Processes* 84: 732–738.

Sturtevant, A.H. (1913). The linear arrangement of six sex-linked factors in Drosophila as shown by their mode of association. *Journal of Evolutionary Zoology* 14: 43–59.

Teer, J.K. and Mullikin, J.C. (2010). Exome sequencing: the sweet spot before whole genomes. *Human Molecular Genetics* 19: R145–R151.

Van Dijk, E.L., Auger, H., Jaszczyszyn, Y., and Thermes, C. (2014). Ten years of next-generation sequencing technology. *Trends in Genetics* 30: 418–426.

Vitaterna, M.H., King, D.P., Chang, A.-M. et al. (1994). Mutagenesis and mapping of a mouse gene, *Clock*, essential for circadian behavior. *Science* 264: 719–725.

Wahlsten, D. (2012). The hunt for gene effects pertinent to behavioural traits and psychiatric disorders: from mouse to human. *Developmental Biology* 54: 475–492.

Wang, Z., Gerstein, M., and Snyder, M. (2015). RNA-Seq: a revolutionary tool for transcriptomics. *Nature Reviews Genetics* 10: 57–63.

Xie, C. and Tammi, M.T. (2009). CNV-seq, a new method to detect copy number variation using high-throughput sequencing. *BMC Bioinformatics* 10: 80.

Zehring, W.A., Wheeler, D.A., Reddy, P. et al. (1984). P-element transformation with period locus DNA restores rhythmicity to mutant, arrhythmic *Drosophila melanogaster*. *Cell* 39: 369–376.

Zhan, S., Zhang, W., Niitepold, K. et al. (2014). The genetics of monrach butterfly migration and warning colouration. *Nature* 514: 317–321.

Zhu, H., Gegear, R.J., Casselman, A. et al. (2009). Defining behavioral and molecular differences between summer and migratory monarch butterflies. *BMC Biology* 7: 14.

Ziller, M.J., Hansen, K.D., Meissner, A., and Aryee, M.J. (2015). Coverage recommendations for methylation analysis by whole-genome bisulfite sequencing. *Nature Methods* 12: 230–232.

12

Nature-Nurture in the Twenty-First Century

Nina Wedell[1], John Hunt[1,2] and David J. Hosken[1]

[1] Centre for Ecology & Conservation, University of Exeter, Penryn Campus, Penryn, TR10 9EZ, UK
[2] School of Science and Health & Hawkesbury Institute for the Environment, Western Sydney University, Hawkesbury, NSW 2793, Australia

Having reached the end of the book, we trust that one of our initial aims has been met and that the long-standing dichotomy that pits nature against nurture when thinking about behaviour has been put to rest. The dichotomy does not exist. Instead, the evidence discussed here reveals a far more nuanced, interesting and rich pattern of interactions between genes and the environment that shape the behaviour of animals, including humans. The chapters throughout this book highlight a variety of topics and behavioural phenotypes ranging from optimality and genetic approaches to the study of behaviour from fly mating to human culture, and provide extensive discussion on the relative importance of genes and environment for all of them. As noted, one general pattern to emerge from behavioural research is that behaviours are largely polygenic and usually affected by the statistical interaction between genes and environment. Importantly, as has been highlighted in several chapters, environments are not just the physical environment but also the biotic and social environments provided by other individuals, which have emerge as key factors shaping behaviour and the effects of behaviour (Tregenza et al. 2003). As these latter environments are highly dynamic, continually changing and evolving, this form of gene-environment interaction (GxE) is also always shifting and thus contributes to the substantial variability observed in behaviour.

The notion that the effect of a gene is inherently dependent on the 'social environment' in which they are found has long been recognized. Dawkins (1976) specifically argued that genes are selected only in relation to other genes in the genome, which represent their immediate social environment. The impact of a gene in isolation usually has little meaning, much as the fitness of an individual can only be quantified in relation to the other individuals in the same population. However, theory has now formalized the role of the social environment and described its impact and interaction with genes (Moore et al. 1997). The social environment is also important in shaping behaviour from another perspective, as it itself can be inherited and therefore directly exert a large impact on genetic evolution (see Chapters 3 and 4). The far-reaching role of the social environment and culture is discussed in full by Barrett and Stulp (Chapter 10), and they make the case that for humans, the socio-economic environment is a major evolutionary driver of behaviour. Clearly, genes and the environment are intrinsically linked in terms of shaping behaviour, and effects also involve multiple feedbacks (see Figure 1.2 in Chapter 1). Things can be rather complicated, but at least understandable in principle.

While our ability to identify genes influencing behaviour and our understanding of how their expression is modified by the environment have increased enormously – largely because of the development, accessibility and reduction in cost of new genomic techniques (see Chapter 11) – there is still much we do not know. One challenge will be to identify the precise genetic basis contributing to behavioural phenotypes and, if that is not challenging enough, fully mapping the genotype-phenotype landscapes across environments. There is an added challenge because genes frequently operate in gene networks, influence the expression of other genes, and are themselves not always expressed. Unravelling the function of such behavioural gene networks will require new approaches and probably a combination of technologies. New gene editing techniques to verify the functional role of candidate genes will also have to be employed, and this is often a laborious and costly process, especially for non-model organisms with interesting behavioural repertoires.

Technological challenges aside, we still have some way to go when it comes to defining behaviours – that is, to creating phenotype maps – and combining the phenotypic and genetic approaches to studying behavioural variation. However, this combined approach has the potential to connect current functional and phylogenetic hypotheses regarding the evolution of behaviour (see Chapter 2), a point that has been made previously (Roff 1994).

We have also begun to make genuine inroads into the functional basis of complex behaviours by unravelling the physiological and neurological underpinnings of behavioural variation. Sometimes new advances come from unexpected research areas. For example, many microbes and parasites have provided novel insights into how nervous systems work and regulate behaviour more generally, as they frequently invade the nervous system of their host and have evolved to target specific neurological pathways to alter key behaviours to their own benefit (Adamo 2013). The extent to which selfish genes and symbionts are responsible for shaping individual behaviour more generally and the extent to which hosts have evolved countermeasures to manipulation remain to be established (see Chapter 8). However, it is increasingly clear that non-self genes can have profound impacts on behaviour above and beyond the classic examples that might spring to mind, such as predator genes altering prey behaviour. We are increasingly aware that effects are far more subtle, nefarious and widespread.

One area only fleetingly covered in the book (see Chapter 8) is the well-documented impact of parasites and pathogens on behaviour (Schmidt-Hempel 1998). There have been recent excellent reviews of the diverse and fascinating ways in which parasites manipulate host behaviour to serve their own ends (Moore 2002; Hughes et al. 2012; Barber et al. 2016). In some cases, the underlying mechanisms of host manipulation are also known, and these frequently involve hijacking the host's nervous system. Classic examples include inducing suicidal behaviour of the host to promote the spread of the parasite, such as fungal parasites that turn ants into mummified spore-sacks by inducing them to crawl to places that maximize the spread of the fungus (Fredericksen et al. 2017), and nematode worms that cause crickets to drown themselves as the worms need water to reproduce (Thomas et al. 2002), to name but two.

However, it is only in the last decade that we have realized the dramatic impact that the microbiome (gut flora) can have in regulating a variety of behaviours in a range of animals from insects to mammals, including humans. This is a rapidly expanding area where new sequencing techniques coupled with dramatic cost reductions have enabled researchers to quantify the microbiome to determine how it affects behaviours, but also to experimentally manipulate part or all of the microbiome to test for resulting behavioural changes. This area promises to provide a fascinating new field with potential links to health and diet, and how the social environment (including mum and dad) affect gut flora and its impact on behaviours. In addition, it also has the potential to provide exciting applied therapeutic value by controlling the risk of chronic diseases in humans simply through modifying gut flora by dietary changes (Sing et al. 2017).

In contrast, an area where we have made substantial inroads into quantifying the impact of genes and the environment on behaviours comes, not surprisingly, from studies where one or a few genes of large effect impact on behavioural phenotypes and can therefore be examined with relative ease. An excellent example involves selective breeding/domestication of the red fox for the fur industry in Russia. Foxes were initially selected for their friendly behaviour (docility and lack of aggression around humans), and became 'tame' after only two to three generation of selection. Simultaneously, they also showed correlated changes in coat colour (became darker) and other physical characteristics (i.e. shortened legs, tail, snout, upper jaw, widened skull, and droopy ears), and produced bigger litters. It is suggested that these dramatic and rapid changes came about through alteration of the brain neurochemistry, which affected hormones and neurotransmitters (Trut et al. 2009). This finding suggests a direct link between gene(s) regulating the expression levels of these neurochemicals and the various behavioural, morphological and even life-history traits. Domestication of dogs has resulted in similar evolution and gene duplications have increased amylase production to facilitate adaptation to a domesticated diet (Axelsson et al. 2013).

An example of a single gene with large effect on multiple traits is the detoxification gene *Cypg6g1* in the fly *Drosophila melanogaster* where upregulation (caused by a transposable element insertion) confers insecticide resistance. In addition to DDT resistance, it also has a dramatic effect on female fecundity (increasing it: McCart et al. 2005), male morphology (smaller body size) and behaviour (reduced male courtship activity and aggression) even when controlling for genetic background (Smith et al. 2011; Rostant et al. 2017). Both the fox and fly examples reveal genes with pleiotropic effects on behaviour – the same gene (or tightly linked genes) affecting more than one trait such as fur colour and aggressiveness in foxes or resistance and courtship in flies. If the same gene or tightly linked genes, as is the case of supergenes (see Chapter 5), affect a suite of traits, then this can promote rapid evolutionary change, including behavioural divergence (Taylor and Campagna 2016). The relative importance of pleiotropy, genetic linkage (i.e. supergenes) and/or genes within tight networks for facilitating rapid behavioural change remains to be quantified. It is also currently unclear to what extent regulation of expression level of key genes is responsible for rapid behavioural change.

An area of behavioural study that has exploded in the last decade is founded on the realization that animals display different personalities or behavioural syndromes. That is, individuals exhibit consistent behaviours across different contexts and over time and these associations vary between individuals or groups of individuals (Dall et al. 2004; Sih et al. 2004). Individuals can be categorized, as having shy/bold/or reactive/exploratory personalities, for example. This categorization implies that there is an underlying genetic basis to such behavioural types and suggests that there is a selective advantage to behavioural-linkage patterns (van Oers and Mueller 2010; Dingemanse and Wolf 2013). One suggestion is that these suites of linked behaviours are due to underlying differences in activity levels that may be determined by genetic differences in metabolic rate (Reale et al. 2010). However, the evidence for this is limited to date (Lantova et al. 2011; Bouwhuis et al. 2014).

There is also evidence that environmental experiences, especially during early development, determine the observed variation and consistency in personality (Stamps and Groothuis 2010); hence, and as expected based on all we know, both genes and the environment are important in shaping animal personality. DNA methylation has been proposed as a mechanism controlling personality variation in the great tit, for example, through regulating gene expression in response to the ontogenetic environment (Verhulst et al. 2016). This implies that such personality traits are shaped not only by genes but also by the social and physical environment experienced during development (an inherited environment), hence generating the GxE effects we argue are so important in determining behaviours. Any GxE will also help maintain variation in personality as GxEs are excellent mechanisms for the maintenance of variation (Hunt and Hosken 2014), which would help answer the question of how different personalities are maintained within populations given a genetic basis (Wolf and Weissing 2012). Explanations could also include balancing selection ranging from antagonistic pleiotropy and overdominance, to spatial and temporal environmental heterogeneity and frequency-dependent selection (Moran 1992; Wolf and Weissing 2012).

The study of GxE, including indirect genetic effects, has been extended in the last decade to include feedback loops between genes and the environment based on the realization that organisms modify the environment in which they live and that in turn this modification alters selection acting on the organism. Such eco-evolutionary feedbacks reveal how interactions between individuals and their environments drive evolution, and in turn how this generates evolutionary feedbacks. Guppies in Trinidad, for example, modify their environment by consuming resources that cocary with the fishes' phenotype. Differences in guppy phenotype and density can promote ecosystem divergence by modification of the local environment. In turn, this change in ecosystem structure translates into differential selection on the guppies themselves, creating a feedback between ecology and evolution (Bassar et al. 2010). Importantly, it is now recognized that this can happen at a much more rapid scale than previously appreciated – in the case of the guppies, habitat modification occurred in the space of a few weeks (Bassar et al. 2010).

Eco-evolutionary dynamics is becoming an emerging field in evolutionary ecology, especially as this can occur at much shorter time scales than 'classic' evolutionary change (Hendry 2016). Clearly, an effective way in which individuals can alter their environment involves behaviour in some shape or form and this behavioural modification will depend on the social context. Similarly, the social environment can also generate evolutionary feedbacks through indirect genetic effects (Moore et al. 1997). It is suggested that behavioural traits may be particularly likely to promote evolutionary change and diversification, as behaviour tends to be very flexible, changes rapidly and is context dependent. This is particularly the case when the social environment consists of individuals with variable genotypes that strongly shape the behaviour of other interacting individuals, thereby promoting evolutionary feedbacks (Bailey et al. 2018).

What does the realization that behaviour is shaped by genes and environments, and that the environment is probably more complex and fluid than we've appreciated, mean for our understanding of ourselves? As discussed in several chapters, GxEs are not new but the question is whether this realization provides any fundamental insights into what it means to be human. Can the recognition that GxEs interact to shape our behaviour illuminate our notion of self-awareness and theory of mind? Will it tell us something about how we are able to recognize ourselves as separate from others and the environment? Perhaps not, especially if mind is an emergent property itself, or even if it evolves from a necessity to "watch the watcher". However, the recognition that we are a product of a complex interaction between genes and our environment, including our social and socio-economic environment, highlights the need to consider ourselves as being greater than the sum of our genetic and environmental parts, and that we may in fact have the possibility to shape ourselves by affecting the environment in which we act. Even further, the acknowledgement that we shape our social environment, which in turn affects our own behaviours and the feedbacks that this entails, may mean that we may begin to value our environment in its entirety even more than currently. The complexity of the genotype-phenotype link also means that although patterns of, and solutions to, human behavioural issues can be identified and average solutions defined, nuanced more individual-focused approaches may be needed to ultimately address behavioural questions in their entirety. That is, addressing average main effects may be possible, but additional variance will lay in interactions, which are trickier to address. If anything, the complex GxEs that define us really do mean that individuals are individual.

Acknowledgements

Thanks again to all our coauthors.

References

Adamo, S.A. (2013). Parasites: evolution's neurobiologists. *Journal of Experimental Biology* 216: 3–10.

Axelsson, E., Ratnakumar, A., Areendt, M.-L. et al. (2013). The genomic signature of dog domestication reveals adaptation to a starch-rich diet. *Nature* 495: 360–364.

Bailey, N., Lucas, M.O., and Moore, A.J. (2018). Indirect genetic effects in behavioural ecology: does behavior play a special role in evolution? *Behavioral Ecology* 29: 1–11.

Barber, I., Mora, A., Payne, E. et al. (2016). Parasitism, personality and cognition in fish. *Behavioural Processes* 141: 205–219.

Bassar, R.D., Marshall, M.C., Lopez-Sepulcre, A. et al. (2010). Local adaptation in Trinidadian guppies alters ecosystem processes. *Proceedings of the National Academy of Sciences USA* 107 (8): 3616–3621.

Bouwhuis, S., Quinn, J.L., Sheldon, B.C., and Verhulst, S. (2014). Personality and basal metabolic rate in a wild bird population. *Oikos* 123: 56–62.

Dall, S.R.X., Houston, A.I., and McNamara, J.M. (2004). The behavioural ecology of personality: consistent individual differences from an adaptive perspective. *Ecology Letters* 7: 734–739.

Dawkins, R. (1976). *The Selfish Gene*. Oxford: Oxford University Press.

Dingemanse, N.J. and Wolf, M. (2013). Between-individual differences in behavioural plasticity within populations: causes and consequences. *Animal Behavior* 85: 1031–1039.

Fredericksen, M.A., Zhang, Y., Hazen, M.L. et al. (2017). Three-dimensional visualization and a deep-learning model reveal complex fungal parasite networks in behaviorally manipulated ants. *Proceedings of the National Academy of Sciences USA* 114: 12590–12595.

Hendry, A.P. (2016). *Eco-Evolutionary Dynamics*. Princeton: Princeton University Press.

Hughes, D., Brodeur, J., and Thomas, F. (2012). *Host Manipulation by Parasites*. Oxford: Oxford University Press.

Hunt, J. and Hosken, D.J. (2014). *Genotype-By-Environment Interactions and Sexual Selection*. Chichester: Wiley Blackwell.

Lantova, P., Zub, K., Koskela, E. et al. (2011). Is there a linkage between metabolism and personality in small mammals? The root vole (*Microtus oeconomus*) example. *Physiology and Behavior* 104: 378–383.

McCart, C., Buckling, A., and Ffrench-Constant, R.H. (2005). DDT resistance in flies carries no cost. *Current Biology* 15: R587–R589.

Moore, J. (2002). *Parasites and the Behavior of Animals*. Oxford: Oxford University Press.

Moore, A.J., Brodie, E.D. III, and Wolf, J.B. (1997). Interacting phenotypes and the evolutionary process: I. Direct and indirect genetic effects of social interactions. *Evolution* 51: 1352–1362.

Moran, N.A. (1992). The evolutionary maintenance of alternative phenotypes. *American Naturalist* 139: 971–989.

Reale, D., Garant, D., Humphries, M.M. et al. (2010). Personality and the emergence of the pace-of-life syndrome concept at the population level. *Philosphical Transactions of the Royal Society B* 365: 4051–4063.

Roff, D.A. (1994). Optimality modeling and quantitative genetics: a comparison of the two approaches. In: *Quantitative Genetic Studies of Behavioral Evolution* (ed. C.R.B. Boake), 49–66. Chicago: University of Chicago Press.

Rostant, W.G., Bowyer, J., Coupland, J. et al. (2017). Pleiotropic effects of DDT resistance on male behaviour and size. *Behavior Genetics* 47: 449–458.

Schmidt-Hempel, P. (1998). *Parasites in Social Insects*. Princeton: Princeton University Press.

Sih, A., Bell, A., and Johnson, J.C. (2004). Behavioral syndromes: an ecological and evolutionary overview. *Trends in Ecology and Evolution* 19: 372–378.

Singh, R.K., Chang, H.W., Yan, D. et al. (2017). Influence of diet on the gut microbiome and implications for human health. *Journal of Translational Medicine* 15: 73.

Smith, D.T., Hosken, D.J., Rostant, W.G. et al. (2011). DDT resistance, epistasis and male fitness in flies. *Journal of Evolutionary Biology* 24: 1351–1362.

Stamps, J.A. and Groothuis, T.G.G. (2010). Ontogeny of animal personality: relevance, concepts and perspectives. *Biological Reviews* 85: 301–325.

Taylor, S. and Campagna, L. (2016). Avian supergenes. *Science* 351: 446–447.

Thomas, F., Schmidt-Rhaesa, A., Martin, G. et al. (2002). Do hairworms (Nematomorpha) manipulate the water seeking behaviour of their terrestrial hosts? *Journal of Evolutionary Biology* 15: 356–361.

Tregenza, T., Wedell, N., Hosken, D.J., and Ward, P.I. (2003). Maternal effects on offspring depend on female mating pattern and offspring environment in yellow dung flies. *Evolution* 57: 297–304.

Trut, L., Oskina, I., and Kharlamova, A. (2009). Animal evolution during domestication: the domesticated fox as a model. *Bioessays* 31: 349–360.

Van Oers, K. and Mueller, J.C. (2010). Evolutionary genomics of animal personality. *Philosphical Transactions of the Royal Society B* 365: 3991–4000.

Verhulst, E.C., Mateman, A., Zwier, M. et al. (2016). Evidence from pyrosequencing indicates that natural variation in animal personality is associated with DRD4 DNA methylation. *Molecular Ecology* 25: 1801–1811.

Wolf, M. and Weissing, F.J. (2012). Animal personalities: consequences for ecology and evolution. *Trends in Ecology and Evolution* 27: 452–461.

Index

Genes and Behaviour: Beyond Nature-Nurture, First Edition.
Edited by David J. Hosken, John Hunt and Nina Wedell.
© 2019 John Wiley & Sons Ltd. Published 2019 by John Wiley & Sons Ltd.